Medical and Applied
VIROLOGY

Proceedings of the Second International Symposium

Edited by

MURRAY SANDERS, M.D.
Medical Director
Gray Research Foundation, Inc.
Fort Lauderdale, Florida

and

EDWIN H. LENNETTE, M.D., Ph.D.
Chief, Viral and Rickettsial Disease Laboratory
California State Department of Public Health
Berkeley, California

WARREN H. GREEN, INC.
St. Louis, Missouri, U.S.A.

CONTRIBUTORS

BALTIMORE, DAVID, PH.D., *Research Associate, The Salk Institute for Biological Studies.*

BARON, SAM, PH.D., *Medical Officer, Viral Biology Section of the Laboratory of Biology of Viruses, National Institutes of Health.*

BEN-PORAT, TAMAR, PH.D., *Assistant Member, Research Laboratory, Department of Microbiology, Albert Einstein Medical Center.*

BILLIAU, ALFOUR J.D.A., M.D., *Aspirant of the Belgian N.F.W.A.*

BOND, JAMES O., M.D.,* *Director, Florida State Board of Health.*

BORECKY, LADISLAV, M.D., *Institute of Virology, Czechoslovak Academy of Science.*

CANTELL, KARI JUHANI, M.D., M.SC.D., *Professor, State Serum Institute, Helsinki, Finland.*

CARVER, DAVID HAROLD, A.B., M.D., *Associate Professor of Pediatrics, The Johns Hopkins University School of Medicine.*

CHANY, CHARLES, PH.D., *Professor of Microbiology, University of Paris.*

CHOPPIN, PURNELL W., M.D., *Associate Professor, The Rockefeller University.*

COMPANS, RICHARD W., B.A., *Graduate Fellow, The Rockefeller University.*

DE CLERCQ, ERIK D.A., M.D., *Navorsingsstasiair of the Belgian N.F.W.O.*

DE SOMER, P., M.D., *Professor of Microbiology, Universiteitsklinieken Rega Institut, Leuven, Belgium.*

FALCOFF, REBECA, PH.D., *Member, Institut de Recherches sur le Cancer.*

FANTES, KARL HEINZ, B.SC., PH.D., *Head Biochemistry, Virus Unit, Glaxo Laboratories Ltd., Stoke Poges, Bucks, England.*

FINTER, N. B., PH.D., *Virus Research Department, Imperial Chemical Industries, Ltd., Cheshire, England.*

FRANKLIN, RICHARD M., PH.D., *Member, The Public Health Research Institute of the City of New York, Inc.*

GORDON, IRVING, M.D., *Professor and Chairman, Department of Microbiology, University of Southern California School of Medicine.*

GRANBOULAN, NICOLE, M.D., *Chargee de Recherches au Centre National de la Recherche Scientifique, Laboratoire de Microscopie Electronique, Institut de Recherches sur le Cancer.*

HAUGEN, ROGER K., M.D.,* *Chief Pathologist, Holy Cross Hospital, Fort Lauderdale, Florida.*

HENLE, WERNER, M.D., *Professor of Viology, The Children's Hospital of Philadelphia.*

HOLMES, KATHRYN V. (MRS.), A.B., *Graduate Fellow, Rockefeller University.*

HUGHES, PAUL W., M.D.,* *Director, Broward County Health Department, Fort Lauderdale, Florida.*

JOKLIK, W. K., PH.D., *Professor, Department of Cell Biology, Albert Einstein College of Medicine, Bronx, New York.*

KAPLAN, ALBERT S., PH.D., *Head, Research Laboratories, Department of Microbiology, Albert Einstein Medical Center.*

LENNETTE, EDWIN H., M.D., *Chief, Viral and Rickettsial Disease Laboratory, Department of Public Health, State of California.*

LEVY, HILTON B., PH.D., *Department of Health, Education, and Welfare, Public Health Service, National Institutes of Health.*

LEWIS, L. JAMES, PH.D., *Assistant Director, Sterling-Winthrop Research Institute.*

LOCKART, ROYCE Z., JR., A.B., M.S., PH.D., *E.I. du Pont de Nemours & Co., Central Research Department.*

MARCUS, PHILIP I., B.S., M.S., PH.D., *Professor of Microbiology, Department of Microbiology and Immunology, Albert Einstein College of Medicine, Bronx, New York.*

MCCLAIN, MARY E., PH.D., *Research Specialist, Viral and Rickettsial Disease Laboratory, California State Department of Public Health.*

MERIGAN, THOMAS C., JR., M.D., *Assistant Professor of Medicine, Chief, Division of Infectious Disease, Department of Medicine, Stanford University School of Medicine.*

PAUCKER, KURT, PH.D., *Associate Professor of Virology, Department of Pediatrics, University of Pennsylvania School of Medicine. On leave of absence; temporarily with the Statens Seruminstitut, Copenhagen, Denmark.*

OSBORN, JUNE E., B.A., M.D., *Instructor in Pediatrics, University of Wisconsin Medical School.*

RADA, BRETISLAV, PH.D., *National Institutes of Health. Present address, Czechoslovak Academy of Sciences, Bratislava, Czechoslovakia.*

ROBINSON, WILLIAM S., M.S., M.D., *Assistant Professor of Molecular Biology, University of California, Berkeley.*

SANDERS, MURRAY, M.D., *Medical Director, Gray Research Foundation, Inc.*

SHATKIN, AARON JEFFREY, PH.D., *National Institutes of Health, Bethesda, Maryland.*

SMORODINTSEV, ANATOL A., M.D., *Chief, Department of Virology, Institute of Experimental Medicine, Academy of Medical Sciences, USSR.*

SPENDLOVE, REX S., PH.D., *Professor and Department Head, Utah State University Department of Bacteriology and Public Health.*

WALKER, DUARD L., A.B., M.A., M.D., *Professor of Medical Microbiology, School of Medicine, University of Wisconsin.*

YOUNGNER, JULIUS S., SC.D., *Professor and Chairman, Department of Microbiology, School of Medicine, University of Pittsburgh.*

**Committee for local arrangements.*

PROGRAM CO-CHAIRMEN

L. JAMES LEWIS, Ph.D.
*Sterling-Winthrop Research Institute**
Rensselaer, New York

and

SAM BARON, M.D.
National Institute of Allergy and Infectious Diseases
Laboratory of Biology of Viruses
National Institutes of Health
Besthesda, Maryland

---

*Present address: Virus Laboratory, Abbott Laboratories, North Chicago, Illinois.

SUPPORTED

By

GRAY RESEARCH FOUNDATION, INC.

SPONSORED

By

NOVA UNIVERSITY
Fort Lauderdale, Florida

PREFACE

T HE *First Annual Symposium on Applied Virology* was held at Boca Raton, Florida and was organized in response to the belief of many workers that there was a real need for a forum in which the practical and applied aspects of virology could be given the serious consideration they merit. Despite the relative lack of advance publicity, the meeting attracted more workers than had been expected and this reception by the scientific community was most gratifying. The Proceedings of the First Symposium appeared in early 1965 and were equally well received.

With this salutory experience, planning shortly thereafter was begun for a Second International Symposium. This eventuated in the *Second International Symposium on Medical and Applied Virology* held at Ft. Lauderdale, Florida in December 1966. Unlike the First Symposium, which had no specific theme but covered a broad area of medical virology, the theme for this Second Symposium was *Interference, Interferon and Viral Inhibitors.* Although the range of subjects covered was relatively broad, the number of papers presented was purposely kept small in order to permit extensive audience participation in the discussions.

This volume represents the Proceedings of this Second Symposium. The papers contained herein are essentially as they were presented; editing has been concerned primarily with the discussions, with emphasis directed at clarity and relevance, and not at length.

The Editors, who were also the General Co-chairmen of the Symposium, acknowledge with gratitude the dedication and untiring efforts of the Program Chairmen, Doctors L. James Lewis and Samuel Baron, and the valued participation, and skillful guidance as session chairmen, of Doctors Samuel Baron, Werner Henle and Richard Franklin.

Finally, we acknowledge our indebtedness to the Gray Research Foundation, Inc., whose generous financial support made possible this Symposium. Of equal importance was the academic sponsorship of Nova University, Fort Lauderdale.

MURRAY SANDERS, M.D.
EDWIN H. LENNETTE, M.D.

Ft. Lauderdale, Florida
and
Berkeley, California

COMMON ABBREVIATIONS USED IN THE TEXT

ACM	actinomycin
AMV	avian myeloblastosis virus
AVP	antiviral substance
BHK	baby hamster kidney
CE	chick embryo
CEC	chick embryo cells
CPE	cytopathic effect
CAR	cystosine arabinoside
DNA	deoxyribonucleic acid
DNase	desoxyribonuclease
EDTA	ethylene diamine tetraacetic acid
EMC	encephalomyocarditis
FHM	fathead minnow line
FPA	fluorophenylalanine
GMK	green monkey kidney
GH	growth hormone
HEL	helenine
HAD	hemagglutinin-hemadsorption
HDP	hemadsorption-producing particle
HEK	human embryonic kidney
IBT	isatin-β-thiosemicarbazone
ICC	immunofluorescent cell count
IF	interferon
LCMV	lymphocytic choriomeningitis virus
LDH	lactic dehydrogenase virus (Riley's agent)
mRNA	messenger ribonucleic acid
MK	monkey kidney
MM	strain of encephalomyocarditis virus
MR	multiplicity reactivation
MCMV	murine cytomegalovirus
NDV	Newcastle disease virus
PFP	plaque-forming particles
PFU	plaque-forming unit
PVP	polyvinylpyrrolidone
PU	protective unit

Pr	pseudorabies unit
RAV	Rous associated virus
RF	replicative form
RK	rabbit kidney
RNA	ribonucleic acid
RNase	ribonuclease
RSB	reticulocyte standard buffer
RSV	Rous sarcoma virus
SFV	Semliki Forest virus
SV	simian virus
SDS	sodium dodecyl sulfate
TRIC	trachoma-inclusion conjunctivitis agents
THS	thyroid-stimulating hormone
UR	uridine
UV	ultraviolet light (inactivated)
VSV	vesicular stomatitis virus
WEE	Western equine encephalomyelitis

CONTENTS

Medical and Applied
VIROLOGY

Photograph by Walter Bird, London

ALICK ISAACS, M.D., F.R.S.

INTRODUCTORY REMARKS

WERNER HENLE

ONE MAY WONDER WHY THE Program Committee chose to plan a separate session each on "Interference" and "Interferon." Does this imply lingering doubts that not all viral interference phenomena are mediated by interferons and that interference *sine* interferon might also exist? This group of papers will examine this question. While it will not be possible to omit references to interferon completely, the discussions should bring to the fore some types of induced cellular resistance in which this inhibitor does not seem to play a role.

Early reports on interference phenomena between live pairs of related, or even vastly different viruses in animals hardly went beyond factual observations, and interpretations of necessity remained speculative. While it became evident in many of the studies that certain rules of dosage and timing of administration of the two opposing viruses had to be observed in order to repress replication of one, neither the number of cells at risk nor the quantity of interfering virus which had to be attained to be ultimately effective could be ascertained under these conditions. These handicaps were at least partly overcome in studies of interference between inactivated and infective myxoviruses in the chick embryo allantois. In this system, the number of cells lining the allantoic cavity was known within narrow limits, all being readily accessible to the opposing virus particles, and the dose of inactivated, interfering virus remained stationary at the predetermined level. In this system, it became possible to study interference with a fair degree of quantitation of the various reactants. Among the most important findings were the observations that interference was an intracellular event, that no more than one inactive virus particle was required to render a cell resistant, that many hours had to elapse before resistance became fully estab-

---

*Research Career Award 5-K6-AI-22,683.

lished and that the resistance covered a broad spectrum of viruses. The chick embryo allantois was still far from ideal for studies of this kind and therefore has been largely replaced in further work by cell cultures as they became available for general use. Under these conditions, many aspects of interference at the cellular level could be assessed more accurately than was heretofore possible and new facts were revealed which escaped detection by the earlier methods.

Many hypotheses have been entertained to explain various interference phenomena. Some of the more plausible ones are listed here, without critical evaluation of their relevance: a) prevention of spread of the excluded virus by inflammatory host responses called forth by the primary infection; b) exhaustion of cells or of metabolites by the interfering virus; c) destruction or blockade of cell receptors by the first virus, thus preventing attachment of the second; d) inactivation of the excluded virus at the cell surface or prevention of its penetration into the cell; e) competition between the two viruses for cellular constituents, such as a key enzyme, and f) production of antiviral factors as a result of the primary infection.

It is somewhat ironical, that the last hypothesis, which is now amply supported by numerous examples of interferon synthesis, has received either little attention in the early studies, or it could not be verified under the experimental conditions employed. For example, in the analysis of the influenza virus-chick embryo system of interference efforts were made to detect interfering components other than the virus particles. Allantoic fluids which were collected as late as 96 hours after infection of chick embryos and from which the virus was largely removed by centrifugation or adsorption onto red cells, failed to induce resistance in the entodermal layer of the allantois. As is known now, these preparations most likely contained considerable levels of interferon, but the assay method was insufficient for its detection and thus bound to fail.

The demonstration of interferon synthesis in the majority of well-studied examples of viral interference does not preclude that others of the suggested mechanisms listed above may play an additional, if not the only role in some situations. For example, in cell cultures persistently infected by a virus, resistance to superinfection by the same or other viruses is a common observation. If the resistance is broad, including many different types of viruses, it is likely to be mediated by an interferon. In some viral carrier cultures, the resistance is rather restricted, however, in that it extends merely to the homologous agent or to close relatives of it. In these cases, mechanisms other than interferon action are to be suspected, such as cell receptor destruction or blockade. A discussion of persistent viral infections of cell cultures was chosen therefore as the first topic of today's session in order to examine the types of resistance that might be encountered in various situations.

Not every viral infection leads to the establishment of interference or interferon synthesis. A highly virulent virus may disturb cellular functions and destroy the cell too rapidly for any significant interferon production to occur. Yet, there are several reports on non-destructive, abortive viral infections in which *no* cellular resistance develops, but, on the contrary, an enhanced susceptibility to other viruses. Discussion of viral enhancement has been chosen as the second topic in the expectation that clarification of its mechanisms might contribute indirectly to the understanding of interference inasmuch as analyses of "abnormal" responses have often served to clarify normal functions of cells or whole organisms.

The last of this group of topics is concerned with alterations in cellular functions during establishment and the duration of interference. A detailed, step-by-step analysis of events taking place in growing cell cultures from initiation to final loss of resistance might reveal, among others, early or late contributions of the interfering agent *per se,* that is the interferon inducer to the protection of cells.

SOME ASPECTS OF PERSISTENT VIRAL INFECTIONS IN CELL CULTURES*

WERNER HENLE**

I HAVE SELECTED TWO observations for discussion which are based upon earlier studies of one type of viral carrier culture, the persistent infection of L cells with Newcastle disease virus (1-6). The first is concerned with changes in the viral population in the course of the persistent infection; and the second with an application of criteria which have evolved from analysis of the L_{NDV} carrier culture system to cell lines derived from Burkitt lymphomas and leukemias which in part harbor as yet unidentified virus particles.

1. Reversible Changes in Newcastle Disease Virus Populations as a Result of Passage in L Cells or Chick Embryos

The basic features of the L_{NDV} carrier system have been described in detail in earlier publications (1-6). It was shown that once the infection has been stabilized the percentage of infected cells remains at all times at a low level (between 1 and 5%), the rest of the cells being protected by interference or interferon induced by incomplete virus particles (5). The protected cells continue to divide, and among their descendants cells appear which are partially or fully susceptible (7). In the competition for these cells, infectious virus particles win on occasion against both the incomplete, interfering units and interferon so that the carrier state is perpetuated (6).

In the course of these studies, it was noted that virus (victoria strain) reisolated after many months from carrier cultures and passed twice in chick embryos (NDV_{L-E-2}) was apparently much more virulent for L cells on primary exposure than the original

*These investigations were supported by grant CA-04568 from the National Cancer Institute and by training grant 5-T1-AI-104 from the National Institute of Allergy and Infectious Diseases, United States Public Health Service.

**Research Career Award 5-K6-AI-22,683.

virus (NDV_O) which had been maintained continuously in chick embryos and was used for initiation of the persistent infection. This difference was examined by Drs. Rodriquez and ter Meulen and the results are reported in detail elsewhere (8).

Infection of L cells at high multiplicities (>5) with NDV_O caused no cytopathic effect and the cultures readily survived as viral carriers. Within 24 hours, all cells had become infected as evident from NDV-specific immunofluorescence, but the infection was of an abortive nature because: (a) only small aggregates of viral antigen were found in the cytoplasm; and (b) only few cells produced infectious virus (<1 pfu per 1000 infected cells). During the same period, large amounts of interferon were synthesized. In contrast, NDV_{L-E-2} destroyed the cultures, all cells revealed within 24 hours large amounts of viral antigen which was dispersed throughout the cytoplasm, up to 200 pfu of progeny were produced, but little, if any interferon.

On further passages of NDV_{L-E-2} in chick embryos, the behavior of the virus populations following exposure of L cells at high multiplicities gradually changed back toward that of NDV_O. From the 4th passage on, the yields of infectious viral progeny declined progressively, interferon synthesis became detectable and rose to maximal levels within the next 8 passages, and the cultures survived with increasing ease, all being viral carriers. Yet, by the 17th passage (NDV_{L-E-17}), the last studied, the yields of infectious progeny were still well above those obtained after infection with NDV_O. Virus populations reisolated in chick embryos from more than 3 months-old carrier cultures initiated with NDV_{L-E-16} again had the properties of NDV_{L-E-2}.

According to these observations, the evident gain or loss of "virulence" of NDV populations for L cells at high multiplicities of infection is reversible by passages either in chick embryos or in L cell cultures. Efforts to determine the nature of these changes in virulence revealed that they were caused not so much by a selection of viral variants in the two host systems, but rather by noninfectious, interferon-inducing components which accumulate in the viral populations during serial passages in chick embryos and which are temporarily lost or reduced in reisolates from L_{NDV}

carrier cultures. This conclusion is based upon two sets of experiments.

In the first set, L cells were exposed to serially decreasing doses of virulent (NDV_{L-E-2}) or avirulent (NDV_O or NDV_{L-E-16}) virus populations and the period of observation was restricted to the first infectious cycle (20 hr). With NDV_{L-E-2}, the yields of infectious virus particles (pfu) per infected cell were, as expected, of the same high order even at a multiplicity of infection of 100, and no trace of interferon became detectable in any of the cultures. The yields of infectious progeny per infected cell were also normal (in excess of 50) when L cells were exposed to NDV_O or NDV_{L-E-2}, as long as the multiplicities of infection were sufficiently low, i.e., not exceeding 0.005 and 0.1, respectively. With serial 10-fold increments in the doses of virus above these levels, the numbers of pfu produced per infected cell declined steadily. With the first significant reduction in viral yields, interferon synthesis became detectable and it attained the same (maximal) levels whether the multiplicity of infection was 1 or 100. Over this range of multiplicites the yields of pfu per infected cell decreased further, however, to reach nearly nil in the case of NDV_O and less than 1 per cent of the standard amount with NDV_{L-E-16}.

These results indicated that the two avirulent virus populations contained a non-infectious component in addition to, and in excess of infectious virus units which was responsible for induction of interferon synthesis and interfered with viral replication in cells simultaneously invaded by both types of particles.

This interpretation was borne out by the second set of experiments in which cloned viral populations derived from NDV_O and NDV_{L-E-16} by 4 serial-picked plaque passages in chick embryo fibroblasts, or 4 serial-limiting dilution passages in the chick embryo allantois were used for exposures of L cells at saturation multiplicities of infection. All of the 6 clones studied were, in contrast to the parent populations, virulent for L cells, produced standard or nearly standard numbers of pfu per infected cell, and induced relatively little or no detectable interferon synthesis. Thus, some of the clones behaved exactly like NDV_{L-E-2} and the remaining ones differed from it only insofar as they failed to destroy the L cell cultures completely and caused the formation of small amounts of interferon. Whether these differences denote that some of the cloned populations contained small amounts of the interfering component, or whether they reflect segregation of viral variants, remains to be ascertained.

From these experiments, it has become evident that infectious NDV particles apparently do not induce significant interferon synthesis. The nature of the non-infectious, interferon-inducing components has not been established. Not all of them appear to be endowed with hemagglutinating activity since the pfu/Ha or EID_{50}/HA ratios of avirulent populations hardly differed from those of the virulent ones. In NDV_O, they were reduced at most by a factor of 10, yet calculations based on the results obtained in L cells at decreasing multiplicities of infection indicated that the ratios between non-infectious and infectious particles must be many times higher than apparent from the pfu/HA ratios. That the interferon-inducing components in NDV_0 are in fact viral in nature was ascertained by their complete neutralization by an antiserum to cloned NDV_{L-E-2}.

It should be pointed out that the virulence of NDV_{L-E-2} for L cells in the sense discussed in demonstrable only under limited conditions. As has been reported previously (5), cells initially infected by this type of virus produce a mixed viral progeny composed of infectious and at least 10 times the number of non-infectious particles. Thus, if all cells are infected initially, the culture does not survive. However, at lower multiplicities of infection, the infectious and non-infectious virus particles of the first cycle progeny complete for the remaining cells. Those invaded by only infectious virus again succumb; those adsorbing only non-infectious virus survive, become resistant and synthesize interferon; and in those which are entered by both types of particles, the infectious process is aborted and these cells also seem to survive and possibly yield interferon. As stated earlier, partly or fully susceptible descendants of the surviving cells may then be invaded on occasion by infectious virus to maintain the viral carrier state (6, 7).

It is thus apparent that the same type of mechanism is responsible for protection and survival of L cells following initial exposure to avirulent virus preparation and for maintenance of the viral carrier state whether initiated by virulent or avirulent virus populations. In all instances, non-infectious interfering and interferon-inducing viral components either derived from infected allantoic or L cells offer the key to the protection of a large part of the cells.

2. Interference and Interferon Synthesis in Cell Lines Derived from Burkitt's Lymphoma or Leukemias.

The majority of cell lines derived from Burkitt's lymphoma were found on electron microscopic examination to harbor as yet unidentified herpes-like virus particles in a small proportion of the cells (9-11). Infected cells were shown to be detectable also by immunofluorescence techniques (12). Efforts were made by Dr. G. Henle, B. Zajac and myself to determine whether the viral carrier state could be recognized by evaluation of certain biological activities of the cultures. These experiments were patterned after the earlier work with L cells persistently infected with Newcastle disease virus (NDV) (1-6). These studies had shown that: (a) only a small proportion of the cells in carrier cultures were infected at any time; (b) L_{NDV} cultures were markedly, but not absolutely resistant to superfection with vesicular stomatitis virus (VSV); (c) this resistance was readily transferred by 10^5-10^6 carrier culture cells to recipient L cell stock populations; (d) low levels of an interferon were detectable in the media of L_{NDV} cultures; (e) carrier populations produced additional, though small amounts of interferon on appropriate viral stimulation, and (f) addition of extraneous murine interferon protected the small VSV-susceptible fraction of carrier culture cells against VSV. The same types of analyses were applied to cell lines derived from Burkitt tumors or leukemic patients.

With the first 3 Burkitt cell lines studied (EB-1, EB-2 and SK-1), results were obtained which matched in every respect those obtained with L_{NDV} cultures. About 1 per cent of the cells harbored the herpes-like agent; the cells were markedly resistant to VSV; they transmitted resistance to human embryonic kidney (HEK) or amnion, but not to murine or rabbit cells; an interferon-like inhibitor was detectable in the culture media; the cultures produced additional small quantities of interferon on exposure to

NDV; and the VSV-susceptible cells were rendered resistant by a human interferon. It appeared to be justified, therefore, to ascribe these results to the viral carrier state which was detected by electron microscopy and immunofluorescence (13, 14).

As more Burkitt cell as well as leukemic cell lines became available, the situation became more complex. Several other Burkitt cell lines behaved like those just described, but two (EB-3 and Kudi), which harbor herpes-like virus particles and are resistant to VSV, were found to be incapable of transmitting their resistance to HEK cells on mixed cultivation. The culture media failed to reveal interferon, and the one tested (EB-3) produced little if any interferon on exposure to NDV and could not be solidly protected by a human interferon. These cells thus do not appear to possess significant capacities to produce interferon nor to be readily protectable by interferon. Similar differences in interferon synthesis and protection have been noted to occur among other types of closely related cells, i.e., sublines of HeLa cells (15), and continuous cultures of polyoma virus-induced hamster tumor cells (16).

Among 2 Burkitt cell lines free of indigenous virus, one (Raji) was also found to be resistant to VSV (17), but the other (EB-4) was highly susceptible. Both lines failed to protect other human cells and no interferon-like inhibitor was detectable in the culture media. The Raji line yielded high titers of interferon on stimulation by NDV and was readily protectable by human interferon. The EB4 cells have not yet been subjected to these tests. Finally, the Ogun line, containing only few cells which harbor virus (<0.1%), exhibited good resistance to VSV and produced somewhat more of the interferon-like inhibitor than any of the others.

None of the 7 cell lines (18-21) derived from leukemic patients and thus far studied were known to harbor herpes-like virus particles and all were highly susceptible to VSV. Yet, on mixed cultivation, 4 of these lines (SK-L1, SK-L2, LK-57 and LK-60) rendered HEK cells resistant to VSV and produced interferon-like inhibitors without apparent extraneous stimulation. They synthesized additional or *de novo* titers of interferon when exposed to NDV and were solidly protectable against VSV by human interferon.

From these various results, it has become evident that the resistance of the Burkitt cell lines to VSV and the detection of the interferon-like inhibitor in the culture media of some of them cannot be ascribed to the presence of herpes-like virus particles. It rather appears that: (a) the cultured lymphoblasts derived from Burkitt tumors belong to the few types of cells which may show an appreciable degree of innate resistance to VSV, and (b) that cells of the leukocyte series may produce an interferon-like inhibitor without obvious stimulation. The latter conclusion has been reached also by Benyesh-Melnick and by Deinhardt and Burnside (personal communications). While this inhibitor was found to be resistant to pH 2 and heating at 60°C for 30 minutes, non-sedimentable at 100,000 x g, susceptible to trypsin, and species specifically oriented in its action (protection of human and simian but not of rabbit or murine cells) it remains to be determined whether it truly represents an interferon.

These studies have shown that "it ain't necessarily so," even when, or perhaps especially when, the first results seem to agree too well with the original premise on which the experiments were based. The possibility that interferons might be produced spontaneously by certain cells is, of course, *per se* of considerable interest. It has been shown previously that certain interferons found in the blood stream of animals after injection of endotoxin are apparently preformed because their appearance is not prevented by inhibitors of protein synthesis (22). Cells of the leukocyte series might well be their source.

REFERENCES

1. HENLE, G., DEINHARDT, F., BERGS, V. V., and HENLE, W.: Studies on persistent infections of tissue cultures. I. General aspects of the system. *J. Exper. Med., 108*:537-560, 1958.

2. BERGS, V. V., HENLE, G., DEINHARDT, F., and HENLE, W.: Studies on persistent infections of tissue cultures. II. Nature of the resistance to vesicular stomatitis virus. *J. Exper. Med., 108*:561-572, 1958.

3. DEINHARDT, F., BERG, V. V., HENLE, G., and HENLE, W.: Studies on persistent infections of tissue cultures. III. Some quantitative aspects of host cell-virus interactions. *J. Exper. Med., 108*:573-589, 1958.

4. HENLE, W., HENLE, G., DEINHARDT, F., and BERGS, V. V.: Studies on persistent infections of tissue cultures. IV. Evidence for the production of an interferon in MCN cells by myxoviruses. *J. Exper. Med., 110*:525-541, 1959.

5. RODRIGUEZ, J. E., and HENLE, W.: Studies on persistent infections of tissue cul-

tures. V. The initial stages of infection of L (MCN) cells by Newcastle disease virus. *J. Exper. Med., 119*:895-922, 1964.

6. HENLE, W.: Interference and interferon in persistent viral infections of cell cultures. *J. Immunol., 91*:145-150, 1963.

7. PAUCKER, K., and CANTELL, K.: Quantitative studies on viral interference in suspended L cells. V. Persistence of protection in cells treated with uv-irradiated Newcastle disease virus and interferon. *Virology, 21*:22-29, 1963.

8. RODRIGUEZ, J. E., TER MEULEN, V., and HENLE, W.: Studies on persistent infections of tissue culture. VI. Reversible changes in Newcastle disease virus populations as a result of passage in L cells or chick embryos. *J. Bacteriol.*, In press, 1967.

9. EPSTEIN, M. A., BARR, Y. M., and ACHONG, B. G.: Studies with Burkitt's lymphoma. In, Methodological Approaches to the Study of Leukemias, *Wistar Institute Symposium Monograph No. 4,* 69-79, 1965.

10. STEWART, S. E., LOVELACE, E., WHANG, J. J., and NGU, V. A.: Burkitt tumor: Tissue culture, cytogenetics and virus studies. *J. Nat'l. Cancer Inst., 34*:319-327, 1965.

11. RABSON, A. S., O'CONOR, G. T., BARON, S., WHANG, J. J., and LEGALLAIS, F. Y.: Morphologic, cytogenetic and virologic studies *in vitro* of a malignant lymphoma from an African child. *Internat. J. Cancer, 1*:89-106, 1966.

12. HENLE, G., and HENLE, W.: Immunofluorescence in cells derived from Burkitt lymphoma. *J. Bacteriol., 91*:1248-1256, 1966.

13. HENLE, G., and HENLE, W.: Evidence for a persistent viral infection in a cell line derived from Burkitt's lymphoma. *J. Bacteriol., 89*:252-258, 1965.

14. HENLE, G., and HENLE, W.: Interference in the detection of viral carrier states. In, *Wistar Institute Symposium Monograph No. 4,* 83-90, 1965.

15. CANTELL, K., and PAUCKER, K.: Studies on viral interference in two lines of HeLa cells. *Virology, 19*:81-87, 1963.

16. HENLE, G., and HENLE, W.: Differences in response of hamster tumor cells induced by polyoma virus to interfering virus and interferon. *J. Natl. Cancer Inst., 31*:143-153, 1963.

17. EPSTEIN, M. A., ACHONG, G., BARR, Y. M., ZAJAC, B., HENLE, G., and HENLE, W.: Morphological and virological investigations on cultured Burkitt tumor lymphoblasts (strain Raji). *J. Natl. Cancer Inst., 37*:547-559, 1966.

18. IWAKATA, S., and GRACE, J. T., JR.: Cultivation *in vitro* of myeloblasts from human leukemia. *New York State J. Med., 64*:2279-2282, 1964.

19. MOORE, G. E., ITO, E., ULRICH, K., and SANDBERG, A. A.: Culture of human leukemia cells. *Cancer, 19*:713-723, 1966.

20. ARMSTRONG, D.: Serial cultivation of human leukemic cells. *Proc. Soc. Exper. Biol. & Med., 122*:475-481, 1966.

21. CLARKSON, B. D., STRIFE, A., and DeHARVEN, E.: Continuous culture of seven new cell lines (SK-L1 to 7) from patients with acute leukemia. To be published.

22. YOUNGNER, J. S., STINEBRING, W. R., and TAUBE, S. E.: Influence of inhibitors of protein synthesis on interferon formation in mice. *Virology, 27*:541-550, 1965.

REPLICATION OF THE PARAINFLUENZA VIRUS SV5 AND THE EFFECTS OF SUPERINFECTION WITH POLIOVIRUS*

PURNELL W. CHOPPIN, RICHARD W. COMPANS,
AND KATHRYN V. HOLMES

SIMIAN VIRUS 5 (SV5), first isolated in 1956 (Hull *et al.*), is one of the most common viruses found in primary cultures of rhesus monkey kidney cells. It is a member of myxovirus subgroup II, which includes mumps, Newcastle disease and parainfluenza viruses. For several years, our laboratory has been engaged in a study of the properties of the virus particle, its replication in monkey kidney and hamster kidney cells, the role of virus-induced alterations of cell membranes in determining virus virulence and yield, and the response of the SV5-monkey kidney cell system to superinfection by other viruses. This chapter summarizes the results of these studies which are described in detail elsewhere (Choppin, 1964a, b; 1965; Choppin and Stoeckenius, 1964; Choppin and Holmes, 1966a, b; Compans and Choppin, 1966; Compans *et al.*, 1966; Holmes and Choppin, 1966).

VIRUS STRUCTURE

SV5 is morphologically similar to the other members of myxovirus subgroup II (Choppin and Stoeckenius, 1964; Compans *et al.*, 1966). In preparations negatively stained with phosphotungstate, the particles are pleomorphic but in general appear spherical, varying from 120 to 460 mμ in their longest dimension. Long filaments were not seen in these preparations, and the lack of filaments has been considered a characteristic of myxovirus subgroup II (Waterson, 1962). However, despite the absence of filaments in negatively stained preparations, filaments over 1 μ in

*Supported by Research Grant AI 05600 from the National Institute of Allergy and Infectious Diseases, and by Grant GM-577 from the Institute of General Medical Sciences, U.S.P.H.S.

16

length were frequently seen in thin sections of SV5-infected cells (Compans *et al.*, 1966). These filaments were apparently either lost or disrupted during purification, concentration, or negative staining. In addition, some of the large pleomorphic forms filled with nucleocapsid, which are commonly seen, might result from distortion of filaments. The abundance of free nucleocapsid in negatively stained preparations, a characteristic of this subgroup of myxoviruses, could be derived in part from disrupted filaments. These results suggest that the apparent rarity of filamentous forms of subgroup II myxoviruses may be the result of preparative procedures rather than a true characteristic of the viruses.

The SV5 virions possess an envelope, derived from the cell membrane, that is covered with projections approximately 100 Å in length which show knoblike thickening of the distal ends. The internal component or nucleocapsid is a single-stranded helix with a diameter of 150-180 Å, composed of ellipsoidal subunits with a long axis of 55-70 Å and a short axis of $\sim$25 Å. Visualization of extended segments of the helical SV5 nucleocapsid enabled the clear demonstration that it is a single-helical structure (Choppin and Stoeckenius, 1964). The dimensions and organization of the SV5 nucleocapsid indicate that it is markedly similar in structure to tobacco mosaic virus (TMV), although the SV5 nucleocapsid is flexible and longer than the TMV particle. Because of the similarities in structure, and the finding of pieces of SV5 nucleocapsid $\geqslant$ 0.9 μ in length, compared to a TMV length of 0.3 μ which contains RNA with a molecular weight of 2 x 10^6, it was suggested (Choppin and Stoeckenius, 1964) that SV5 and the other subgroup II myxoviruses might contain RNA with a molecular weight several times 2 x 10^6. Subsequently, Duesberg and Robinson (1965) were successful in isolating RNA from Newcastle disease virus with a molecular weight of 7.5 x 10^6.

Recently (Compans and Choppin, 1967), it has been possible to obtain from SV5-infected cells viral nucleocapsid in high yield, to purify it by CsCl density gradient centrifugation, and to begin a study of its physical and chemical properties. Preliminary data suggest a unit length of the nucleocapsid of approximately 1.02 μ, a density in CsCl of 1.297, an RNA content of 4.1%, and the absence of DNA. Further studies on the nucleocapsid and its RNA and protein are in progress.

REPLICATION IN MONKEY KIDNEY CELLS

Early studies (Choppin, 1964) indicated that SV5 multiplies to high titer in primary cultures of rhesus monkey kidney cells with minimal cytopathic effects. However, in spite of the lack of cell destruction, the virus forms plaques in these cells with high efficiency, the plaques consisting of intact cells with diminished neutral red staining.

After inoculation of monkey kidney cells with SV5 at a multiplicity of 40 PFU/cell, the latent period is approximately 6 hours, and the doubling time during the exponential increase in infective virus is approximately 50 minutes. In 24 hours, yields of up to 1500 PFU/cell are obtained, and the cells can continue to produce large amounts of virus for as long as 30 days if lactalbumin hydrolysate medium with 2% calf serum is used and changed daily. Under these conditions, the monolayers remain intact and cytopathic effects are minimal. The changes that do occur usually appear after 3 or 4 days and consist of slight granularity of the cytoplasm, some polykaryocyte formation, and scattered coalescence of rounded cells. After several days small eosinophilic inclusions may appear. If the medium is not changed daily these changes become more extensive.

Infection of monkey kidney cells with SV5 does not significantly inhibit cellular DNA, RNA, or protein synthesis within 48 hours (Holmes and Choppin, 1966), and infected cells can divide. Infected MK cells have been carried through as many as 11 serial passages. The yield of infective SV5 decreased with passage; however, immunofluorescence studies indicated that essentially all cells in the cultures at each passage contained viral antigen.

In summary, SV5-infected primary rhesus monkey kidney cells can survive and produce infective virus for many days. Cellular macromolecular synthesis is not inhibited, and infected cells can divide. SV5 infection of MK cells is thus an example of moderate virus-cell interaction (Dulbecco, 1965).

The replication of SV5 is not inhibited by 5-fluoro-2'-deoxyuridine, 5-bromo-2'-deoxyuridine or 5-iodo-2'-deoxyuridine. Actinomycin D does not inhibit viral growth for 10 hours after addition of the drug, however continued production of SV5 after this time is inhibited by concentrations as low at 0.1 μg/ml (Choppin, 1965). These results suggest that neither DNA synthesis nor DNA-dependent RNA synthesis is required for the synthesis of the virus, and that the inhibition of the continued production of the virus by actinomycin is secondary to effects of the drug on the host cell.

Since actinomycin D does not directly inhibit the replication of SV5 RNA, it was possible to investigate SV5-induced RNA synthesis in actinomycin treated cells (Choppin and Holmes, 1966a b; Holmes and Choppin, 1966) as has been done in many laboratories with a number of other RNA viruses. In MK cells treated with actinomycin D, 5 μg/ml, and given 30-minute pulses of uridine-H³ at intervals, virus-induced RNA synthesis was first detected at 3 hours, reached a maximum at 10 hours, and then declined, but was still continuing at 24 hours. (The term "virus-induced" RNA synthesis is used rather than "viral RNA," since it has not been established how much of this RNA is incorporated into virus particles. As will be described below, however, most of the *nucleocapsid* formed in MK cells does appear to be incorporated into virus.)

The rate of virus-induced RNA synthesis was compared with that of total cellular RNA under similar labeling conditions, but in the absence of actinomycin. At its maximum rate at 10 hours, virus-induced RNA synthesis was < 1% of the total cellular synthesis, e.g., with a 30 minute pulse of uridine-H³, 260 CPM were incorporated into virus-induced RNA and 37,464 into total cellular RNA. Thus, although a high yield of infective virus is produced by MK cells, virus-induced RNA synthesis is only a small fraction of the total cellular synthesis. This is perhaps one of the reasons why these cells can continue to produce virus for long periods, divide, and, as described below, be infected with and produce a normal yield of a second virus.

Experiments were also done in actinomycin treated cells to which uridine-H³ was added at the time of infection and was present throughout, and the cumulative curve depicting virus-induced RNA synthesis compared with that of infective virus. In these experiments, SV5-induced RNA synthesis precedes the production of infective virus by about two hours.

Electron microscopic studies (Compans *et al.,* 1966) of SV5 infection of MK cells indicated that virus particles adsorb to the cell surface and appear to penetrate by a process resembling phagocytosis. The first detectable step in virus morphogenesis is the appearance in the cytoplasmic matrix of the helical nucleocapsid, which then aligns under regions of cell membrane that acquire viral surface projections. Assembly and release of virus particles occurs by a budding process involving incorporation into the viral envelope of a well defined unit membrane which is continuous with, and morphologically identical to, that of the host cell. Both spherical and filamentous virus particles are formed. The filaments frequently contain regularly coiled nucleocapsid throughout their length. Except for the budding process at the cell surface, MK cells show little evidence of infection. In spite of the large amount of infective virus being produced, there is no accumulation of nucleocapsid during the first 24 hours when virus production is at its maximum; only scattered pieces are seen in the cytoplasm. Thus, the rate of formation of nucleocapsid in MK cells appears to be well balanced with its continuous release within mature virus particles at the cell surface.

REPLICATION IN BABY HAMSTER KIDNEY CELLS

The replication of SV5 has been studied in several lines of hamster kidney cells, most extensively in BHK21-F cells (Holmes and Choppin, 1966), a heteroploid variant of the BHK21 line of Macpherson and Stoker (1962). In BHK21-F cells inoculated at a multiplicity of 16 pfu/cell, the latent period is 6-7 hours, similar to that in MK cells, and during the early exponential increase period the doubling time is also similar, though perhaps slightly longer. However, the late events in infection of the two cells differ markedly. In BHK21-F cells, virus production ceases by 12-15 hours, and the total yield of infective virus is only 7-10 pfu/cell. In contrast to the minimal cytopathic effects in MK cells, SV5-infected BHK21-F cells undergo massive cell fusion which begins at about 6 hours and increases with time until, by about 12 hours, nearly all nuclei are in multinucleate cells. By 14-18 hours, the entire monolayer has become a giant syncytium. By 24-30 hours after inoculation, the syncytium usually disintegrates, peeling off the underlying surface of the Petri dish in large pieces. Eosinophilic inclusions are prominent in the cytoplasm at about 12 hours, and become larger and more numerous with time.

Although the yield of infective virus from BHK21-F cells is very low, immunofluorescence studies indicate that, as with MK cells, viral antigen is detectable in the perinuclear areas at about 3 hours, and at 7 hours is found in the cytoplasm of nearly every cell. By 24 hours the cytoplasm appears to be filled with large fluorescent foci and diffuse fluorescence. In addition, electron microscopic studies (Compans *et al.*, 1966) showed that extremely large amounts of viral nucleocapsid accumulate in the cytoplasm of these cells. The eosinophilic cytoplasmic inclusions correspond to these large collections of nucleocapsid. Thus, although the BHK21-F cell appears to synthesize a great deal of nucleocapsid, there appears to be a block in its elaboration within mature virus particles at the cell membrane. The steps in the morphogenesis of SV5 in BHK21-F cells as seen in the electron microscope are in general similar to those in MK cells, except that in BHK21-F cells aberrant budding is frequently found. This consists of markedly pleomorphic particles and budding at the tips of long stalks of cytoplasm. By 22 hours after infection, virus budding at the surface of the BHK21-F cells has ceased. It should be noted that by this time membrane alterations resulting in extensive cell fusion have already occurred.

The incorporation of thymidine-H[3], uridine-H[3] and leucine-H[3] into DNA, RNA, and protein in SV5-infected BHK21-F cells was investigated in experiments using 30-minute pulses of the labeled precursors (Holmes and Choppin, 1966). Cellular biosynthesis was not significantly inhibited until 12-15 hours after infection, but after this time there was a rapid decrease in the synthesis of all three macromolecular species. However, as described above, by 12 hours almost all nuclei are in giant cells, and by 14-18 hours the monolayer consists of a giant syncytium which subsequently degenerates. Thus, SV5 infection of BHK21-F cells does not shut off cellular macromolecular synthesis soon after infection, but does so only after extensive cell fusion has occurred.

Although SV5 infection of BHK21-F cells results in massive cell fusion followed by cell death, a very few scattered cells may survive, and such cells have been harvested and carried through up to 28 serial passages. In each passage, there was extensive giant cell formation which increased with the time the cultures were held, but some cells always survived. Immunofluorescence studies of the passages showed that > 99% of the cells contained viral antigen. The yield decreased after the first passage, but virus production at a low level continued throughout the 28 passages of the cells. These results indicate that SV5-infected BHK21-F cells can divide, and that infection *per se* does not shut off vital cellular processes or invariably result in cell death, although disintegration following fusion is the usual result.

Fusion of BHK21-F cells can be induced rapidly by concentrated virus. After inoculation of 2000-6000 pfu/cell, fusion is detected within 20 minutes, and by 1 hour most of the nuclei are in large polykaryocytes. Such rapid fusion can be produced by UV-inactivated virus, and in the presence of actinomycin D or puromycin, indicating that neither infective virus nor cellular RNA or protein synthesis are required to induce the alterations in the cell membranes which lead to cell fusion (Holmes and Choppin, 1966).

The fusion of BHK21-F cells has been studied with time-lapse photomicrography (Holmes and Choppin, 1966) and several other interesting phenomena have been observed which will be mentioned here only briefly. Giant cells formed by fusion are capable of division; cells containing as many as 16 nuclei have been observed to undergo aberrant mitosis. Frequently, such division results in the formation of extremely large nuclei with many nucleoli, presumably through pooling of material from many of the nuclei in the original giant cell. The nuclei in the syncytia migrate through relatively long distances of cytoplasm, and frequently arrange themselves rather regularly in long lines.

RESUMÉ OF THE EFFECTS OF SV5 ON MONKEY KIDNEY AND BABY HAMSTER KIDNEY CELLS

The high virus yield, minimal cytopathic effects, and long survival of SV5-infected MK cells is in marked contrast to the low yield, extensive fusion, and disintegration of SV5-infected BHK21-F cells. There are, however, similarities in the replication of the virus in the two cell types. Some of the effects of the virus in these cells are summarized in Table I. The available data suggest that the ability of the virus to alter cell membranes may explain the differences in the interactions of SV5 with the two cell types. That the membranes of MK and BHK21-F cells respond differently to the virus is demonstrated clearly, not only by fusion of BHK21-F cells during virus replication, but also by rapid fusion of these cells, in contrast to MK cells, after inoculation of concentrated virus. That death of BHK21-F cells is secondary to alterations of cell membranes is suggested by the findings that infected cells can divide, and that cellular DNA, RNA, and protein synthesis is not inhibited until after there has been massive cell fusion. Disintegration occurs through tearing and detachment of the large syncytium. Since virus-induced RNA synthesis in BHK21-F as well as in MK cells is $< 1\%$ of total cellular RNA synthesis, cell death would not be expected on the basis of diversion of precursors to virus synthesis or preempting the protein synthesizing system of the cell. All of the available evidence is compatible with the hypothesis that whether SV5 behaves as a moderate or virulent virus depends on the response of the cell membrane to the virus.

TABLE I

COMPARISON OF EFFECTS OF SV5 ON PRIMARY RHESUS MONKEY KIDNEY
AND BABY HAMSTER KIDNEY (BHK21-F) CELLS

	Cells	
	Monkey Kidney	*BHK21-F*
Latent period	$\sim$6 hours	$\sim$6 hours
Virus-induced RNA synthesis	$<1\%$ of cell RNA	$<1\%$ of cell RNA
Virus yield	1500 PFU/cell/24 hours	7-10 PFU/cell
Cytoplasmic accumulation of nucleocapsid	Minimal	Extensive
Aberrant budding of virus	No	Yes
Fusion of infected cells	$\pm$	Extensive
Rapid fusion by concentrated virus	No	Extensive
Division of infected cells	Yes	Yes
Shut off of cellular biosynthesis	No	After extensive fusion
Cell death	No	After extensive fusion

The reactions of the cell membranes may also play a significant role in determining the large differences in the amount of infective virus produced by the two cells. A membrane injured as a result of the virus-induced alterations that lead to the fusion of BHK21-F cells might be expected to be less efficient in the elaboration of mature virus through the budding process; furthermore as cells fuse, the total surface area, and therefore the potential area available for budding, decreases. On the other hand, monkey kidney cells, which show little morphological change, continue to produce high yield of virus for days. In addition, two other lines of hamster kidney cells (BHK21/13 and HKCC) which fuse less than BHK21-F cells, produce more virus (Holmes and Choppin, 1966) ; there is thus an inverse relationship between the degree of fusion and virus yield. The electron microscopic observations that budding from the surface of BHK21-F giant cells has ceased by 22 hours, and that there is an extensive accumulation of nucleocapsid within these cells also support the concept that there may be a block in the elaboration of virus at the cell membrane. On the other hand, in monkey kidney cells there appears to be a balance between formation of nucleocapsid and its continuous release within mature virus particles.

LACK OF VIRAL INTERFERENCE AND THE EFFECTS OF SUPERINFECTION OF SV5-INFECTED CELLS

Early studies (Choppin, 1964a) indicated that SV5 infection of MK cells did not significantly interfere with superinfection by a variety of challenge viruses, which included vesicular stomatitis, vaccinia, influenza A2, polio type 2, ECHO 7, and Coxsackie B4. This lack of interference in SV5-infected MK cells was demonstrated in two ways. First, the ability of the various viruses to form plaques on monolayers which were previously infected with SV5 was investigated, and second, the yields of the superinfecting viruses were determined. In both types of experiments, MK cells were employed which were inoculated with SV5 at a multiplicity of 40-70 pfu/cell, and had been producing large amounts of this virus for 2-3 days before inoculation with the challenge virus. In these experiments, the SV5-infected cells were as susceptible to infection with, and plaque formation by, each of the challenge viruses as were control cells, and there was little or no decrease in the yield of the superinfecting virus. In addition to these experiments, MK cells were inoculated with SV5 at multiplicities ranging from 0.0001 to 100 pfu/cell, and the supernatant medium, obtained from such cells at daily intervals for 5 days, was tested for interfering activity against vesicular stomatitis virus plaque formation on fresh monolayers of MK cells; no interference was detected (Choppin, 1964b).

The above results indicate that SV5 infection of rhesus monkey kidney cells under these conditions does not significantly interfere with the replication of a number of superinfecting viruses by interferon production or by any other means.

The ability to superinfect MK cells that were already producing SV5 made it possible to study the effect of the superinfecting virus on the continued replication of SV5. It is known that poliovirus rapidly turns off cellular DNA-dependent RNA synthesis and protein synthesis (Darnell, 1962; Fenwick, 1963; Zimmerman *et al.*, 1963; Holland and Peterson, 1964; Bablanian *et al.*, 1965). It was therefore of interest to determine whether the inoculation of poliovirus at high multiplicities would turn off ongoing SV5 replication, specifically, SV5 RNA-directed RNA and protein synthesis (Choppin and Holmes, 1966a, b).

MK monolayers were inoculated with SV5 at a multiplicity of approximately 40 pfu/cell. Essentially every cell was infected under these conditions, as demonstrated by the appearance in each cell of viral antigen detected by immunofluorescence, by confluent hemadsorption, and by plating of the cells as infectious centers. After 24 hours the SV5-infected cells were inoculated with poliovirus type 1, Mahoney strain, at a multiplicity of 325 pfu/cell. Immunofluorescence studies indicated that essentially all cells were infected with poliovirus, and a typical poliovirus single cycle growth curve was obtained. Under these conditions, there was no inhibition in SV5 production until after 6 hours, when most of the poliovirus had already been made, and the cells were beginning to show the cytopathic effects of this virus. To rule out the possibility that failure of poliovirus to shut off SV5 replication was due to the presence of a large backlog of SV5 components within the cell at the time of inoculation of poliovirus, puromycin and 6-azauridine were added to SV5-infected cells under similar conditions. Inhibition of ongoing SV5 replication was detected within one hour, and replication ceased by about 2 hours with puromycin and by 4 hours with 6-azauridine. Thus, there does not appear to be a large supply of viral components in the MK cell, and continued protein and RNA synthesis is necessary for continued production of SV5.

The above results suggest that poliovirus does not shut off SV5 RNA-directed RNA and protein synthesis. To investigate further this question, the effect of the inoculation of poliovirus on the incorporation of uridine-H^3 into SV5-induced RNA was investigated in actinomycin-treated cells. MK cells were infected with SV5 as above, and 4 hours prior to inoculation with poliovirus, actinomycin D was added at a concentration of 5 μg/ml. The cells were then inoculated with poliovirus, 325 pfu/cell, in the presence of guanidine, 1.1 mM. This system was designed to enable a study of the effects of poliovirus on SV5-induced RNA synthesis in the absence of cellular RNA synthesis, which is inhibited by actinomycin, and of poliovirus RNA synthesis, which is inhibited by guanidine. It has been previously shown in HeLa and HEL cells (Bablanian, *et al.,* 1965; Penman and Summers, 1965) that guanidine does not prevent the shut off of cellular biosynthesis by poliovirus, although it may delay it slightly. To determine if this was the case in the system employed in the present studies, the effect of poliovirus and guanidine on MK cell RNA and protein synthesis was investigated. In pulse-labeling experiments, inhibition of both RNA and protein synthesis was detectable at 2 hours; 50% inhibition of protein synthesis was found at $3\frac{1}{2}$ hours and of RNA synthesis at 5 hours; and by 12 hours protein and RNA synthesis were reduced to 25 and 30% of the controls, respectively. Inhibition was also demonstrated in experiments in which labeled precursors were added at the time of inoculation with poliovirus, and their cumulative incorporation into RNA and protein determined. Thus, inhibition of MK cell RNA and protein synthesis by poliovirus does occur in the presence of guanidine, although the rate of inhibition is slightly slower than it is in HeLa cells.

When virus-induced RNA synthesis in actinomycin-treated SV5-infected cells was investigated after poliovirus inoculation in the presence of guanidine, it was found that not only was there no inhibition of ongoing virus-induced RNA synthesis, but that there was significant stimulation. In an experiment in which uridine-H³ was added at the time of poliovirus inoculation, 3480 CPM were incorporated into RNA in SV5-infected cells at 12 hours, and 7900 into such cells superinfected with poliovirus in the presence of guanidine. In pulse labeling experiments, the rate of virus-induced RNA synthesis in the poliovirus superinfected cells compared to that of control SV5-infected cells, varied from 113% at 2 hours, to 156% at 10 hours. That this increase in actinomycin-resistant RNA synthesis was not due simply to failure of guanidine to inhibit poliovirus RNA synthesis was shown by control experiments done simultaneously in cells not previously infected with SV5; no detectable poliovirus RNA synthesis occurred in these cells. The remaining possibilities include: 1) that the increase represents a stimulation in SV5 RNA synthesis, and 2) that it represents utilization of SV5 RNA polymerase by poliovirus RNA. The available evidence suggests that the first possibility is more likely. There is a slight increase in the yield of infective SV5 virus for the first 5 hours after the addition of poliovirus and guanidine, and there is no evidence of replication of infective poliovirus. The precise explanation of the observed increase in actinomycin-resistant virus-induced RNA synthesis in SV5-infected cells after inoculation with poliovirus and guanidine must await further experiments which are in progress, including the isolation and characterization of the RNA.

The above studies indicate that although poliovirus inhibits cellular DNA-dependent RNA synthesis, superinfection with poliovirus does not shut off ongoing SV5 RNA-dependent RNA synthesis. Similarly, SV5 protein synthesis is not inhibited by poliovirus under conditions in which cellular protein synthesis *is* inhibited. Willems and Penman (1966) have recently reported that the inhibition of host cell protein synthesis involves an inactivation of host cell messenger RNA, which apparently becomes incapable of reassociating with ribosomes. From the present results, it is clear that whatever the precise mechanism is by which poliovirus inhibits cellular protein synthesis, this mechanism can distinguish between protein synthesis directed by cellular messenger RNA and that directed by SV5-viral messenger. These results thus provide an example of a distinction between cellular and viral messenger RNA. In a sense, the SV5-infected cell super-infected with poliovirus is the complement of the virus-interferon systems described in detail elsewhere in this volume; whereas interferon treatment of cells results in inhibition of viral protein synthesis while cellular protein synthesis continues, in the system described above, viral protein synthesis is permitted but cellular synthesis is inhibited.

SUMMARY

SV5-infected primary rhesus monkey kidney (MK) cells produce large amounts of virus continuously for many days with little evidence of cell damage. Cellular macromolecular synthesis is not inhibited and infected cells can divide. In contrast, in baby hamster kidney (BHK21-F) cells SV5 infection is cytocidal, but inhibition of cellular macromolecular synthesis and cell death occur only after extensive cell fusion; little infective virus is produced although the cytoplasm contains much viral antigen. In both cell types, the rate of SV5-induced RNA synthesis, at its maximum, is $< 1\%$ of total cellular RNA synthesis. Electron microscopic studies indicate that in MK cells there is a balance between synthesis of nucleocapsid and its continuous release within mature virus particles. In BHK21-F cells, in which membrane alterations have resulted in cell fusion, large aggregates of nucleocapsid accumulate in the cytoplasm, suggesting a block in virus maturation at the cell surface. These findings suggest that the virulence and yield of SV5 may depend on the response of the cell membrane to the virus.

SV5 infection does not interfere with superinfection by a number of challenge viruses. In cells producing SV5, superinfection with poliovirus at high multiplicity does not inhibit ongoing SV5 replication until 6 hours, when the cells show cytopathic effects of poliovirus. In actinomycin-treated SV5-infected cells, superinfection with poliovirus in the presence of guanidine leads to a stimulation of virus-induced RNA synthesis and a slight enhancement of the production of infective SV5. These results indicate that poliovirus infection does not shut off SV5 RNA-directed RNA synthesis or protein synthesis under conditions in which cellular DNA-dependent RNA synthesis and cellular protein synthesis are inhibited.

ACKNOWLEDGMENTS
The authors wish to thank Miss Cathleen O'Connell and Miss Linda Ferrer for excellent technical assistance.

REFERENCES

1. BABLANIAN, R., EGGERS, H. J., and TAMM, I.: Studies on the mechanism of poliovirus-induced cell damage. I. The relation between poliovirus-induced metabolic and morphological alterations in cultured cells. *Virology, 26:*100-113, 1965.
2. CHOPPIN, P. W.: Multiplication of a myxovirus (SV5) with minimal cytopathic effects and without interference. *Virology, 23:*224-233, 1964a.
3. CHOPPIN, P. W.: Unpublished experiments, 1964b.
4. CHOPPIN, P. W.: Effect of actinomycin D and halogenated deoxyuridines on the replication of simian virus 5. *Proc. Soc. Exper. Biol. & Med., 120:*699-702, 1965.
5. CHOPPIN, P. W., and HOLMES, K. V.: Replication of simian virus 5 and effect of superinfection with poliovirus. *Bacteriol. Proc.,* 128, 1966a.
6. CHOPPIN, P. W., and HOLMES, K. V.: Replication of simian virus 5 RNA and the effects of superinfection with poliovirus. Manuscript in preparation, 1966b.
7. CHOPPIN, P. W., and STOECKENIUS, W.: The morphology of SV5 virus. *Virology, 23:*195-202, 1964.
8. COMPANS, R. W., and CHOPPIN, P. W.: Isolation and properties of the helical nucleocapsid of the parainfluenza virus SV5. *Proc. Natl. Acad. Sci. U.S., 57:* 949-956, 1967.
9. COMPANS, R. W., HOLMES, K. V., DALES, S., and CHOPPIN, P. W.: An electron microscopic study of moderate and virulent virus-cell interactions of the parainfluenza virus SV5. *Virology, 30:*411-426, 1966.
10. DARNELL, J. E.: Early events in poliovirus infection. *Cold Spring Harbor Symp. Quant. Biol., 27:*149-158, 1962.
11. DUESBERG, P. H., and ROBINSON, W. S.: Isolation of the nucleic acid of Newcastle disease virus (NDV). *Proc. Natl. Acad. Sci., 54:*794-800, 1965.
12. DULBECCO, R.: Characteristics of virus-cell complexes. *Am. J. Med., 38:*669-677, 1965.
13. FENWICK, M. L.: The influence of poliovirus infection on RNA synthesis in mammalian cells. *Virology, 19:*241-249, 1963.
14. HOLLAND, J. J., and PETERSON, J. A.: Nucleic acid and protein synthesis during poliovirus infection of human cells. *J. Mol. Biol., 8:*556-573, 1964.
15. HOLMES, K. V., and CHOPPIN, P. W.: On the role of the response of the cell membrane in determining virus virulence. Contrasting effects of the parainfluenza virus SV5 in two cell types. *J. Exper. Med., 124:*501-520, 1966.
16. HULL, R. N., MINNER, J. R., and SMITH, J. W.: New viral agents recovered from tissue cultures of monkey kidney cells. I. Origin and properties of cyto-

pathogenic agents SV1, SV2, SV4, SV5, SV6, SV11, SV12 and SV15. *Am. J. Hyg., 63:*204-215, 1956.

17. MACPHERSON, I., and STOKER, M.: Polyoma transformation of hamster cell clones — an investigation of genetic factors affecting cell competence. *Virology, 16:*147-151, 1962.

18. PENMAN, S., and SUMMERS, D.: Effects on host cell metabolism following synchronous infection with poliovirus. *Virology, 27:*614-620, 1965.

19. WATERSON, A. P.: Two kinds of myxovirus. *Nature, 193:*1163-1164, 1962.

20. WILLEMS, M., and PENMAN, S.: The mechanism of host cell protein synthesis inhibition by poliovirus. *Virology, 30:*355-367, 1966.

ENHANCEMENT OF ANIMAL VIRUS GROWTH BY ANOTHER VIRUS*

KARI CANTELL

WHEN ANIMAL CELLS ARE exposed to two different viruses at one time, there are a number of possible results. The yield of each virus may be unaffected, reduced or increased by the double infection, and virus particles with genotypic and/or phenotypic properties of the two parent viruses may be found in various proportions among the progeny. Interference is the oldest object of intensive study among these multiplex phenomena. This may partly be due to technical reasons; when only rough methods were available, it was easier to detect interference than hybridization or enhancement. The interest in interference phenomena may also reflect the general attitude that we are concerned to suppress virus infections and not to favor them.

Most of the chapters in this book are concerned with inhibition of virus multiplication. The present chapter deals with enhancement of virus growth. Various situations in which one virus favors the growth of another will be surveyed. The terms "helper virus" and "helper system" will be used in a broad sense for all such situations, regardless of the mechanisms involved. Special consideration will be given to the evidence that the helper virus may act by suppressing the production and/or action of interferon (IF).

HELPER SYSTEMS AMONG ANIMAL VIRUSES

Table I is a compilation of the helper systems known among animal viruses. In most cases, the helper virus and the virus which is helped show considerable similarities to each other. The helper systems are divided into three groups to make this point clear. In the first group, the similarity between the two partners involved in the helper system is obvious. All the members of the second

*This work was supported in part by the Sigrid Jusélius Foundation and the Finnish National Research Council for Medical Sciences.

group are ether-resistant DNA viruses of cubic symmetry. Most members of the last group are ether-sensitive RNA viruses. The last two helper systems do not fit into the general pattern, since in these cases the two viruses are quite different. There is no doubt that more helper systems involving quite different pairs of viruses (e.g., RNA — DNA, lipid — nonlipid) will be found when a more intensive search is made. However, generally speaking, the motto of the table may be true and new helper systems are perhaps more likely to be found among similar than among dissimilar viruses.

TABLE I
HELPER SYSTEMS AMONG ANIMAL VIRUSES

"Similia similibus curentur"

Pox (vaccinia, rabbitpox, ectromelia, myxoma etc.)	—	Pox	see 1-3
Herpes simplex	—	Herpes simplex	4,5
Entero (polio, echo, Coxsackie)	—	Entero	6-11
SV_{40}, adeno-SV_{40} hybrid	—	Adeno	12-16
Adeno	—	Adeno-SV_{40} hybrid	17,18
SV_{40}, adeno, polyoma	—	K	19,20
SV_{15}, adeno	—	AAV	21-24
RAV	—	RSV	25-27
Mouse leukemia	—	MSV	28,99
LCM	—	Rabies	29,30
Mumps, Sendai	—	Sindbis	31-33
Mumps, Sendai, influenza B	—	VSV	34-37
Influenza	—	Influenza	see 38
NDV	—	NDV	39
Parainfluenza 3, hog cholera, virus diarrhea	—	NDV	40-46
Parainfluenza 3	—	Polio RNA	41
Parainfluenza 3	—	Pseudorabies	41

The mechanisms operating in many of the helper systems are not yet known. However, several different points of helper virus action have been disclosed, as schematically presented in Figure 1. The helper effect of the lymphocytic choriomeningitis virus (LCMV) on the rabies virus is suspected to occur at the cell surface, which may become more "penetrable" by rabies virus particles in the presence of LCMV (30). The nongenetic reactivation of pox viruses is due to uncoating of the reactivable virus by the enzymes (s) induced by the reactivating virus (3). Under certain conditions, enteroviruses may fail to multiply because they can not synthetize the RNA polymerase needed for replication of viral RNA, but they can be rescued by other enteroviruses producing the essential enzyme (s) (8-11). The helper effect of SV_{40} on adenovirus takes place after uncoating of the adenovirus but the precise mechanism is not clear (15, 16). There seem to be some examples of the rescue of animal viruses by complementation (4, 5, 47, 98) and there are many examples of a partial rescue of animal viruses by genetic recombination (2, 38, 48). In the RAV — RSV system, the helper virus supplies the defective Rous sarcoma virus with structural proteins or perhaps with the whole viral coat (25-27). Many other mechanisms by which the helper virus exerts its action are undoubtedly still awaiting discovery. The possibility of interplay between different mechanisms must also be kept in mind.

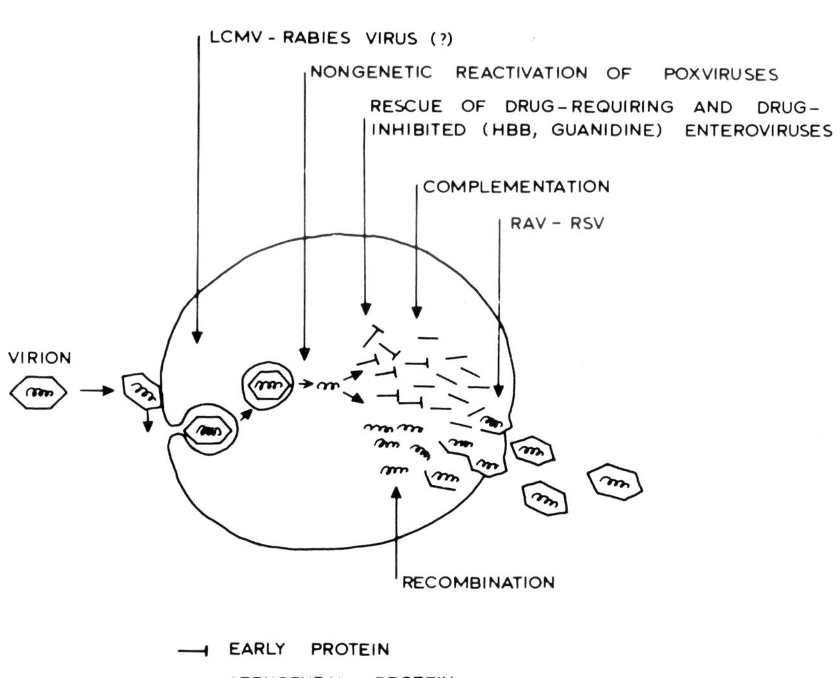

Figure 1.

INHIBITION OF INTERFERON AND ENHANCEMENT OF VIRUS GROWTH

The infectivity of a virus particle is relative and depends on the system of assay (49). When a virus fails to multiply in a cell, the virus may be called "noninfectious" or the cell "insusceptible." On the other hand, the same situation can also be considered as a success of cellular defense. Interferon (IF) is thought to represent a cellular defense mechanism against viruses (50). Accordingly, it is reasonable to expect that various agents possessing anti-IF activity will be capable of enhancing viral replication under suitable conditions. A collection of agents which are known to be able to inhibit the production and/or action of IF (51, 52), and which have also been found to enhance virus infections in cell cultures, is given in Table II. It is pointed out that many different agents showing anti-IF activity *can* enhance virus growth. There is indirect evidence that in some cases, at least, the enhancement is due to the anti-IF action. Increase of Sindbis virus plaque size has been induced with several different carcinogenic hydrocarbons (75-77) and steroid hormones (71) which inhibit the synthesis of IF. Chemically related substances which do not affect the produc-

TABLE II

ENHANCEMENT OF VIRUS GROWTH BY AGENTS CAPABLE OF INHIBITING
THE PRODUCTION AND/OR ACTION OF INTERFERON (IF)

Anti-IF agent	Enhancement *Virus*	*Criterion*	*References*
Actinomycin D	Several arbo	Number of infected cells, latent period, yield	53-58
	Mumps, Sendai, NDV	Latent period, yield	58-62
	Measles rubella	Latent period, yield	63-65
	Polio, reo	Yield	66,67
	Bittner, herpes-like	Electron microscopic particles	68,69
Mitomycin C	Measles	Yield, plaque size	70
Puromycin	Mumps, Sendai	Yield	61,62
Steroids	Sindbis	Plaque size	71
	Influenza B	Yield	72
	Polio	Number of infected cells	73
	Vaccinia	Number of infected cells, yield	74
Carcinogenic hydrocarbons	Sindbis, vaccinia	Plaque size	75-77
Triethylene-melamine	Semliki Forest	Plaque size	78
UV-irradiation	Sindbis	Plaque size	79
	Rubella	Yield, plaque size	65

tion of IF have not been found to increase virus plaque size (71, 75-77).

Observations concerning the anti-IF activity of viruses are presented in Table III. It can be seen that different viruses may inhibit the production of IF, the antiviral action of IF or both. It must be emphasized that the anti-IF effect may be manifested only by the given virus-cell combination and under appropriate conditions. The viruses on the list may be good inducers of IF after inactivation, in other cells, etc. On the other hand, the cells of the list may produce large amounts of IF when stimulated with other inducers or under other conditions.

TABLE III
INHIBITION OF PRODUCTION AND/OR ACTION OF INTERFERON (IF) BY VIRUSES

Virus	Cell	Inhibition	References
Influenza A	Chick chorioallantois		80
Fowl plaque, NDV	Chick fibroblast		81
Sendai	HeLa	Production of IF	33
VSV	Krebs-2		82
Friend	Mouse		83
Mouse cytomegalo	Mouse spleen		84
Sendai	Chick fibroblast, HeLa		36,33
Mumps	Chick fibroblast, human amnion	Action of IF	32
Tick-borne encephalitis	Chick fibroblast, L		85-87
Parainfluenza 3	Calf kidney	Production and action of IF	40,41
Bovine virus Diarrhea	Calf kidney		46

The mechanisms involved in the anti-IF action of viruses are poorly understood. In some cases, anti-IF activity appears to be associated with infectious virus particles (40, 80, 86), whereas in other systems inhibition of IF can also be caused by inactivated virus (82). The inhibition is not necessarily caused by the virus particles themselves. Indeed, recent findings indicate that it can be mediated by substances, presumably proteins, which are produced in virus-infected cells but which can be separated from the virus particles (Table IV). More information will surely soon become available about the nature and interrelationship of these factors. The interpretation of experimental data is complicated by the fact that virus preparations may contain IF, substances enhancing (95) or blocking its production and substances inhibiting its action (see Table). It is sufficient here to point out that all the substances exhibiting anti-IF activity have already been shown to be capable of enhancing virus multiplication (see Table).

TABLE IV

INHIBITION OF INTERFERON (IF) AND ENHANCEMENT OF VIRUS INFECTION
BY "NONVIRAL" FACTORS FROM VIRUS-INFECTED CELLS

Factor	*Source*	*Inhibition*	*Activity Enhancement*	*References*
Blocker	Chick embryos infected with some myxoviruses	Production of IF	Plaque number of Chikungunya virus	88
Enhancer	— " —	Action of IF	Yield of some myxoviruses	89-92
Stimulator	Rat embryo cells infected with adeno, polyoma, SV_{40}	Action of IF	Plaque number and yield of K virus	19,20
Anti-interferon	Chick cells infected with NDV or cowpox virus, HeLa cells infected with poliovirus	Action of IF	Yield of NDV	93,94

The virus-cell systems in which the inhibition of IF has been described are again listed in Table V. In addition, Table V shows that enhancement of the multiplication of another virus has been observed in many of these systems. It seems probable that, at least in some instances, the enhancement is actually due to inhibition of IF. First, the autointerference and persistent infection of NDV in calf kidney cells appear to depend upon IF. The autointerference can be suppressed and the persistent infection activated by parainfluenza 3 virus, which inhibits both the production and action of IF in these cells (40, 41, 96). Second, the plaque number of vesicular stomatitis virus (VSV) in chick cells is increased by infection with Sendai virus, which inhibits the action of IF. As chick embryos develop, their cells become increasingly resistant to VSV and more sensitive to IF and, at the same time, the helper activity of Sendai virus is gradually strengthened (36). No enhancement has been observed in some of the virus-cell systems in which the inhibition of IF has been described (Table V). This is not very surprising, since such enhancement has not been carefully sought either.

TABLE V

ENHANCEMENT OF VIRUS GROWTH BY ANOTHER VIRUS INHIBITING
THE PRODUCTION AND/OR ACTION OF INTERFERONS (IF)

Anti-IF virus	Cell	Enhancement		
		Virus	Criterion	
Influenza A	Chick chorioallantois	—		80
Fowl plaque, NDV	Chick fibroblast	—		81
Sendai	— " —	VSV	Plaque number	35,36
Sendai	HeLa	NDV	Plaque number and size, yield	33
Sendai	— " —	Sindbis	Yield	33
Mumps	Chick fibroblast	Sindbis	Plaque size	32
Parainfluenza 3	Calf kidney	NDV	Yield	40
Parainfluenza 3	— " —	Pseudorabies	Yield	41
Parainfluenza 3	— " —	Polio RNA	Yield	41
Bovine virus diarrhea	— " —	NDV	Yield	41
VSV	Krebs-2	—		82
Tick-borne encephalitis	Chick fibroblast	—		87
Tick-borne encephalitis	L	—		85,86
Friend	Mouse	—		83
Mouse cytomegalo	Mouse spleen	NDV	Yield	84

In order to demonstrate enhancement, it may be necessary to study number of infected cells, length of latent period, size of plaques, rate of production and yield of infectious virus in experiments with several different viruses. It is probable that new helper systems will be found when virus multiplication is analyzed in this fashion in virus-cell systems exhibiting anti-IF effect. Such virus-cell combinations are likely to be found among virus infections which do not lead to detectable IF formation.

Most of the known helper systems which seem to be based on inhibition of IF have been found among myxo- and arboviruses (Table I). These viruses are considered to be the best inducers of IF (97), which implies that IF has become an effective cellular defense mechanism against these viruses. The role of IF in many helper systems among other groups of viruses has not yet been adequately studied. IF is not virus-specific. Therefore, a virus paralyzing the function of IF might be thought to assist the multiplication of related and unrelated viruses equally well. Yet, there might be some degree of virus-specificity in the anti-IF action of viruses, based, for example, on the relationship between intracellular topography and virus multiplication; the action of IF could be inhibited only in the proximity of the helper virus.

CONCLUDING REMARKS

Double virus infections of man may be not infrequent because of the common occurrence of subclinical and latent infections. Enhancement of virus growth is one of the many possible outcomes of double infection. Many of the helper systems listed above are unlikely to be encountered in nature. However, the mechanisms which have been demonstrated in the laboratory may well operate in virus infections of man.

A defective virus which can *only* grow with the aid of another virus could survive by intimately associating with its helper or by making use of a host which is persistently infected with the helper virus. A looser partnership with the helper virus appears to suffice for the propagation of the adeno-associated viruses. Their large number in relation to adenovirus may enable them to follow adenoviruses during laboratory passages, but it is less likely that they could do so during transmission of adenovirus from man to man. The chances of a defective virus finding a helper are probably better in the laboratory than in nature. Defective viruses are therefore more likely to be found among old laboratory strains than among fresh isolates.

REFERENCES

1. FENNER, F.: Studies on the reactivation of pox-viruses. *In: Perspectives in Virology, Vol. 3,* Edited by M. Pollard. New York, Harper & Row, 1963, pp. 68-86.
2. FENNER, F., and SAMBROOK, J. F.: The genetics of animal viruses. *Ann. Rev. Microbiol., 18:*47-94, 1964.
3. JOKLIK, W. K.: The molecular basis of the viral eclipse phase. *Progr. Med. Virol., 7:*44-96, 1965.
4. ROIZMAN, B., and AURELIAN, L.: Abortive infection of canine cells by herpes simplex virus. I. Characterization of viral progeny from co-operative infection with mutants differing in capacity to multiply in canine cells. *J. Molec. Biol., 11:*528-538, 1965.
5. ROIZMAN, B.: Abortive infection of canine cells by herpes simplex virus III. The interference of conditional lethal virus with an extended host range mutant. *Virology, 27:*113-117, 1965.
6. DRAKE, J. W.: Interference and multiplicity reactivation in polio viruses. *Virology, 6:*244-264, 1958.
7. LEDINKO, N., and HIRST, G. K.: Mixed infection of HeLa cells with polioviruses types 1 and 2. *Virology, 14:*207-219, 1961.
8. CORDS, C. E., and HOLLAND, J. J.: Replication of poliovirus RNA induced by heterologous virus. *Proc. Nat. Acad. Sci. (Wash.), 51:*1080-1082, 1964.
9. WECKER, E., and LEDERHILGER, G.: Curtailment of the latent period by double-infection with polio-viruses. *Proc. Nat. Acad. Sci. (Wash.), 52:*246-251, 1964.
10. IKEGAMI, N., EGGERS, H. J., and TAMM, I.: Rescue of drug-requiring and drug-inhibited entero-viruses. *Proc. Nat. Acad. Sci. (Wash.), 52:*1419-1426, 1964.
11. AGOL, V. I., and SHIRMAN, G. A.: Interaction of guanidine-sensitive and guanidine-dependent variants of poliovirus in mixedly infected cells. *Biochem. Biophys. Res. Commun., 17:*28-33, 1964.
12. RABSON, A. S., O'CONOR, G. T., BEREZESKY, I. K., and PAUL, F. J.: Enhancement of adenovirus growth in African green monkey kidney cell cultures by SV$_{40}$. *Proc. Soc. Exp. Biol. (N.Y.), 116:*187-190, 1964.
13. BEARDMORE, W. B., HAVLICK, M. J., SERAFINI, A., and MCLEAN, I. W., JR.: Interrelationship of adenovirus (type 4) and papovavirus (SV-40) in monkey kidney cell cultures. *J. Immunol., 95:*422-435, 1965.
14. ROWE, W. P.: Studies of adenovirus - SV$_{40}$ hybrid viruses, III. Transfer of SV$_{40}$ gene between adenovirus types. *Proc. Nat. Acad. Sci. (Wash.), 54:*711-717, 1965.
15. REICH, P. R., BAUM, S. G., ROSE, J. A., ROWE, W. P., and WEISSMAN, S. M.: Nucleic acid homology studies of adenovirus type 7 - SV$_{40}$ interaction. *Proc. Nat. Acad. Sci. (Wash.), 55:*336-341, 1966.
16. FELDMAN, L. A., BUTEL, J. S., and RAPP, F.: Interaction of a papovavirus and adenoviruses I. Induction of adenovirus tumor antigen during abortive infection of simian cells. *J. Bact., 91:*813-818, 1966.

17. Butel, J. S., Melnick, J. L., and Rapp, F.: Detection of biologically active adenovirions unable to plaque in human cells. *J. Bact.*, *92*:433-438, 1966.

18. Rowe, W. P., and Baum, S. G.: Studies of adenovirus SV_{40} hybrid viruses II. Defectiveness of the hybrid particles. *J. Exp. Med.*, *122*:955-966, 1965.

19. Brailovsky, C., and Chany, C.: Un facteur produit par l'adénovirus 12 en culture cellulaire, stimulant la multiplication du virus K du Rat. *C. R. Acad. Sci. (Paris)*, *260*:2634-2637, 1965.

20. Chany, C., and Brailovsky, C.: Personal communication, 1965.

21. Atchison, R. W., Casto, B. C., and Hammon, W. Mc.D.: Adenovirus-associated defective virus particles. *Science, 149*:754-756, 1965.

22. Hoggan, M. D., Blacklow, N. R., and Rowe, W. P.: Studies of small DNA viruses found in various adenovirus preparations: physical, biological, and immunological characteristics. *Proc. Nat. Acad. Sci. (Wash.), 55*:1467-1474, 1966.

23. Archetti, I., Bereczky, E., and Bocciarelli, D. S.: A small virus associated with the simian adenovirus SV 11. *Virology, 29*:671-673, 1966.

24. Mayor, H. D., and Melnick, J. L.: Small deoxyribonucleic acid-containing viruses (Picodnavirus group). *Nature, 210*:331-332, 1966.

25. Hanafusa, H., Hanafusa, T., and Rubin, H.: The defectiveness of Rous sarcoma virus. *Proc. Nat. Acad. Sci. (Wash.), 49*:572-580, 1963.

26. Hanafusa, H.: Analysis of the defectiveness of Rous sarcoma virus III. Determining influence of a new helper virus on the host range and susceptibility to inteference of RSV. *Virology, 25*:248-255, 1965.

27. Vogt, P. K.: A heterogeneity of Rous sarcoma virus revealed by selectively resistant chick embryo cells. *Virology, 25*:237-247, 1965.

28. Hartley, J. W., and Rowe, W. P.: Production of altered foci in tissue culture by defective Moloney sarcoma virus particles. *Proc. Nat. Acad. Sci. (Wash.), 55*:780-786, 1966.

29. Koprowski, H., Wiktor, T. J., and Kaplan, M. M.: Enhancement of rabies virus infection by lymphocytic choriomeningitis virus. *Virology, 28*:754-756, 1966.

30. Wiktor, T. J., Kaplan, M. M., and Koprowski, H.: Rabies and lymphocytic choriomeningitis virus (LCMV) infection of tissue culture; enhacing effect of LCMV. *Ann. Med. Exp. Fenn., 44*:290-296, 1966.

31. Frothingham, T. E.: Enhancement of Sindbis virus plaque size in cell cultures treated with mumps virus-infected egg fluid. *Virology, 19*:583-586, 1963.

32. Frothingham, T. E.: Further observations on cell cultures infected concurrently with mumps and Sindbis viruses. *J. Immunol., 94*:521-529, 1965.

33. Maeno, K., Yoshii, S., Nagata, I., and Matsumoto, T.: Growth of Newcastle disease virus in a HVJ carrier culture of HeLa cells. *Virology, 29*:255-263, 1966.

34. Henle, W.: Personal communication, 1965.

35. Valle, M., and Cantell, K.: The ability of Sendai virus to overcome cellular resistance to vesicular stomatitis virus, I. General characteristics of the system. *Ann. Med. Exp. Fenn., 43*:57-60, 1965.

36. Cantell, K., and Valle, M.: The ability of Sendai virus to overcome cellular resistance to vesicular stomatitis virus, II. The possible role of interferon. *Ann. Med. Exp. Fenn., 43*:61-64, 1965.

37. Cantell, K., and Valle, M.: Unpublished observations.

38. Kilbourne, E. D.: Influenza virus genetics. *Progr. Med. Virol., 5*:79-126, 1963.

39. GRANOFF, A.: Studies on mixed infection with Newcastle disease virus III. Activation of non-plaque-forming virus. *Virology, 14:*143-144, 1961.
40. HERMODSSON, S.: Inhibition of interferon by an infection with parainfluenza virus type 3 (PIV-3) . *Virology, 20:*333-343, 1963.
41. HERMODSSON, S.: Action of parainfluenza virus type 3 on synthesis of interferon and multiplication of heterologous viruses. *Acta Path. Microbiol. Scandinav., 62:*224-238, 1964.
42. KUMAGAI, T., SHIMIZU, T., and MATUMOTO, M.: Detection of hog cholera virus by its effect on Newcastle disease virus in swine tissue culture. *Science, 128:* 366, 1958.
43. KUMAGAI, T., SHIMIZU, T., IKEDA, S., and MATUMOTO, M.: A new *in vitro* method (END) for detection and measurement of hog cholera virus and its antibody by means of effect of HC virus on Newcastle disease virus in swine tissue culture I. Establishment of standard procedure. *J. Immunol., 87:*245-256, 1961.
44. KUMAGAI, T., SHIMIZU, T., IKEDA, S., and MATUMOTO, M.: A new *in vitro* method (END) for detection and measurement of hog cholera virus and its antibody by means of effect of HC virus on Newcastle disease virus in swine tissue culture IV. Repraisal of effect of resum in culture medium and time of challenge with ND virus. *Nat. Inst. Anim. Hlth. Quart., 4:*135-144, 1964.
45. INABA, Y., OMORI, T., and KUMAGAI, T.: Detection and measurement of non-cytopathogenic strains of virus diarrhea of cattle by the END method. *Arch. ges. Virusforsch., 13:*425-429, 1963.
46. DIDERHOLM, H., and DINTER, Z.: Interference between strains of bovine virus diarrhea virus and their capacity of suppressing interferon of a heterologous virus. *Proc. Soc. Exp. Biol. (N.Y.), 121:*976-980, 1966.
47. COOPER, P. D.: Rescue of one phenotype in mixed infections with heat-defective mutants of type 1 poliovirus. *Virology, 25:*431-438, 1965.
48. MC CLAIN, M. E., and GREENLAND, R. M.: Recombination between rabbitpox virus mutants in permissive and nonpermissive cells. *Virology, 25:*516-522, 1965.
49. ISAACS, A.: Particle counts and infectivity titrations for animal viruses. *Advanc. Virus Res., 4:*111-158, 1957.
50. BARON, S.: The biological significance of the interferon system. *In: Interferons.* Edited by N. B. FINTER. Amsterdam, North-Holland Publ. Co., 1966, pp. 268-293.
51. BURKE, D. C.: The production of interferons by viruses: mechanisms. *In: Interferons.* Edited by N. B. FINTER. Amsterdam, North-Holland Publ. Co., 1966, pp. 55-86.
52. SONNABEND, J. A., and FRIEDMAN, R. M.: Mechanism of interferon action. *In: Interferons.* Edited by N. B. FINTER. Amsterdam, North-Holland Publ. Co., 1966, pp. 219-221.
53. HELLER, E.: Enhancement of Chikungunya virus replication and inhibition of interferon production by actinomycin D. *Virology, 21:*652-656, 1963.
54. WAGNER, R. R.: Interferon control of viral infection. *Trans. Ass. Amer. Phycns., 76:*92-101, 1963.
55. ZHDANOV, V. M., GAIDAMOVICH, S. YA., and VAGZHANOVA, V. A.: Acceleration of reproduction of Venezuelan equine encephalitis virus by actinomycin D. *Acta Virol., 8:*378-379, 1964.

56. CARVER, D. H., and MARCUS, P. I.: Enhanced production of interferon with aging of chick embryo cells in vitro. *Bact. Proc.*, 124, 1966.

57. STOLLAR, V., STEVENS, T. M., and SCHLESINGER, R. W.: Effect of actinomycin D and 6-azauridine on replication of type 2 dengue virus. *Bact. Proc.*, 113, 1966.

58. FRIEDMAN, R. M.: Role of interferon in viral interference. *Nature, 201*:848-849, 1964.

59. WHEELOCK, E. F.: Intracellular site of Newcastle disease virus nucleic acid synthesis. *Proc. Soc. Exp. Biol. (N.Y.), 114*:56-60, 1963.

60. BUKRINSKAYA, A. G., and ZHDANOV, V. M.: Shortening by actinomycin D of latent period of multiplication of Sendai virus. *Nature, 200*:920-921, 1963.

61. WHITE, D. O., and CHEYNE, I. M.: Stimulation of Sendai virus multiplication by puromycin and actinomycin D. *Nature, 208*:813-814, 1965.

62. NORTHROP, R. L., and WALKER, D. L.: A virus related body (VRB) of a persistent mumps virus infection. *Fed. Proc., 25*:312, 1966.

63. ANDERSON, C. D., and ATHERTON, J. G.: Effect of actinomycin D on measles virus growth and interferon production. *Nature, 203*:671, 1964.

64. MATUMOTO, M., ARITA, M., and ODA, M.: Enhancement of measles virus replication by actinomycin D. *Jap. J. Exp. Med., 35*:319-329, 1965.

65. VAHERI, A.: Personal communication, 1966.

66. COOPER, P. D.: The inhibition of poliovirus growth by actinomycin D and the prevention of the inhibition by pretreatment of the cells with serum or insulin. *Virology, 28*:663-678, 1966.

67. LOH, P. C., and SOERGEL, M.: Growth characteristics of reovirus type 2: actinomycin D and the preferential synthesis of viral RNA. *Proc. Soc. Exp. Biol. (N.Y.), 122*:1248-1250, 1966.

68. LYONS, M. J., LASFARGUES, E. Y., and CAME, P. E.: Effect of actinomycin D and halogenated deoxyuridines on the replication of Bittner virus *in vitro. Nature, 212*:100-101, 1966.

69. YOUNG, B. G., and ENGELS, B.: Effect of DNA and RNA inhibitors on replication of herpes-like virus in tissue cultured lymphoma cells. *Fed. Proc., 25*:477, 1966.

70. RAPP, F., and VANDERSLICE, D.: Enhancement of the replication of measles virus by mitomycin C. *J. Immunol., 95*:753-758, 1965.

71. REINICKE, V.: Increase of size of Sindbis virus plaques due to the effect of steroid hormones. *Acta Path. Microbiol. Scandinav., 65*:554-558, 1965.

72. KILBOURNE, A. D., and TATENO, I.: *In vitro* effects of cortisone on multiplication of influenza B virus. *Proc. Soc. Exp. Biol. (N.Y.), 82*:274-277, 1953.

73. HELLMAN, A., KALTER, S. S., and DAYE, G. T., JR.: The influence of hydrocortisone on cellular susceptibility to infection with virulent and avirulent poliovirus. *Acta Virol., 9*:224-229, 1965.

74. HOLDEN, M., and ADAMS, L. B.: The influence of hydrocortisone on vaccinia virus growth in L cells. *J. infect. Dis., 110*:268-277, 1962.

75. DE MAEYER, E., and DE MAEYER-GUIGNARD, J.: Inhibition by 20-methylcholanthrene of interferon production in rat cells. *Virology, 20*:536-539, 1963.

76. DE MAEYER, E., and DE MAEYER-GUIGNARD, J.: Inhibition by 3-methylcholanthrene of interferon formation in rat embryo cells infected with Sindbis virus. *J. Nat. Cancer Inst., 32*:1317-1331, 1964.

77. DE MAEYER-GUIGNARD, J., and DE MAEYER, E.: Effect of carcinogenic and non-

carcinogenic hydrocarbons on interferon synthesis and virus plaque development. *J. Nat. Cancer Inst., 34:*265-276, 1965.

78. DE MAEYER-GUIGNARD, J., and DE MAEYER, E.: Inhibition of interferon synthesis and stimulation of virus plaque development in mammalian cell cultures after uv-irradiation. *Nature, 205:*985-987, 1965.

79. DE MAEYER, E., and DE MAEYER-GUIGNARD, J.: Personal communication.

80. LINDEMANN, J.: Interferon und inverse Interferenz. *Z. Hyg. Infekt.-Kr., 146:* 287-309, 1960.

81. ISAACS, A.: Production and action of interferon. *Cold Spr. Harb. Symp. Quant. Biol., 27:*343-349, 1962.

82. WAGNER, R. R., and HUANG, A. S.: Inhibition of RNA and interferon synthesis in Krebs-2 cells infected with vesicular stomatitis virus. *Virology, 28:*1-10, 1966.

83. WHEELOCK, E. F.: The effects of nontumor viruses on virus-induced leukemia in mice: reciprocal interference between Sendai virus and Friend leukemia virus in DBA/2 mice. *Proc. Nat. Acad. Sci. (Wash.), 55:*774-780, 1966.

84. OSBORN, J. E., and MEDEARIS, D. N., JR.: Suppression of interferon and antibody and multiplication of Newcastle disease virus in cytomegalovirus infected mice. *Proc. Soc. Exp. Biol. (N.Y.),* In press.

85. VILČEK, J., and STANČEK, D.: Unresponsiveness to the action of interferon developed in persistently infected L cells. *Life Sci., 2:*895-901, 1963.

86. STANČEK, D.: The role of interferon in tick-borne encephalitis virus-infected L cells. IV. Origin of insusceptibility of persistently infected L cells to the action of exogenous interferon. *Acta Virol., 9:*409-415, 1965.

87. LIBÍKOVÁ, H.: Lowered susceptibility to exogenous interferon during the first stages of virus infection of cells. *Acta Virol., 9:*279-281, 1965.

88. ISAACS, A., ROTEM, Z., and FANTES, K. H.: An inhibitor of the production of interferon ("blocker"). *Virology, 29:*248-254, 1966.

89. KATO, N., OKADA, A., and OTA, F.: A factor capable of enhancing virus replication appearing in parainfluenza virus type 1 (HVJ) -infected allantoic fluid. *Virology, 26:*630-637, 1965.

90. KATO, N., OKADA, A., and OTA, F.: Production of a viral growth enhancing factor (enhancer) in eggs infected with influenza virus (PR8). *Arch ges. Virusforsch., 17:*631-640, 1965.

91. KATO, N., and OTA, F.: Effect of enhancer on the replication of Newcastle disease virus in chorioallantoic membranes. *Arch ges. Virusforsch., 18:*116-118, 1966.

92. KATO, N., OHTA, F., and OKADA, A.: Counteraction between interferon and enhancer. *Virology, 28:*785-788, 1966.

93. GHENDON, YU. Z.: On the ability of certain viruses to block the effect of interferon. *Acta Virol., 9:*186-187, 1965.

94. GHENDON, YU. Z., BALANDIN, I. G., and BABUSHKINA, L. M.: On the mechanism of the ability of certain viruses to block the action of interferon. *Acta Virol., 10:*268-270, 1966.

95. PAUCKER, K.: Personal communication, 1966.

96. HERMODSSON, S.: Rôle of interferon in the autointerference of Newcastle disease virus (NDV). *Acta Path. Microbiol. Scand., 62:*133-144, 1964.

97. HO, M.: The production of interferons. *In: Interferons.* Edited by N. B. FINTER. Amsterdam, North-Holland Publ. Co., 1966, pp. 21-54.

98. Burge, B. W., and Pfefferkorn, E. R.: Complementation between temperature-sensitive mutants of Sindbis virus. *Virology, 30:*214-223, 1966.
99. Huebner, R. J., Hartley, J. W., Rowe, W. P., Lane, W. T., and Capps, W. I.: Rescue of the defective genome of Moloney sarcoma virus from a noninfectious hamster tumor and the production of pseudotype sarcoma viruses with various murine leukemia viruses. *Proc. Nat. Acad. Sci. (Wash.), 56:*1164-1169, 1966.

VIRAL INTERFERENCE IN AGED CULTURES OF CHICK EMBRYO CELLS

ROYCE Z. LOCKART, JR.

INTRODUCTION

THIS CHAPTER IS CONCERNED, at least in part, with the cellular changes associated with viral interference. It is important to understand the nature of the changes that are able to transform what is usually a lethal infection into one with much milder consequences. It has long been known that newborns are more sensitive to the effects of a number of viruses than adults. Both mice and chickens show this relationship to Western equine encephalomyelitis (WEE) virus. A number of associates and I have worked with WEE virus for the last 10 years. All of our work has been in tissue cultures of mouse and chicken cells. On several occasions, titrations of Western equine encephalomyelitis (WEE) virus were not carried out until two or three days after the chick-embryo (CE) monolayers prepared for the purpose had been made. Frequently, the resulting plaques were much smaller and the resultant virus titers were lower than those usually obtained. Further investigations showed that aged CE cultures also produced significantly lower yields of WEE virus. Cultures which had been aged, and which made low titers of virus, frequently showed little or no cytopathic effects and instead could be maintained for prolonged periods as carrier cultures. Incubation with actinomycin D or the reduction of the cell concentration a day before infection permitted full yields of virus and complete cellular destruction. As our thinking has been oriented towards interferon for the past few years, and interferon has been repeatedly implicated in the maintenance of the virus carrier condition, we attempted to see what relationship might exist between culture age and interferon production. Data are presented which suggest that the inhibition of virus production by the aged cultures might be due to the earlier production of interferon.

MATERIALS AND METHODS

Cell Cultures. Monolayer cultures of cells were prepared from chicken embryos 11 days of age. The cells were prepared according to the method of Dulbecco and Vogt (1954) and grown in Eagle's medium containing 3% calf serum (EC medium).

Virus. Two strains of Western equine encephalomyelitis (WEE) virus were used. On chick-embryo (CE) monolayer cultures, they produced plaques with average sizes of about 2.5 mm for the small plaque variant and about 7.0 mm for the large-plaque variant. Virus suspensions for use as stocks were prepared on CE tissues and dispensed in 1-ml quantities which were kept frozen until used. Titers were between 1 and 3 x 10^9 plaque-forming unit (PFU) per ml. The procedures used for virus titrations have already been described (Lockart, 1963).

Interferon. Interferon was prepared in the same manner as virus except that the CE cultures were incubated two or three days prior to infection. The assay method for interferon has already been described (Lockart and Sreevalsan, 1963). A protective unit (PU) of interferon is defined as the reciprocal of the dilution of interferon which prevents the appearance of any cytopathic effects (CPE) when monolayers of CE cells containing about 2 x 10^6 cells are incubated overnight with 2 ml of interferon and then challenged with 10 or more PFU/cell of WEE virus.

Solutions. Phosphate buffered saline (Dulbecco and Vogt, 1954) to which was added 0.1% bovine serum albumin, fraction V (Mann Research Laboratories, New York, N.Y.) was used to wash the cell cultures and to dilute virus suspensions for titrations. It is referred to as PBSA. PBSA and medium were warmed to 37° before use. Actinomycin D was generously provided by Merck, Sharp and Dohme, West Point, Pennsylvania.

RESULTS

Onset of Suppression of Viral Reproduction. Monolayer cultures of CE cells were prepared in 100 mm petri dishes. Each day for 5 days, three cultures were dispersed with trypsin and the number of cells in each were determined. Two other cultures were inoculated by the addition of 1.6x10^8 PFU of WEE virus. Those

TABLE I
EFFECT OF CULTURE AGE ON VIRUS PRODUCTION AND CPE

Days of Incubation	Avg. No. Cells Per Monolayer[1] $(x10^7)$	Avg. Total[2] Yields PFU/culture	PFU/cell	CPE[3] (48 hrs p.i.)
1	1.3	$1.15x10^{10}$	821	+ + + +
2	1.6	$1.0x10^{10}$	625	+ + + +
3	1.6	$2.3x10^9$	144	+ +
4	1.8	$7.5x10^8$	42	±
5	2.3	$6.0x10^8$	26	—
1	1.9	$9.0x10^9$	473	+ + + +
2	2.1	$1.3x10^{10}$	619	+ + + +
3	2.2	$2.9x10^9$	132	+ + + +
4	2.0	$4.1x10^9$	205	+
5	2.4	$2.6x10^9$	108	±

1. Averaged from counts made on three separate cultures.
2. Average virus titers in the fluids pooled from two cultures 24 hours after their inoculation.
3. CPE: — = none; ± = very slight; + = slight but definite; + + + + = complete; p.i. = post-inoculation.

inoculated were incubated at 37° for 1 hour to permit virus adsorption to occur. The cultures were then washed three times with PBSA and 5 ml of EC medium were added to each. Fluids from the two infected cultures were collected and pooled 24 hr post-inoculation (p.i.). The samples were kept frozen until titrated. Five ml of fresh EC medium were added to each of the cultures and they were viewed microscopically for cytopathic effects (CPE). At daily intervals, the cultures were scanned and the amount of CPE estimated. The results of two such experiments are given in Table I. Little cell multiplication occurred during the 5 days of incubation. In the first experiment the number of cells increased 1.8 fold, while they increased only 1.4 fold in the second experiment. Virus yields were markedly reduced after the third day as was the degree of CPE which resulted. Those cultures showing little or no CPE could be kept as long as 10 days after infection without showing additional CPE. Minor variations were noted between the two experiments but in both, clear-cut reductions in viral yields and CPE were evident when infection of the cultures were delayed until 5 days after their preparation.

Comparison of One-step Growth Curves in Cultures after One and Eleven Days of Incubation. Twenty-two cultures prepared the previous day and a similar number prepared 11 days previously were washed and then infected by the addition of 1 ml of PBSA containing 6.7x10⁷ PFU of the large plaque-forming variant of WEE virus. They were incubated at 36° for 1 hr, washed two times with PBSA and each culture was given 5 ml of EC medium. Samples of medium were removed from two separate cultures at various times for 14 hr after the addition of virus and frozen. The

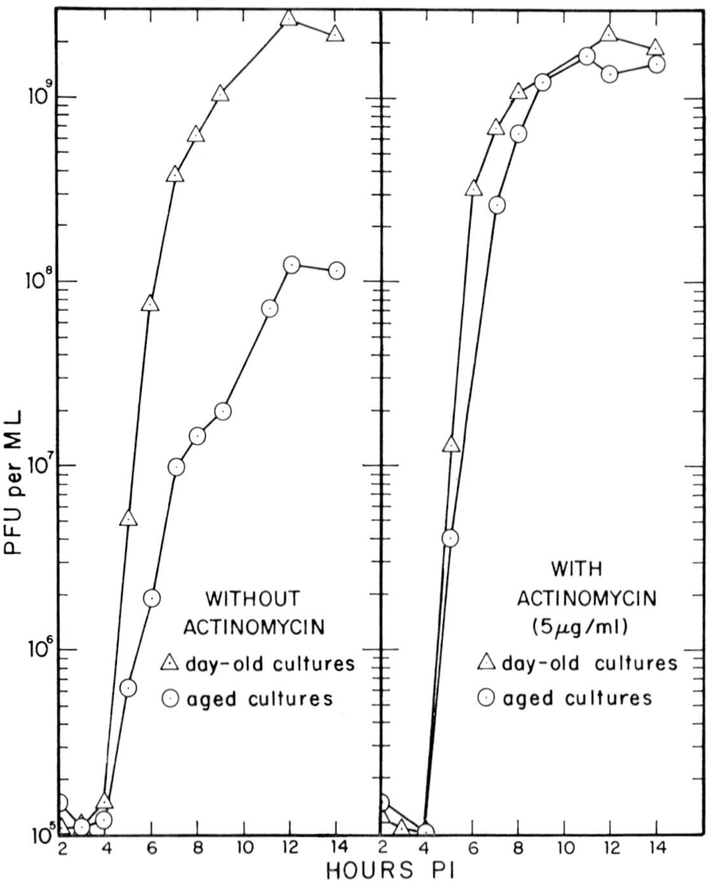

Figure 1. Comparison of one-step growth curves in cultures after one and eleven days of incubation. Cultures of CE tissues prepared one and eleven days previously were infected by the addition of >5 PFU per cell of WEE virus. The virus inoculum for one half of the cultures contained 5 μg/ml of actinomycin. Samples of medium were removed at various times and titrated for virus by the plaque technique. Triangles — one-day cultures; circles — eleven-day cultures.

virus titers of the samples were determined by the plaque technique. The results are plotted in Figure 1A. Virus production commenced between 4 and 5 hr p.i. in both sets of cultures and continued until 12 hours p.i. The aged cultures produced only 5% as much virus as the day-old cultures and it was produced at a much slower rate.

Reversal of the Reduced Virus Production of Aged Cultures by Actinomycin. This experiment was part of the same experiment described above. In this part of the experiment, the virus was diluted in PBSA containing actinomycin D at a concentration of 5 μg per ml. Cultures, one day and 11 days of age, received 1 ml of virus as before. Incubation, washing, and sampling were done at the same times as before. The results are plotted in Figure IB. There was no difference in the kinetics of virus production or the amount of virus produced by aged or day-old cultures. Furthermore, both curves are essentially identical with that obtained from the day-old cultures which had not received actinomycin as depicted in Figure 1A. The same type of experiment was repeated twice more. Similar results were obtained each time. In the absence of actinomycin, some variation was seen in the kinetics of virus production in the aged cultures. In a second experiment with the large plaque-forming variant, virus production did not commence until 6 hr p.i. But, as before, the highest titer was reached at 12 hr p.i. and represented a level only 3% of that of the day-old cultures. In the experiment with the small plaque-forming variant of WEE virus, production commenced at the same time as in the day-old cultures but stopped at 7 hr p.i. The amount of virus present at 12 hr p.i. was about 5% of that of the day-old cultures or aged cultures which had received actinomycin.

Lack of interferon in the medium. Medium was removed from CE cultures which had been incubated 8 days. It was dialyzed overnight against 50 volumes of Eagle's medium without serum. Day-old cultures of chick-embryo tissues were incubated overnight with the dialyzed medium and then checked for virus reproduction. The titers of virus produced were always as high as those from cultures of a similar age incubated with the original medium or fresh medium. It was concluded that the inhibition of virus production did not result from free interferon.

Inability to Reverse Interferon Action by Actinomycin D. The complete restoration of virus-producing ability of the aged cultures by a one-hour period of incubation with actinomycin strongly suggests that the inhibition was not a result of previous interferon action. Cultures of CE cells incubated overnight with interferon were only slightly affected in their virus-producing capacity by actinomycin. Data to illustrate this point are presented in Table II. Cultures of CE cells were incubated overnight with sufficient interferon to decrease virus production by approximately one and two logs. The cultures were washed and one-half of them were incubated for an hour with actinomycin (5 μg/ml). They were again washed and infected with WEE virus at a high multiplicity. Samples of fluid were removed from the various cultures 23 hours later and titrated.

TABLE II
NON-REVERSIBILITY OF INTERFERON ACTION BY ACTINOMYCIN

	Interferon *(PU*)*	*Virus Yields* *(PFU/ml at 23 hr p.i.)*		*Ratio*
		—Act.	*+Act*	*(+Act)* *(—Act)*
	None	7.5×10^8	8.4×10^8	1.12
Exp. 1	4	4.5×10^6	2.8×10^7	6.2
	1	1.6×10^8	2.8×10^8	1.8
	None	1.3×10^9	1.2×10^9	0.9
Exp. 2	4	6.0×10^6	8.4×10^6	1.4
	1	1.7×10^8	2.4×10^8	1.4

*PU $=$ protective units (1 PU $=$ approx. 6 PDD 50 units).

Formation of Plaques on Aged Cultures of CE Tissues by Vaccinia and Newcastle's Disease Viruses. Both Newcastle disease and vaccinia viruses produce plaques on CE tissues. Neither the number or the size of the plaques produced by these two viruses were affected by the age of the cultures.

The Possibility of Interferon Action. Two observations tend to exclude the possibility that the inhibition noted in aged cultures was a result of previous interferon action. They are the reversion to full susceptibility resulting from incubation with actinomycin and the lack of a lessened response by NDV and vaccinia viruses. It is possible, however, that interferon produced during the infection was responsible for the inhibition. If such were the case, one might expect the aged cultures, after infection, to produce interferon sooner, faster, in greater quantities or a combination of the above, or to respond to interferon more rapidly than non-aged cultures. A series of experiments were done to compare day-old and aged (8-11 day) cultures with respect to the following: 1) time at which interferon could first be detected in the medium; 2) time at which one PU of interferon was found in the medium, and 3) the final titer of interferon. In each experiment, aged and day-old cultures were infected by the addition of 2.5×10^7 PFU of the large plaque-forming variant of WEE virus. After adsorption, washing, and replacement of medium, the fluids from three of each of the day-old and aged cultures were removed at hourly intervals until 9 hr p.i. and at 24 hr p.i .The fluids were pooled into their respective types and later assayed on day-old CE cultures by the CPE protection test to determine how much interferon was present. The results of 6 such experiments are shown in Table III.

TABLE III

INTERFERON PRODUCTION BY AGED CULTURES

Exp. No.	1	2	3	4	5	6	Avg.
A. Time at which interferon was first detected in fluids (hr. p.i.)							
Aged cultures	3	4	4	5	3	4	3.8
day-old cultures	6	6	7	8	7	7	6.8
B. Time at which one PU of IF was found in fluids (hr. p.i.)							
Aged cultures	6	5	8	9	10	9	7.8
day-old cultures	9	7	9	9	9	9	8.7
C. Interferon titers at 24 hr p.i. (PU/ml)							
Aged cultures	128	128	16	32	16	16	56
day-old cultures	32	16	8	16	16	16	17

Interferon was found in the supernatant fluids of aged cultures earlier in all of the experiments. On the average, it was found at 3.8 p.i., which amounted to an average of 3 hr earlier than when it could first be detected in the day-old cultures. There did not appear to be a significant difference in the time at which one PU of interferon could first be found in the cultures, thus one cannot surmise that aged cultures made interferon at a much faster rate than the day-old cultures. Fluids from the aged cultures, however, usually contained more interferon at 24 hr p.i. The aged cultures produced more interferon in 4 of the 6 experiments and equal amounts in the other two.

Comparison of the Rates of Onset of Virus Inhibition Resulting from Interferon on Day-old and Aged CE Cultures. Monolayer cultures of chick embryo tissues which had been incubated for 5 days and cultures prepared the previous day were incubated for periods of time varying from 1 to 5 hours with 1 PU of interferon. At each hour, the cultures were washed and inoculated with about 10 PFU/cell of WEE virus. Fluids were collected

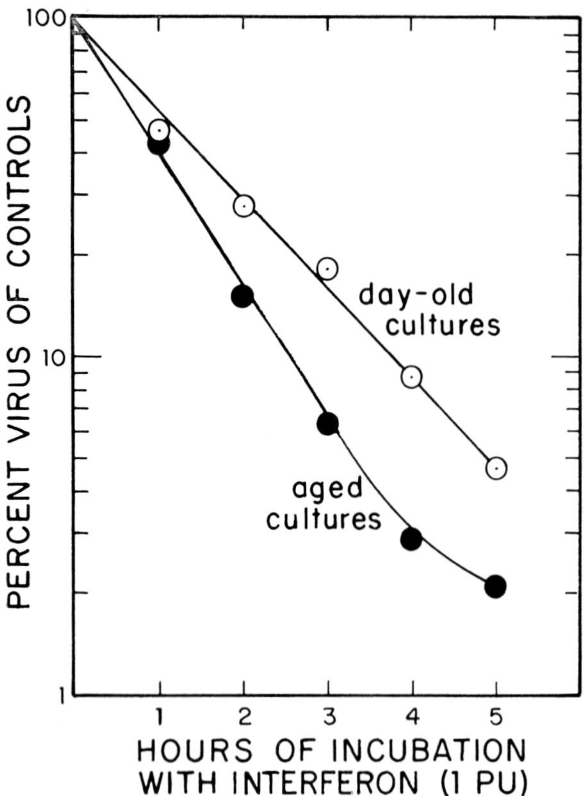

Figure 2. Comparison of the rates of onset of virus inhibition resulting from interferon on day-old and aged cultures of CE tissues. Cultures were incubated for periods between one and five hours with 1 PU of interferon. At each hour, the cultures were washed and inoculated 20 hours later and titers for virus. The above points which indicate the degree of inhibition of virus yields represent the over-age of 5 experiments.

20 hr p.i. and later titrated. The results from 4 experiments were averaged and are plotted in Figure 2. The rate at which the onset of the inhibition of virus synthesis was brought about by interferon was somewhat faster in the aged cultures.

Reversal of Aging Effect by Dilution of the Culture. The CE monolayers used were always started as confluent cell layers. It occurred to us that cell contact effects might play a role in the aging phenomenon. To investigate this possibility, monolayer cultures of CE cells were prepared in 100 mm petri dishes. Four days later, several of the cultures were dispersed by incubation for 5 minutes in 0.05% trypsin solution. The suspended cells were recultured at both the original concentration and at one tenth their formed concentration. In addition, cultures from fresh embryos were prepared. The following day, one of each of the cultures was again suspended by means of trypsinization and cell counts were made. The other cultures were inoculated with WEE virus. Virus titers of the fluids collected at 23 hr p.i. were determined. The results of this experiment are shown in Table IV. Another experiment in which only part of the above was done is shown also in Table IV. Cells which had been suspended, diluted and allowed to incubate overnight were able to produce virus yields as great as cells which had been removed from embryos the previous day. Only when the cells were diluted did they regain their complete virus-producing capacity, showing that exposure to trypsin was not responsible for the regained susceptibility.

TABLE IV

RESTORATION OF VIRUS-PRODUCING CAPACITY BY CELL DILUTION

		Cells		Virus Yields (24 Hrs. p.i.)	
Exp. No.	Age (Days)	Treatment	Cells/Culture	PFU/Culture	PFU/Cell
1	1	none	8.3×10^6	2.45×10^{10}	2960
	5	none	1.7×10^7	1.1×10^{10}	647
	5	trypsinized at 4 days	2.4×10^7	1.5×10^{10}	625
	5	trypsinized and diluted at 4 days	1.6×10^6	9.5×10^9	5900
2	5	none	9.7×10^6	8.0×10^9	824
	5	trypsinized and diluted at 4 days	2.4×10^6	6.5×10^9	2700

1.5×10^7 PFU added to infect the plates.

DISCUSSION

The nature of cell susceptibility or, *vice versa,* cellular resistance to infection by viruses is not totally understood. During the course of our routine work, we became aware that cultures of chick-embryo cells became increasingly resistant to infection with WEE virus as a result of prolonged incubation prior to infection. Such cultures displayed no increased resistance to vaccinia or Newcastle disease viruses, however. WEE virus production by cells from the aged cultures could be returned to levels usually obtained from cells in day-old cultures by the addition of actinomycin D at the time of infection or by the reduction of the cell population the day before infection. These results seem to rule out the possibility that the resistance was due to preformed interferon. Also, a number of attempts to find interferon in the medium from the aged cultures failed unless the medium was concentrated about 100-fold. While the mechanism of the resistance cannot be adduced with certainty, there is some evidence to indicate that interferon was responsible. The aged cultures of CE cells were found to produce interferon much earlier than day-old cultures and to respond to its action somewhat faster. These two factors together might account for the resistance seen. The rapid production of resistance in cells after infection might explain why some viruses are produced in higher titers in the presence of actinomycin (Heller, 1963). The lack of resistance to vaccinia and Newcastle disease viruses is also consistent with this interpretation since both of these viruses cause the production of little or no interferon in chick-embryo cells in tissue culture.

No reason can be given why the incubation of the CE cultures should heighten their ability to produce interferon and increase the rate at which they respond to it. Ho and Enders (1959) very early noted that CE cultures produced higher yields of interferon when they had been incubated two weeks prior to infection with a CE-adapted RMC strain of type 2 poliovirus. The ability to reverse the effect by dilution of the cells suggests the possibility that the biochemical consequences of cell contact might play a role. It will be recalled that the cultures initially contained quite high cell concentrations and that very little cell multiplication was observed. Results of essentially a similar nature have been obtained by Carver and Marcus (1966). The full manuscript of their findings was kindly made available to me prior to this presentation.

The results strongly point out the importance of standardized conditions for those working with interferons, especially when the highly sensitive arboviruses are used in the assays.

ACKNOWLEDGEMENTS

This work was done at the University of Texas. It was supported by Grants AI-03536 and 5-K3-AI-19,385 issued by The National Institutes of Health, Department of Health, Education, and Welfare. I wish to acknowledge the excellent technical assistance provided by Mrs. Barbara Horn.

REFERENCES

DULBECCO, R., and VOGT, M.: One step growth curve of Western equine encephalomyelitis virus on chicken embryo cells grown *in vitro* and analysis of virus yields from single cells. *J. Exper. Med., 99:*183-199, 1954.

LOCKART, R. Z., JR., and SREEVALSAN, T.: The effect of interferon on the synthesis of viral nucleic acid in *Viruses, Nucleic Acids and Cancer,* 17th Annual Symposium on Fundamental Cancer Research at the University of Texas M. D. Anderson Hospital and Tumor Institute, Houston, Texas, Baltimore, Williams and Wilkins, 1963, pp. 447-461.

LOCKART, R. Z., JR.: Production of an interferon by L cells infected with Western equine encephalomyelitis virus. *J. Bacteriol., 85:*556-566, 1963.

HELLER, E.: Enhancement of chikungunya virus replication and inhibition of interferon production by actinomydin D. *Virology, 21:*652-656, 1963.

CARVER, D. H., and MARCUS, P. I.: Enhanced production of interferon with aging of chick embryo cells *in vitro. Bact. Proc.,* 124, 1966.

EFFECT OF NUCLEOSIDE ANALOGUES ON CELLS INFECTED WITH PSEUDORABIES VIRUS*

ALBERT S. KAPLAN AND TAMAR BEN-PORAT

INTRODUCTION

IN THE PAST FEW YEARS, a certain measure of success has been achieved in the therapy of herpetic infections by local application of so-called antiviral agents. Striking results have been obtained in the control of herpes keratitis by the nucleosides BUDR,** IUDR and Ara-C. Whether these analogues inhibit the multiplication of cells and virus to the same degree, or whether they are selectively antiviral, has been a subject of controversy (Hanna and Wilkinson, 1965; Smith, 1963; Deinhardt, 1965; Kaufman, 1965).

Ideally, drugs effective in the control of viral diseases should be inhibitory to viral growth only. Since the DNA of the viruses analyzed to date do not possess unusual bases, analogues of the derivatives of DNA will, in principle, be as effective in interfering with cellular as they are with viral nucleic acid synthesis. However, infection of cells with the herpes viruses induces changes in the level of activity of a variety of enzymes concerned with the synthesis of DNA. Some of these enzymes may be structurally different from the enzymes performing the same function in the noninfected cells; this difference might permit the selective inhibition by certain agents of the synthetic processes of the infected

*This investigation was supported by grants from the U. S. Public Health Service (Al-02432 and Al-03362), and from the National Science Foundation (GB-1386), and by a U. S. Public Health Service Career Program Award (5-K3-AI 19, 335) from the National Institute of Allergy and Infectious Diseases.

**Abbreviations:* Pr, pseudorabies virus; RK, rabbit kidney; PFU, plaque-forming unit; Ara-C, arabinosyl cytosine; BUDR, 5-bromo-2'-deoxyuridine; IUDR, 5-iodo-2'-deoxyuridine; FUDR, 5-fluoro-2'-deoxyuridine; CMP, cytidine monophosphate; dCMP, deoxycytidine monophosphate; CDP, cytidine diphosphate; dCDP deoxycytidine diphosphate; TTP, thymidine triphosphate; TMP, thymidine monophosphate; DNase, deoxyribonuclease.

cells only. Although this possibility does exist, studies on differences in the response to inhibitory agents of the enzymes present in infected and noninfected cells have not been observed to the present time.

In addition to possible qualitative differences between the enzymes present in infected and noninfected cells, the quantitative differences in the level of activity of these enzymes may also provide an approach to the chemotherapy of viral diseases. The experiments presented in this paper show the importance of these quantitative differences in the response of infected and noninfected cells to analogues of nucleosides. These experiments are concerned mainly with the mode of antiviral action of IUDR and Ara-C in RK cell cultures infected with Pr virus. Some experiments relating to the search for new antiviral substances will also be discussed.

RESULTS AND DISCUSSION

Mode of Antiviral Action of IUDR

IUDR is a potent inhibitor of the growth of DNA-containing viruses (see review by Eggers and Tamm, 1966). Evidence has been obtained in various laboratories that it is incorporated into viral DNA and that in infected cells exposed to IUDR, viral DNA and viral antigens accumulate (Smith and Dukes, 1964; Kaplan *et al.*, 1965; Kaplan and Ben-Porat, 1966). However, despite the accumulation of the viral components within the infected cells their assembly into viral particles does not occur. The lack of assembly of viral particles in IUDR-treated cells could be due to the presence of the iodine atom in the viral DNA, thereby altering the structure of the DNA molecule which then would be unable to fold properly in order to fit into viral capsids. On the other hand, some protein normally involved in virus assembly may be functionally defective in IUDR-treated cells.

To differentiate between these two alternatives, we attempted to create conditions which would allow IUDR-containing viral DNA molecules to be coated, and thus to eliminate the first possibility. Cells infected with Pr virus were permitted to synthesize normal viral DNA either before or after the synthesis of IUDR-

DNA. This procedure should provide the infected cells with the proper information for the synthesis of the proteins normally involved in the process of virus assembly. Since viral DNA forms within the infected cell a precursor pool from which DNA is withdrawn at random to form mature virions (Ben-Porat and Kaplan, 1963), some of the IUDR-containing DNA should become encapsidated under these conditions, provided this DNA presents no structural impediment to the coating process. The following two experiments were performed:

1) Monolayer cultures of RK cells were incubated with either thymidine or IUDR for the first 4 hours after infection; an excess of IUDR was then added. The DNA was labelled with ^{14}C-8-adenine and the amount of label in total DNA and in DNase-insensitive DNA was determined.

2) In another experiment viral DNA containing IUDR was allowed to accumulate in the cells by incubating the Pr virus-infected cultures with ^{3}H-IUDR for the first 7 hours after infection. The labelled IUDR was then replaced with an excess of either unlabelled thymidine or unlabelled IUDR, and the amount of radioactive IUDR-containing DNA which became DNase-insensitive was tested.

In both cases, similar results were obtained. The presence of thymidine in the cultures either before or after IUDR-DNA had accumulated favored the accumulation of some DNase resistant IUDR-DNA (Kaplan and Ben-Porat, 1966).

In order to determine whether double-stranded IUDR-containing DNA can be enclosed in viral particles, the buoyant density of the DNase-insensitive DNA in density gradients of CsCl was analyzed. Cells were infected and incubated with IUDR for the first 8 hours of the infective process. The IUDR was removed and was replaced with an excess of thymidine. At the end of the virus growth cycle (16 hours), the cells and the virus were harvested and the buoyant density of the DNA present in these samples, before and after DNase treatment, was determined in gradients of CsCl in the analytical centrifuge. In both cases, three peaks of viral DNA appeared in the gradients, one with a density of 1.794 g/cm^3, representing IUDR-DNA, one with a density of 1.732 g/cm^3, representing normal viral DNA, and a peak with an intermediate density (1.763 g/cm³) , which corresponds to the density of a hybrid with one strand of IUDR-DNA and one of thymidine-DNA and which resulted from the semi-conservative replication of IUDR-DNA (Kaplan and Ben-Porat, 1964) . The presence in the DNase-treated sample of DNA with a buoyant density of 1.794 g/cm³ showed that DNA containing IUDR in both strands can indeed be coated.

The relative amounts of viral DNA containing thymidine, of viral DNA containing IUDR, and of hybrid DNA present in the sample can be determined by measuring the areas under each peak. Table I shows that there was approximately as much viral DNA containing IUDR in both strands as normal viral DNA in the total DNA extracted from the cultures. The DNase-insensitive DNA, however, contained a significantly greater proportion of IUDR-DNA than it did of thymidine-DNA. On the basis of random withdrawal of DNA from a precursor pool, in which IUDR-DNA has as much chance to be coated as thymidine-DNA, this result is to be expected, because the IUDR-DNA molecules were present throughout the period of virus maturation, whereas the thymidine-DNA started accumulating when maturation had begun.

TABLE I

EXTENT OF INCORPORATION OF IUDR-DNA INTO A DNAASE-INSENSITIVE FORM

Sample analyzed	Thymidine-DNA $p = 1.732\ g/cm^3$	Hybrid DNA $p = 1.763\ g/cm^3$	IUDR-DNA $p = 1.794\ g/cm^3$
Total DNA	90*	64	101
DNase-insensitive DNA	26	20	40
% of total DNA that is DNase-insensitive	29	31	40

*Relative units. From Kaplan and Ben-Porat (1966) .

These results show clearly that the presence of the iodine atom in viral DNA was no barrier to the coating process and that molecules containing IUDR can be assembled into a form in which they no longer are sensitive to the action of DNase, provided the infected cells are supplied with thymidine either before or after IUDR-DNA has accumulated in the infected cells. Thus, it seems that if thymidine is supplied to the cells at some time during the growth cycle the cells synthesize thymidine-containing viral DNA and the formation of functional proteins that are involved in virus assembly can then occur. As a result, some of the IUDR-DNA is encapsidated. If this interpretation is correct, it would appear that IUDR-DNA, although incapable of correctly coding for these proteins, can replicate if thymidine is supplied to give rise to DNA that can perform this function normally. In the next set of experiments, we tested whether IUDR-DNA can initiate the infective process and can replicate to give rise to normal virus, i.e., whether viral particles containing IUDR-DNA are infectious.

Formation of Infectious Viral Particles Containing IUDR-DNA

RK cells in stationary phase were incubated for 16 hours with FUDR and then infected at a multiplicity of 1 PFU/cell. Unadsorbed virus was removed by repeated washings and the cultures were treated as described in Table II. Twelve hours after infection the amount of infectious virus produced by the cultures was determined. The yield of infectious virus from cultures incubated with IUDR throughout the infective process (sample C) was 0.37% that of the control cultures (sample A). In contrast, cultures incubated with thymidine for the first 4 hours after infection and thereafter incubated with IUDR (sample B), yielded approximately 9% as much infectious virus as the control cultures.

TABLE II

EFFECT OF INCUBATION OF INFECTED CELLS WITH THYMIDINE FOR FOUR HOURS AFTER INFECTION AND THE SUBSEQUENT ADDITION OF IUDR ON THE YIELD OF INFECTIOUS VIRUS

Sample	Nucleoside Present in Medium Between		Virus titer PFU/ml $(x10^{-5})$	Infectivity (%)
	0 and 4 Hr after Infection	*4 and 12 Hr after Infection*		
A	thymidine, 2 μg/ml	thymidine, 100 μg/ml	1250	100
B	thymidine, 2 μg/ml	IUDR, 100 μg/ml	112	9.0
C	IUDR, 10 μg/ml	IUDR, 100 μg/ml	4.7	0.37

From Kaplan and Ben-Porat (1966).

In order to test whether the infectious virus produced by sample B consisted, at least partially, of viral particles containing IUDR-DNA, its radiation sensitivity was determined, since the presence of IUDR or BUDR in the DNA of a variety of organisms is known to increase their sensitivity to radiation (Greer and Zamenhof, 1957; Kozinski and Szybalski, 1959; Djordjevic and Szybalski, 1960; Sauerbrier, 1961; Stahl *et al.*, 1961; Opara-Kubinska *et al.*, 1961; Howard-Flanders *et al.*, 1962; Kaplan *et al.*, 1962; Tanooka, 1964; Rupp and Prusoff, 1964). The survival curves of the virus present in samples A, B and C after exposure to gamma rays are shown in Figure 1. The virus present in samples B and C was about twice as sensitive to radiation as that present in sample A.

Since the virus in samples C and B exhibited the same degree of sensitivity to radiation, we concluded that the increased amount of infectious virus in sample B consisted of viral particles containing IUDR-DNA and that, therefore, IUDR-DNA can initiate the infective process and give rise to normal thymidine-containing DNA.

Is IUDR a Specific Antiviral Agent?

It is quite evident from the experiments described above that IUDR is an effective inhibitor of virus multiplication because it is incorporated into viral DNA. This DNA controls the formation of non-functional proteins normally involved in the assembly of viral particles. Since survival of the virus is dependent upon the orderly process of normal virus assembly, virus multiplication may be especially sensitive to the replacement of thymidine by IUDR in viral DNA. However, some cellular function essential for cell multiplication may be equally sensitive to this interchange of nucleosides in cellular DNA.

There have been contradictory reports on the question of selective antiviral action of IUDR, i.e., whether the multiplication of the virus is more affected than that of the host cells by the drug. A comparison of the effect of different concentrations of IUDR on the multiplication of virus and of cells is relatively difficult, since the cycle of virus multiplication is short compared to the generation time of the cells and a large number of virions is produced during each cycle. The two cannot therefore be tested under identical conditions. Furthermore, if the virus is allowed to undergo several cycles of growth, mutants resistant to IUDR may appear in the drug-treated cultures.

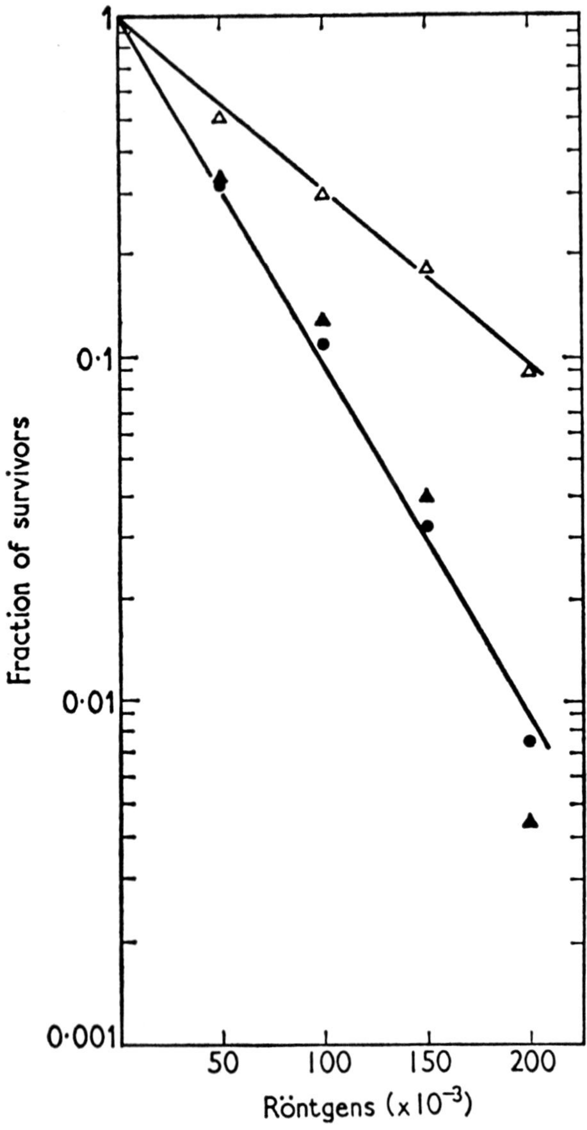

Figure 1. Sensitivity to ^{60}Co irradiation of Pr virus obtained from cultures treated with thymidine or IUDR. Cultures were treated as described in the text and in Table II. Sample A, open triangles; sample B, filled circles; sample C, filled triangles. From Kaplan and Ben-Porat (1966).

Infection of cells with viruses of the herpes group results in an increase in the level of activity of several enzymes, including thymidine kinase. The possibility that the thymidine kinase in infected cells may have a greater affinity for IUDR than the enzyme present in noninfected cells has been considered. This was found not to be the case (Prusoff *et al.*, 1965). However, even though differences in the affinity of the thymidine kinases of infected and noninfected cells for IUDR have not been observed, the difference in the level of activity of these enzymes in the cells may affect the susceptibility of these cells to the drug. The relatively high level of thymidine kinase in infected cells may elevate the level of phosphorylated IUDR which would then compete for incorporation into DNA with the thymidine derivatives synthesized by the cell. If this were indeed the case, a greater substitution of IUDR for thymidine would be obtained in the DNA of infected cells than of noninfected cells.

TABLE III

RELATIONSHIP BETWEEN CONCENTRATION AND SUBSTITUTION OF IUDR FOR THYMIDINE IN DNA

Concentration of IUDR (μg/ml)	IUDR Substituted in DNA		IUDR Substituted Per IUDR-containing Strand*	
	Infected	Noninfected	Infected	Noninfected
	(%)	(%)	(%)	(%)
0.5	62.7†	0	62.7	0
1	75.4	0	75.4	0
5	93.2	35.0	93.2	70.0
10	98.3	43.3	98.3	86.6
100	100	49.1	100	98.2

*Calculated on the basis that IUDR was substituted for thymidine in both strands of viral DNA but only in one strand of cellular DNA.
†Calculated from the buoyant density in CsCl. (W. D. Rupp, personal communication.)

The degree of substitution of IUDR for thymidine in the DNA of infected and noninfected cells treated with various concentrations of IUDR was determined from the buoyant density of the DNA in gradients of CsCl. Table III shows that a considerable degree of substitution of IUDR for thymidine occurred in viral DNA, even at a concentration of the drug in the growth medium of 1 μg/ml or less. In contrast, at this low concentration of IUDR there was in the noninfected cells little or no DNA with a buoyant density significantly different from that of normal cellular DNA. However, since the banding pattern of the DNA from the non-infected cells treated with low concentrations of IUDR appeared to have a slight skew on the side of the higher density, we considered the possibility that there had been in these samples some incorporation of IUDR into DNA and that the peaks of the DNA

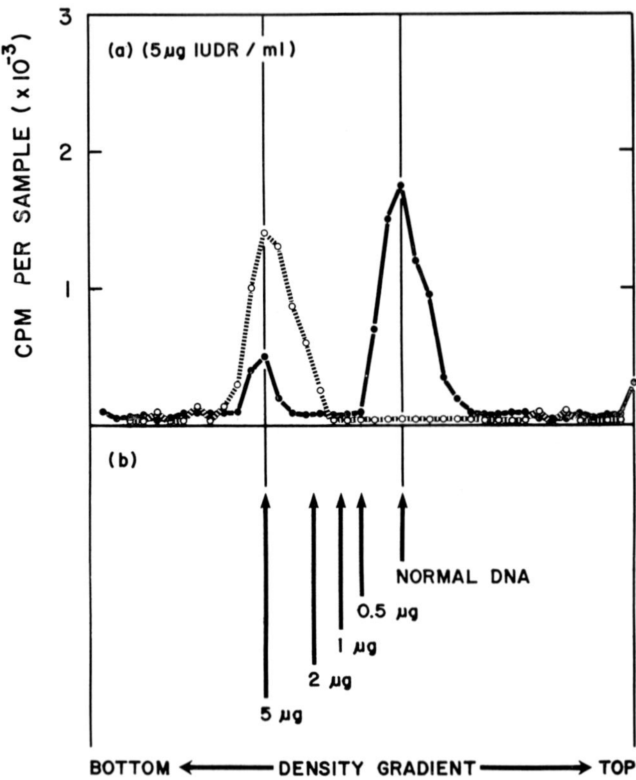

Figure 2. Distribution in CsCl gradients of DNA extracted from cells incubated with various concentrations of IUDR. RK cells were incubated with Eagle's medium with 0.04 μc/ml thymidine-2-[14]C (specific activity 0.12 mc/mg) for 3 hours. The cultures were washed and further incubated in Eagle's medium for 18 hours. Different amounts of IUDR, as well as 1 μc/ml adenine.[3]H (specific activity 3.5 mc/mg), were added to the cultures. Twelve hours thereafter the cultures were harvested, and the DNA was extracted and centrifuged to equilibrium in a CsCl solution (average density, 1.780 g/cm³). Samples were collected dropwise and the radioactivity in the DNA was determined. ^{14}C-labelled DNA, filled circles; ^{3}H-labelled DNA, open circles.

containing IUDR and that of normal, thymidine-containing DNA overlapped. In order to test this possibility and to distinguish between normal cellular DNA and the cellular DNA containing IUDR, we turned to the use of radioactive tracers, a more sensitive method than ultraviolet light absorption for this purpose.

The experiment described above was repeated. This time, however, normal cellular DNA was labelled with carbon-14 prior to the addition of IUDR, while the DNA synthesized during treatment of the cells with IUDR was labelled with tritium. The DNA extracted from the cells was centrifuged to equilibrium in a gradient of CsCl in a Spinco preparative centrifuge and the distribution in the gradient of radioactive DNA was determined. The results of this experiment are given in Figure 2 (a and b) and Table IV.

TABLE IV

REPLACEMENT OF THYMIDINE BY IUDR IN THE DNA
OF NONINFECTED CELLS DETERMINED FROM THE POSITION OF
RADIOACTIVE DNA IN DENSITY GRADIENTS OF CsCl

Concentration of IUDR (μg/ml)	Replacement (%)
0.5	21
1	31
2	45
5	70

See text for details.

Figure 2a gives the positions of the DNA extracted from cells which had been incubated with 5 μg of IUDR/ml. The position of the C^{14}-labelled DNA corresponds to the position of normal DNA containing thymidine. The peak of DNA containing both carbon-14 and tritium consists of DNA containing one strand of normal DNA and one strand of DNA which has been synthesized while IUDR was present in the cultures. From the experiment summarized in Table III, we know that noninfected cells incubated with 5 μg/ml IUDR synthesize DNA in which 70% of the thymidine is replaced by IUDR. Thus, we have two markers, one consisting of normal thymidine-containing DNA and one of DNA in which 70% of the thymidine is replaced by IUDR. From the position of other DNA preparations relative to these markers we can calculate the degree of substitution of IUDR for thymidine in these DNA preparations. Figure 2b indicates the positions in the gradient of DNA synthesized in the presence of various concentrations of IUDR. The approximate degree of substitution of thymine for IUDR in these preparations was calculated and is given in Table IV.

It is clear from these experiments (see Tables III and IV) that considerable differences exist between the degree of substitution in the DNA of infected and noninfected cells when the cells are supplied with low concentrations of IUDR (0.5 μg to 2 μg/ml). This difference disappears at higher concentrations of the drug.

If one wishes to extrapolate from the results of these experiments done in cell culture to what can be expected to occur in the animal, one might predict that it should be possible to secure effective antiviral action and relatively low toxicity by using throughout the treatment constant levels of low concentrations of IUDR.

Mode of Antiviral Action of Ara-C

Ara-C is as effective as IUDR in the control of herpes keratitis, although it is slow compared to IUDR in clearing the initial infection of herpes keratitis (Kaufman, 1965; Underwood *et al.*, 1965). Moreover, this analogue is considerably more toxic than IUDR to the host cells (Kaufman *et al.*, 1964). We studied the effect of Ara-C on infected and noninfected cells in order to ascertain whether results obtained in cell culture would enable us to predict the effects of the drug in animals.

The effect of various concentrations of Ara-C on the incorporation of ^{3}H-cytidine into the DNA of infected and noninfected RK cells was tested. It is clear from the results shown in Figure 3 that the incorporation of cytidine into DNA is less affected in infected cells than in noninfected cells by Ara-C. Essentially the same results were obtained when the incorporation of labelled adenine or thymidine was tested.

In order to determine the reason for the differential effect of Ara-C on the synthesis of DNA in infected and noninfected cells, it is necessary to understand the mode of action of Ara-C on DNA synthesis. Unfortunately, however, this is still unclear.

The inhibitory effect of Ara-C on DNA synthesis is reversed

competitively by deoxycytidine. Furthermore, Ara-C inhibits the formation of dCMP but not of CMP from uridine, an observation that led to the conclusion that the drug prevents the reduction of CDP to dCDP (Chu and Fischer, 1962). However, when the effect of Ara-C was tested on CDP reductase *in vitro,* no inhibitory effect of the drug on the activity of this enzyme could be demonstrated (see review by Cohen, 1966).

In order to determine whether Ara-C inhibits the formation of dCDP in RK cells, the effect of Ara-C on the incorporation into DNA of cytidine and deoxycytidine was compared. If Ara-C inhibits the activity of CDP reductase, the size of the intracellular pool of dCDP which normally dilutes the radioactive deoxycyti-

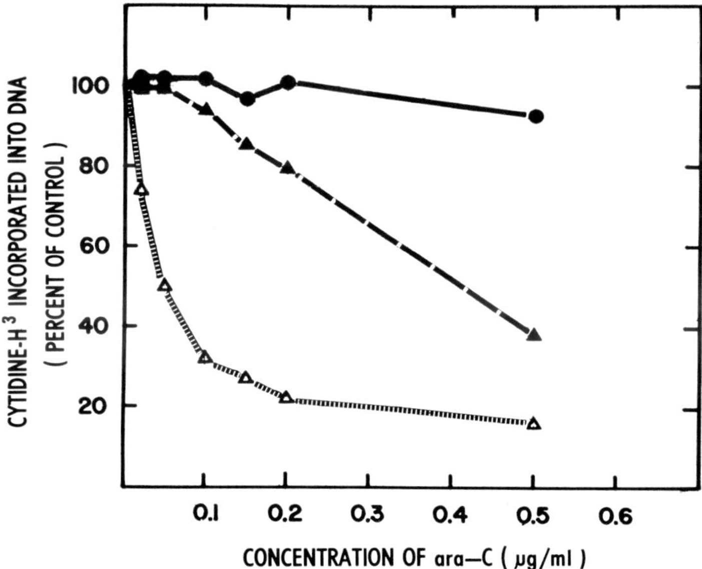

Figure 3. Effect of various concentrations of Ara-C on the incorporation of cytidine into the DNA of infected and noninfected cells. Cultures were infected with an adsorbed multiplicity of 10. Unadsorbed virus was removed and the cultures were incubated for 3 hours with Eagle's medium containing various concentrations of Ara-C. Mock infected cultures were treated identically. Cytidine-³H, 1 μc/ml (specific activity, 5 mc/mg), was added to the cultures which were harvested 4 hours later and the radioactivity in the DNA was determined. Infected, stationary-phase cultures, filled circles; infected, log-phase cultures, filled triangles; noninfected, log-phase cultures, open triangles.

dine that is supplied externally will be reduced. The specific activity of the deoxycytidine incorporated into DNA should therefore be higher, and this will be reflected in the specific activity of the DNA. Figure 4 shows the results of this experiment. It is clear that there is an increased incorporation of deoxycytidine-^{3}H into the DNA of noninfected cells incubated with 0.1 μg/ml or less of Ara-C. The same concentration of the drug, on the other hand, inhibits by approximately 70% the incorporation of cytidine into DNA. In infected cells, a concentration of 0.1 μg/ml of Ara-C has no effect on the incorporation of either deoxycytidine or of cytidine. At higher concentrations of Ara-C the incorporation of both nucleosides decreases.

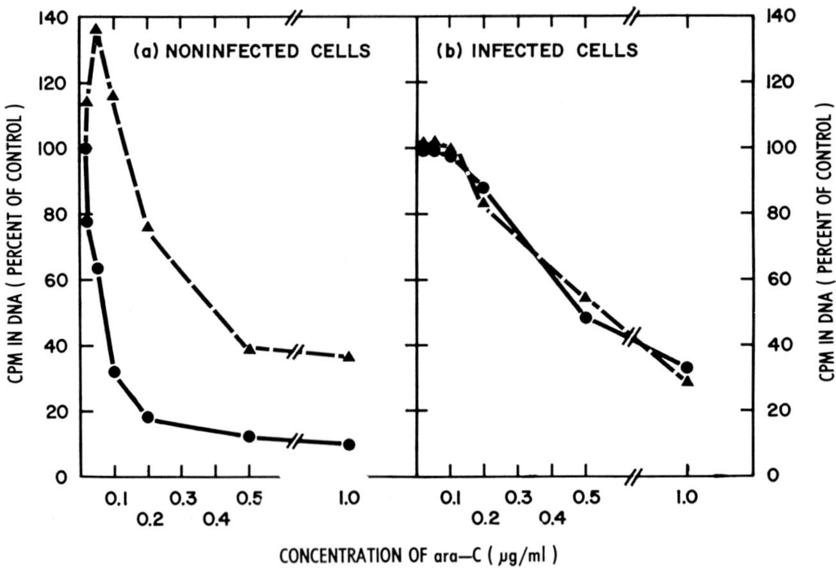

Figure 4. Effect of various concentrations of Ara-C on the incorporation of deoxycytidine and cytidine into the DNA of infected and noninfected cells. The experiment was performed as described in the legend to Figure 3. Deoxycytidine-^{3}H or cytidine-^{3}H, 1 μc/ml, (specific activity, 0.5 mc/mg) was added. Deoxycytidine, filled triangles; cytidine, filled circles.

At concentrations above 0.1 μg/ml, Ara-C inhibits the incorporation of cytidine, as well as deoxycytidine, both in infected and noninfected cells. No effect of Ara-C on the incorporation of the same nucleosides into RNA could be detected, indicating that penetration and phosphorylation of these compounds had occurred.

This second type of inhibition can be reversed partially, if the nucleosides are added to the cultures at a 20:1 ratio of deoxycytidine to Ara-C. These findings are in general agreement with the results of others working with different systems (Cohen, 1966). It is not known, however, whether high concentrations of deoxycytidine reverse the inhibitory action of Ara-C merely by preventing the phosphorylation of the drug (Chu and Fisher, 1962) or whether, in addition to interfering with CDP reductase, Ara-C also affects one of the kinases required for the phosphorylation of deoxycytidine.

The effect of Ara-C on the activity of some of the enzymes involved in the synthesis of DNA in extracts of infected cells was tested and the following results were obtained: Ara-C does not affect the phosphorylation of deoxycytidine, dCMP, or dCDP. At high concentrations, the drug does affect the activity of DNA polymerase.

Thus, so far we have been unsuccessful in clarifying the mechanism by which Ara-C at high concentrations inhibits DNA synthesis. It is clear, however, from our experiments that at relatively low concentrations Ara-C inhibits CDP reductase and that this inhibition does not occur in infected cells. This may be due to the fact that the enzyme present in infected cells is less susceptible to the inhibitory actions of Ara-C. On the other hand, Ara-C may be less effective in infected cells because of the presence of a greater amount of reductase in these cells than in noninfected cells.

The almost complete resistance of infected stationary phase cells to Ara-C could be explained on the basis of the inability of these cells to phosphorylate the drug. In spite of active DNA synthesis, these cells incorporate very little deoxycytidine into DNA.

Studies on CDP Reductase in Infected and Noninfected Cells

The fact that Ara-C seemed to be effective in inhibiting the CDP reductases present in infected but not in noninfected cells was of some interest, and we attempted to determine whether it was due to a qualitative difference between the enzymes or whether the level of activity of the enzyme was greatly increased after infection of the cells with Pr virus.

We attempted to assay for CDP reductase by the technique described by Reichard (1962) and, although this enzyme was measurable, in our hands, by this technique when assayed in extracts of mouse embryo, we have been unable to detect its activity in extracts of either infected or noninfected RK cells.

We therefore decided to obtain more information about this enzyme by indirect methods. CDP reductase is reported to be inhibited by TTP; the inhibition of the reductase is the basis for the inhibition of DNA synthesis in cultures exposed to high concentrations of thymidine (Morris *et al.*, 1963; Reichard *et al.*, 1961).

In the next experiment, the effect of thymidine on the incorporation of adenine-8-^{14}C and deoxycytidine-^{3}H into the DNA of infected and noninfected cells was tested. Figure 5 shows that in noninfected cells the result expected on the basis of an inhibition of CDP reductase was obtained, i.e., a decrease in the incorporation of adenine and an increase in the incorporation of ^{3}H-deoxycytidine into DNA. This increase in the incorporation of deoxycytidine is due presumably to the inhibition of the formation of dCDP from CDP and a consequent reduction of the endogenous pool of deoxycytidine derivatives that normally dilutes the radioactive deoxycytidine supplied to the cells.

High concentrations of thymidine are ineffective in suppressing the incorporation of either adenine or deoxycytidine into DNA of infected cells (Fig. 5). No depletion of deoxycytidine derivatives could be detected in the infected cells incubated with thymidine, even though the intracellular pool of TMP and TTP was approximately the same in these cells as in noninfected cells treated identically.

Thus, again conditions were realized in which inhibition of CDP reductase was obtained in noninfected cells but not in infected cells. We cannot conclude from these results whether a different enzyme is present in the infected cells or whether infection induces high levels of activity of this enzyme.

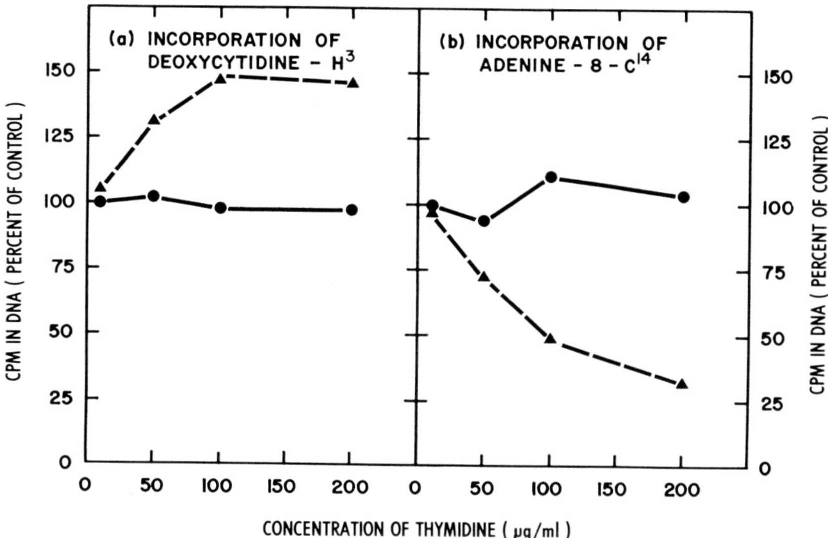

Figure 5. Effect of various concentrations of thymidine on the incorporation of deoxycytidine and adenine into the DNA of infected and noninfected cells. Logarithmically-growing cultures were infected as described earlier and incubated with various concentrations of thymidine for 3 hours. Deoxycytidine-^{3}H, 1 μc/ml, (specific activity, 0.5 mc/mg) or adenine-^{3}H, 1 μc/ml, (specific activity 3.5 mc/mg) was then added. The cultures were harvested 4 hours later and the radioactivity in the DNA determined. Infected cultures, filled circles; noninfected cultures; filled triangles.

We are now attempting to adapt the established method of assay for CDP reductase to the RK cell system, and we hope that we will then be able to answer this question.

Comparison of Incorporation of Various Nucleosides into the DNA of Infected and Noninfected Cells

It has been established in various laboratories that infection of cells with viruses of the herpes group will increase the level of activity of some of the enzymes involved in the synthesis of DNA. These results were obtained by assaying *in vitro* the activity of the enzymes present in extracts of infected and noninfected cells. Another approach that may be used to detect changes in the metabolic activity of cells after virus infection is to determine the relative rates of incorporation of various nucleosides into DNA. This method allows one to determine the relative degree of utilization of the different metabolic pathways by which the phosphorylated deoxynucleotides necessary for DNA synthesis are obtained by the infected cells.

The incorporation of various nucleosides into infected and noninfected cells was determined. Table V gives the results of an experiment in which the incorporation of the nucleosides, as well as of ^{32}P, into the DNA of noninfected log-phase cells and infected log-phase cells was determined.

TABLE V
COMPARISON OF INCORPORATION OF VARIOUS NUCLEOSIDES INTO THE DNA OF
INFECTED AND NONINFECTED CELLS

Radioactive Compound Added	Noninfected Log-phase Cultures	Infected Log-phase Cultures
	(mμmoles incorporated/sample x 10²)	
Cytidine	26.6	72.0
Deoxycytidine	65.0	16.7
Guanosine	66.5	99.5
Deoxyguanosine	27.2	73.4
Adenosine	82.1	27.5
Deoxyadenosine	66.5	22.2
Thymidine	98.5	89.8
Phosphorus-32	30.3	32.5

Cells were infected with an adsorbed multiplicity of 10. Four hours after infection the radioactive compounds were added. Five hours thereafter the cultures were harvested and the number of counts incorporated into the DNA were determined. Radioactive compounds added: 1 μc/ml (specific activity, 5 μc/μmole) of ^{3}H-labelled nucleosides; 1 μc/ml (carrier-free) of ^{32}P.

In another experiment, the incorporation of nucleosides into RNA and DNA of infected and noninfected cells was determined. The data are, in this case, presented as per cent incorporation of the nucleoside into the DNA of infected cells as compared to its incorporation into the DNA of noninfected cells (Table VI).

Definite differences between the degree of incorporation of the various compounds into DNA were found. Both guanosine and deoxyguanosine incorporation was higher in infected than in noninfected cells. On the other hand, the incorporation of adenosine and deoxyadenosine into the DNA of infected cells was lower. These results can be readily understood in terms of the differences in base composition between viral and cellular DNA. Viral DNA contains approximately 37 moles % of guanine and 13 moles %

of adenine. In host cell DNA, on the other hand, the guanine content is 22 moles % while the adenine content is 28 moles %. On the basis of these differences alone, one would expect a difference of up to 50% in the incorporation of adenine and guanine derivatives between infected cells and noninfected cells. The differences observed are, however, greater than expected on this basis.

TABLE VI

EFFECT OF INFECTION ON UPTAKE OF NUCLEOSIDES INTO DNA AND RNA

Radioactive Compound Added	Incorporation into DNA	Incorporation into RNA
	(Per Cent of Noninfected Cultures)	
Guanosine	260	16
Deoxyguanosine	365	15
Adenosine	36	11
Deoxyadenosine	29	8
Cytidine	420	16
Deoxycytidine	27	13
Thymidine	51	—
Uridine	320	11

This experiment was performed as described in the legend to Table V.

An interesting point emerges when the incorporation of cytidine and deoxycytidine into the DNA of infected and noninfected cells is compared. While deoxycytidine is more readily incorporated than cytidine into the DNA of noninfected cells, the reverse situation is true for infected cells. The latter incorporate approximately four times more cytidine into their DNA than do noninfected cells; the incorporation of deoxycytidine is reduced by approximately the same factor.

In the previous section, we have shown that DNA synthesis in virus-infected cells is not affected by agents that inhibit CDP reductase. The fact that cytidine is more readily incorporated into the DNA of infected cells than into the DNA of noninfected cells indicates that an increase in the level of activity of this enzyme has taken place after infection of the cells with Pr virus. (No difference in cytidine kinase activity was detected.)

The relatively high level of incorporation of cytidine into the DNA of infected cells as compared to noninfected cells may provide another approach to viral chemotherapy. Analogues of cytidine which can be reduced and incorporated into DNA should become incorporated into the DNA of infected cells to a greater extent than into the DNA of noninfected cells. The danger of using compounds which may be incorporated into cellular DNA and cause mutation or cancer has been pointed out repeatedly. However, if the quantitative difference between the incorporation of these analogues into the DNA of infected and noninfected cells is sufficiently great, the dosage can be manipulated to allow incorporation of the drugs into viral DNA and keep their incorporation into cellular DNA at a low level. These analogues may then be useful in controlling viral disease without being appreciably toxic to the host. We have tested some analogues of cytidine (generously supplied by Dr. S. S. Cohen) and of those tested, only bromocytidine possessed antiviral activity. The antiviral activity of bromocytidine was relatively low compared to IUDR; it was less effective by about a factor of 10^2. Nevertheless, while the antiviral activity of bromocytidine and bromodeoxycytidine was approximately the same, the latter was about twice as inhibitory to cellular growth.

Although these results are far from satisfactory, in view of the large difference in the degree of their incorporation into the DNA of infected and noninfected cells, it seems that analogues of cytidine which are capable of being reduced and incorporated into DNA may be useful as antiviral agents.

CONCLUDING REMARKS

The experiments presented here show that model systems studied in cell culture are helpful in predicting the usefulness of a drug as an antiviral agent. It is clear from these data that IUDR can, in principle, be classified as a selective antiviral agent, since it is incorporated to a greater extent into the DNA of infected cells than into the DNA of noninfected host cells. Ara-C, on the other hand, is more effective in inhibiting DNA synthesis in noninfected cells than in infected cells, and doses of this drug that are effective in suppressing virus growth will therefore be strongly toxic to the noninfected host cells.

Both of these effects — the greater susceptibility of infected cells to IUDR and their greater resistance to Ara-C — are probably due to the changes in the levels of activity of some enzymes which occur after infection of the cells. These differences in the levels of enzymatic activity between infected and noninfected cells seem, therefore, to provide a means by which certain aspects of the metabolism of the infected cells can be inhibited specifically.

We have shown that the incorporation of cytidine into the DNA of infected cells is much greater than it is in noninfected cells. The contrary is true for the incorporation of deoxycytidine. On the basis of this information, one may assume that analogues of cytidine which can be incorporated into DNA will be more effective in suppressing virus growth than in affecting the host cells. However, for such compounds to be effective it is essential that they should not be deaminated and enter a pool of precursors which are used effectively by the infected and the noninfected cells.

It is evident, of course, that results obtained in cell culture may not reflect the situation as it prevails in an animal and the dangers of extrapolating from these results are many. Nevertheless, knowledge of the physiology of virus-infected cells does provide a rational basis for predicting the kind of drug that may be effective as an antiviral agent.

REFERENCES

1. BEN-PORAT, T., and KAPLAN, A. S.: *Virology, 20:*310, 1963.
2. CHU, M. Y., and FISCHER, G. A.: *Biochem. Pharmacol., 11:*423, 1962.
3. COHEN, S. S.: *Progr. Nucleic Acid Research and Molec. Biol. 5:*1, 1966.
4. DEINHARDT, F.: *Ann. New York Acad. Sci., 130:*218, 1965.
5. DJORDJEVIC, B., and SZYBALSKI, W.: *J. Exper. Med., 112:*509, 1960.
6. EGGERS, H. J., and TAMM, I.: *Ann. Rev. Pharmacol., 6:*231, 1966.
7. GREER, S., and ZAMENHOF, S.: *Abstr. 131st Meeting Amer. Chem. Soc.* p. 3c, 1957.
8. HANNA, C., and WILKINSON, K. P.: *Exper. Eye Res., 4:*31, 1965.
9. HOWARD-FLANDERS, P., BOYCE, R. P., and THERIOT, L.: *Nature, 195:*51, 1962.
10. KAPLAN, A. S., and BEN-PORAT, T.: *Virology, 23:*90, 1964.
11. KAPLAN, A. S., and BEN-PORAT, T.: *J. Molec. Biol., 19:*320, 1966.
12. KAPLAN, A. S., BEN-PORAT, T., and KAMIYA, T.: *Ann. New York Acad. Sci., 130:* 226, 1965.
13. KAPLAN, H. S., SMITH, K. C., and TOMLIN, P. A.: *Radiation Res., 16:*98, 1962.
14. KAUFMAN, H. E.: *Ann. New York Acad. Sci., 130:*168, 1965.
15. KAUFMAN, H. E., CAPELLA, J. A., MALONEY, E. D., ROBBINS, J. E., COOPER, G. M., and UOTILAI, M. H.: *Arch. Ophthalmol., 72:*535, 1964.
16. KOZINSKI, A. W., and SZYBALSKI, W.: *Virology, 9:*260, 1959.
17. MORRIS, N. R., REICHARD, P., and FISCHER, G. A.: *Biochem. Biophys. Acta, 68:* 93, 1963.

18. Opara-Kubinska, A., Lorkiewicz, A., and Szybalski, W.: *Biochem. Biophys. Res. Commun., 4:*288, 1961.
19. Prusoff, W. A., Bakhle, Y. S., and Sekeley, L.: *Ann. New York Acad. Sci., 130:* 135, 1965.
20. Reichard, P.: *J. Biol. Chem., 237:*3513, 1962.
21. Reichard, P., Canellakis, Z. N., and Canellakis, E. S.: *J. Biol. Chem., 236:* 2514, 1961.
22. Rupp, W. D., and Prusoff, W. A.: *Nature, 202:*1288, 1964.
23. Sauerbrier, W.: *Virology, 15:*465, 1961.
24. Smith, K. O.: *J. Immunol., 91:*582, 1963.
25. Smith, K. O., and Dukes, C. B.: *J. Immunol., 92:*550, 1964.
26. Stahl, F. W., Craseman, J. M., Okun, L., Fox, E., and Laird, C.: *Virology, 13:* 98, 1961.
27. Tanooka, H.: *Radiation Res., 21:*26, 1964.
28. Underwood, G. E., Elliott, G. A., and Buthala, D. A.: *Ann. New York Acad. Sci., 130:*151, 1965.

INTRINSIC INTERFERENCE: A NEW TYPE OF VIRAL INTERFERENCE[1]

PHILIP I. MARCUS[2] AND DAVID H. CARVER[3]

INTRODUCTION

THE HEMADSORPTION-NEGATIVE plaque test (18) was developed as a means of accurately assaying rubella virus as a nonhemadsorbing, noncytopathic virus. The test has proved to be highly reliable, and is based on the fact that cells infected with rubella virus are completely refractory to superinfection by Newcastle disease virus (NDV), while remaining susceptible to many other viruses. Cells refractory to NDV may be scored as: 1) hemadsorption-negative (HAD⁻) plaques standing out against a background of NDV-susceptible, HAD⁺, cells; 2) individual HAD⁻ cells in a population of widely dispersed single HAD⁺ cells, or 3) as cell survivors (colonies). In each case, the principal characteristic used to score the rubella virus-infected cell is its absolute resistance to NDV. This state of refractoriness is termed *intrinsic interference* (18). This communication extends these studies to show that the capacity to induce intrinsic interference is not restricted to rubella virus but may involve several kinds of seemingly unrelated noncytopathic viruses and also possibly some virulent types. This paper also presents evidence that intrinsic interference is: 1) an intrinsic property of the virus-infected cell which does not extend to uninfected cells in the culture; 2) induced under conditions that rule out completely interferon-mediated interference, and 3) presumably brought about the action of a protein (s) en-

[1]Some aspects of this investigation were reported at the 66th Annual meeting of the American Society for Microbiology, Los Angeles, Calif., May 3, 1966 (*Bacteriol. Proceed.* p. 123, 1966), and have appeared in *J. Virology, 1:*334, 1967.

[2]Recipient of a U. S. Public Health Service Research Career Development Award (2-K3-GM-15,461-06) from the National Institute of Allergy and Infectious Diseases. On leave-of-absence, Sept. 1967 — Sept. 1968, The Salk Institute for Biological Studies, La Jolla, Calif.

[3]Present address: Department of Pediatrics, The Johns Hopkins Hospital, Baltimore, Maryland 21205.

coded for by the genome of the inducing virus. Intrinsic interference is defined as: a viral genome induced cellular state of resistance to NDV, coexistent with a state of susceptibility to a broad spectrum of other viruses.

MATERIALS AND METHODS

Cell Cultures and Medium. Primary green monkey kidney (GMK) cells were obtained from Microbiological Associates (Md.) or Baltimore Biological Labs. (Md.) as confluent monolayers. These were trypsinized (0.05% trypsin + 0.005 M EDTA) and used in hemadsorption-negative (HAD−) plaque tests (18) as confluent monolayers on 60 mm plastic petri dishes (Falcon, Calif.) or plated as single cells on 35 mm dishes. Other cell types were used in a similar way. Their specific designations are noted below along with the individuals to whom we are grateful as initial suppliers: Baby hamster kidney, BHK21/WI-2 (A. Vaheri); L-cells (W. Joklik) ; Fathead minnow line, FHM (7) (S. Silverstein, L. Sturman) ; V5 line of Rauscher leukemia virus carrying mouse cells (R. Bases, B. Wright). GMK, BHK, and L-cells were grown in NCI solution (Grand Island Biol. Lab., N. Y.) supplemented with 6 per cent calf serum (= Attachment Solution (18)) and 3 per cent fetal bovine serum. The concentration of fetal bovine serum was raised to 10% for cultures of FHM cells, and incubation for growth was always at 34°C (7) for this poikilothermic line. V5 cells were grown in F10 medium (8) with 15% fetal bovine serum. Just prior to, and during, challenge with Newcastle disease virus (NDV) all cell cultures were treated similarly, i.e., rinsed, and incubated at 37°C in Attachment Solution (18) .

Virus Stocks and Preparation. Rubella virus-F8 (M. R. Balsamo) stocks were prepared as described previously (18) and in BHK cells to give 1-10 x 10⁵ HAD− PFP/ml (18) . Sindbis virus (J. F. Enders) was grown in primary chick embryo cell (CEC) monolayers incubated in Simpson-Hirst medium as reported (5) to give titers of about 1 x 10⁹ PFP or HAD− PFP/ml. NDV stocks of 1-5 x 10⁹ PFP/ml were prepared from chick allantoic fluid. The California strain of NDV was used in most experiments, but three other strains behaved similarly; Beaudette, Mass. —HIK, and a vaccine strain. Vaccinia virus (7N) was harvested from HeLa cell monolayers or the chorioallantoic membrane of eggs. Poliovirus-MEF1, type 2 (J. F. Enders) was grown in chick embryo or GMK cell monolayers. Vesicular stomatitis virus, VSV (S. Silverstein) was propagated in CEC monolayers to give stocks of about 1×10^8 PFP/ML. Influenza B virus (R. Simpson) was propagated in eggs, and encephalomyocarditis virus, EMC (R. Bases) in Krebs ascites cells. Lactic dehydrogenase virus, Riley's agent, LDH virus (R. Bases, V. Riley) was used as supplied in high titer by Dr. Riley and Mr. Loveless. A noncytopathic hemadsorbing simian virus (BTV) was isolated as a contaminant from one lot of primary GMK cells and was assayed by hemadsorption foci (10) using bovine or turkey erythrocytes. Virulent poliovirus, type 1, was supplied by Dr. D. Summers.

Hemadsorption Producing Particle (HDP) Assay. The hemadsorption-negative (HAD$^-$) plaque technique has been described in detail (18). The procedure used to detect hemadsorption producing particle (HDP) activity of a virus preparation on single cells has been described less thoroughly (17). One ramification of this latter technique is its use to detect individual cells that manifest intrinsic interference. This may be accomplished as follows: cells, usually in monolayer culture, are exposed to the intrinsic interference inducing virus and then incubated under various test conditions as reported in the text. To determine the fraction of cells refractory to NDV, i.e., showing intrinsic interference, the infected cell monolayer first is challenged with NDV at a multiplicity, m, $=10$ PFP (in typical experiment 1×10^8 NDV-PEP in a volume of 0.3 ml are added to a monolayer containing 5×10^6 cells and adsorbed for 30-60 minutes at 37°C). Nonadsorbed NDV is then washed off and any surface-situated nonengulfed virus is neutralized with NDV-antiserum to eliminate any background of spurious red blood cell binding in the single cell hemadsorption test (18). Immediately following the adsorption of NDV, the monolayers are trypsinized to obtain monodisperse cells. These cells are plated at low densities (usually 1×10^5 cells/35 mm Petri dish), incubated for about 15 hours to allow NDV hemagglutinin incorporation into the plasma membrane (17), and differentiated as to HAD$^+$ or HAD$^-$ cells by the single cell hemadsorption test as illustrated in Figure 1. The steps taken to detect intrinsic interference in individual cells are listed below and will be referred to collectively as the *intrinsic interference test*.

1) Infect cells with intrinsic interference inducing virus.
2) Incubate virus-cell complexes to develop state of intrinsic interference.
3) Challenge infected cells with NDV at m = 10 PFP.

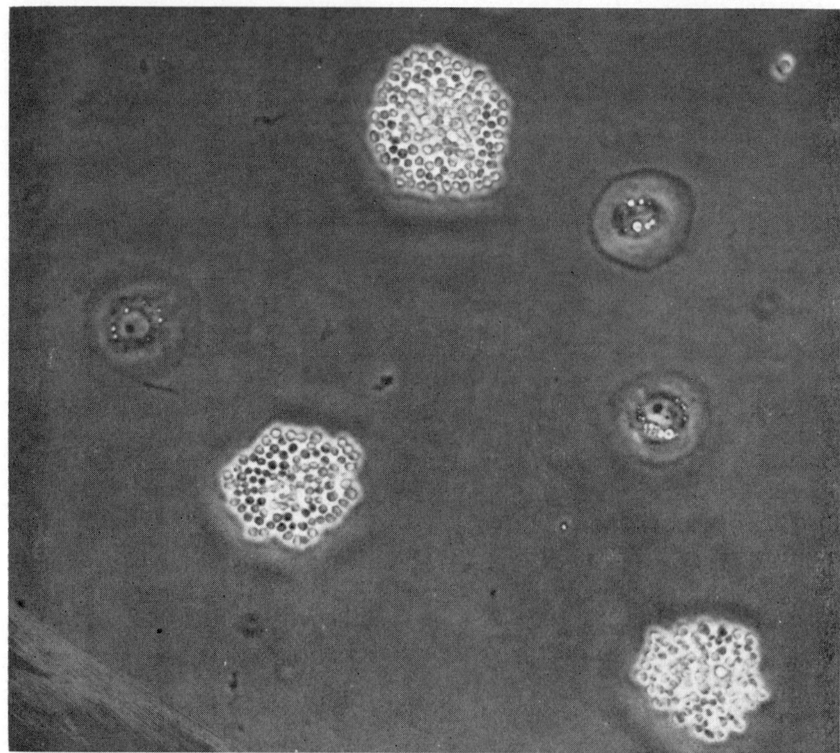

Figure 1. Single cell hemadsorption test to detect individual cells rendered refractory to NDV, i.e., hemadsorption-negative (HAD$^-$), by an intrinsic interference inducing virus. Green monkey kidney cell monolayers were infected with Sindbis virus (m = 5 PFP) and 11 hours later challenged with NDV (m = 10 PFP), trypsinized and plated as single cells, incubated an additional 15 hours and tested for their hemadsorption reaction. Cells susceptible to NDV, i.e., not manifesting intrinsic interference are represented by the 3 HAD$^+$ cells covered entirely with bovine erythrocytes. Cells demonstrating intrinsic interference are refractory to NDV and appear as 3 HAD- cells. Bovine erythrocytes have a diameter of about 6μ.

4) Trypsinize challenged cells and plate as single cells.

5) Run single cell hemadsorption test and determine microscopically the fraction of cells manifesting intrinsic interference by counting HAD$^-$ (NDV-refractory cells) and HAD$^+$ (NDV-susceptible) cells.

Ultraviolet Light Inactivation of Sindbis and Rubella Viruses. Ultraviolet light survival curves for Rubella and Sindbis viruses were carried out under conditions where T2 phage and NDV could be used as biological actinometers (18). Rubella virus survivors were assayed as HAD$^-$ —PFP and Sindbis virions as PFP on young CEC monolayers grown in Simpson-Hirst medium (5). All survival curves showed 1-hit inactivation kinetics. The 37% survival dose, D_o, defined as the dose of UV light required to reduce the number of infectious particles to e^{-1} (0.37 survivors = 1 lethal hit) is the same for both viruses, 110 ergs mm^{-2}. For comparison, D_o values for NDV and T2 phage irradiated in the same medium (Attachment Solution) are 42.5 and 25 ergs mm^{-2}, respectively. UV-inactivated virus, usually at 5×10^{-5} survival level, was used immediately after irradiation.

Infective Center Assay. Control cells, and cells infected with intrinsic interference inducing viruses were tested for their capacity to act as infective centers for NDV. NDV was absorbed to cell monolayers (m=10 PFP), the free virus was washed away, the surface-situated virions neutralized with NDV-antiserum, and the cells dispersed with trypsin, all within a 90 minute period at 37°C. Monodisperse cells were counted and plated as infective centers as described previously (19). Triplicate plates were run for each assay, and dilutions of virus-cell complexes were adjusted to obtain 50 to 150 plaques per plate. The results were expressed as the per cent of maximum NDV plaque count relative to control cultures.

RESULTS

Sindbis virus can be obtained readily in high titer compared to the other noncytopathic viruses used here; it is easy to assay, and its reactions are typical of viruses that induce intrinsic interference. Consequently, Sindbis virus has been used in most of the experiments reported here. All viruses listed as possessing intrinsic interference inducing capacity (Table I) have been tested in experiments similar to those described below for Sindbis virus, and have been shown to react in the same manner unless specifically noted. For obvious technical reasons this statement does not apply to some of the viruses for experiments involving high multiplicities.

TABLE I

VIRUSES THAT INDUCE INTRINSIC INTERFERENCE

Virus	*Designation*	*Host Cell*
rubella	F-8 strain and fresh isolates	GMK, L-cell, Human embryo kidney
Sindbis	Grp A arbovirus	GMK, FHM[c], HeLa
West Nile	Grp B arbovirus	GMK
poliovirus	MEF, 1, type 2	GMK
poliovirus[a]	Type 1	HeLa
Lactic dehydrogenase virus	LDH virus, Riley's agent	Primary mouse embryo cells
JLS-V5 mouse cell agent[b]	JLS-V5 line carrying Rauscher murine leukemia virus	JLS-V5 mouse cell line
Mouse cell agent[b]		Present in 3 of 5 different lots of primary mouse embryo cell cultures

[a]Cytopathic, see text for special test conditions.
[b]Cells and supernatants tested by J. Loveless and V. Riley and found to be free of LDH virus. Possible identity of all agents in mouse cell cultures has not been ruled out.
[c]This line of cells is infected with a *Mycoplasma sp.*, but shows 100% HAD+ cells when challenged with NDV.

Scoring Intrinsic Interference (NDV-refractory, HAD⁻ cells). The fraction of NDV-refractory cells is determined as described in Materials and Methods. Figure 1 shows the appearance of HAD+ and HAD⁻ GMK cells that have been infected with Sindbis virus at m = 5 PFP, incubated for 11 hours, and then challenged with NDV as in the 5-step procedure outlined above. After the 15-hour incubation period allowed for NDV hemagglutinin incorporation into the surface of infected cells essentially all of the HAD+ cells are saturated with respect to the number of adsorbed red blood cells (Fig. 1) as in control preparations, and the contrast between HAD+ and HAD⁻ cells is maximal. Thus, hundreds of cells may be scanned rapidly and the fraction of NDV-refractory (HAD⁻) cells ascertained to a high degree of accuracy. This experimental procedure has been used to define several characteristics of intrinsic interference. These are documented below.

Basic Characteristics of Intrinsic Interference

Rate of Induction — A Function of Multiplicity. The rate at which intrinsic interference is induced and the dependence of rate on multiplicity was determined as follows: GMK cell monolayers were infected with various multiplicities of Sindbis virus, challenged with NDV at different intervals, and the fraction of NDV-refractory cells determined. The results from such an experiment are illustrated in Figure 2, where each point represents a count of at least 400 cells. The fraction of HAD⁻ cells is seen to increase linearly from time zero and at a rate dependent upon the multiplicity of the inducing virus, the higher the multiplicity the shorter the time of induction to interference. For example, the time required to induce intrinsic interference in 50% of the cells at m = 1,5,25, and 100 PFP is 18, 11, 8, and 5 hours, respectively. Although no lag is apparent in the induction process under the con-

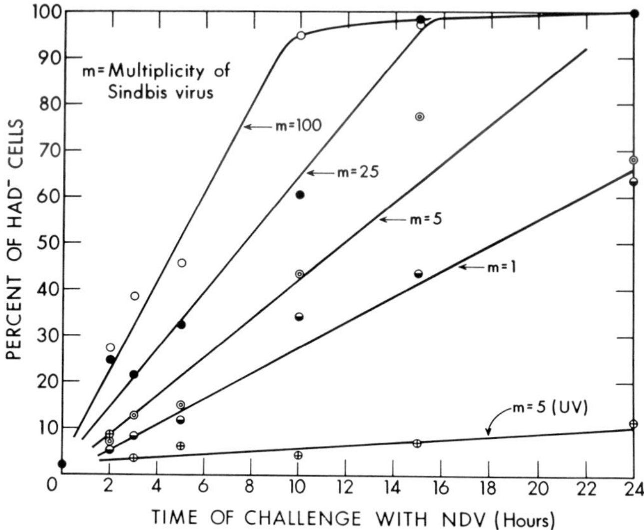

Figure 2. Dependence of rate of induction of intrinsic interference on viral multiplicity. Monolayers of GMK cells were infected at various multiplicities (m) of Sindbis virus, challenged with NDV (m = 10 PFP) at different intervals, and the fraction of HAD- cells determined 15 hours later by the intrinsic interference test.

ditions of virus attachment used, e.g., 0.3 ml virus/60 ml plate for 30 minutes at 37°C, when Sindbis virus is adsorbed at 4°C and the temperature then raised to 37°C, all curves extrapolate to a lag of about $\frac{1}{2}$ hour. When cells infected at m = 1 PFP were incubated for a few hours beyond 24 hours (not shown), the fraction of HAD$^-$ cells increased only slightly above the 62 per cent noted at 24 hours — a value consistent with the fraction of cells calculated from the Poisson distribution to receive 1 PFP. The 3% point at zero time represents a baseline of control cells physiologically noninfectable at the time of NDV challenge, an observation reported previously (16). The fraction of noninfectable cells varies from 1 to 10% with different lots and kinds of cells and is determined as a control in each experiment.

TABLE II
VIRUSES AND THEIR CYTOPATHIC EFFECT ON CELLS
MANIFESTING INTRINSIC INTERFERENCE

Virus	Cytopathicity on NDV Refractory Cells[a]
Echo - 11	+[b]
Encephalomyocarditis virus	+
Influenza-B	+
Poliovirus	+
Vaccinia	+
Vesicular stomatitis virus	+
Simian virus, noncytopathic, hemadsorbing	+[c]
Newcastle disease virus:	
California	—[d]
Beaudette	—
Mass.-HiK	—
Blacksburg, vaccine	—

[a]All cell types listed in Materials and Methods were susceptible to NDV, and most cell types were susceptible to the majority of the test viruses. Notable exceptions are referred to in the text.
[b]+ denotes complete cell destruction within 24 hours, indistinguishable from control cultures similarly infected at a high multiplicity.
[c]For this noncytopathic, hemadsorbing virus, + denotes infection scored by hemadsorption.
[d]— denotes the complete absence of a cytopathic effect and cell appearance indistinguishable from uninfected controls.

Half-life of the Interference. Once the NDV-refractory state is reached in 100% of the cells it may persist indefinitely if the cell monolayer is not disrupted and is maintained in a healthy state with adequate changes of medium. For example, rubella virus-infected monolayers of GMK cells remained refractory to NDV for the entire length of a 2-month test period. At the end of this time, the NDV-resistant cells were as susceptible to the other test viruses (Table II) as were control cells not previously infected with rubella virus. However, the NDV-refractory state has a measurable half-life under certain conditions. If GMK cell monolayers rendered 100 per cent refractory to NDV by prior infection with rubella or Sindbis virus are dispersed with trypsin and plated at lesser cell densities in fresh medium, the HAD⁻ state is lost. The kinetics of this return to NDV-susceptibility may be ascertained by challenging the cells with NDV at daily intervals and determining the number of HAD⁻ and HAD⁺ cells. Results of this type are presented in Figures 3 and 4 for rubella and Sindbis

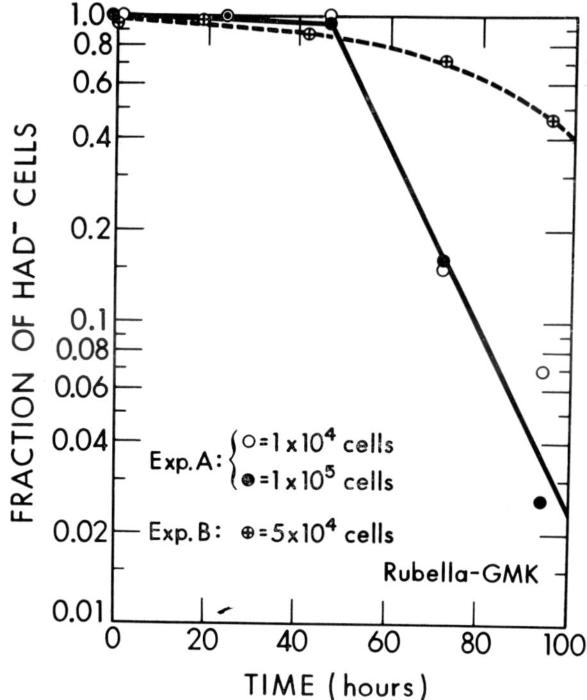

Figure 3. Loss of intrinsic interference following subculture of rubella virus-infected cells. Monolayers of GMK cells infected with rubella virus and maintained in a 100% NDV-refractory state for 1 week were subcultured, plated at various cell densities in 60 mm dishes, challenged with NDV at approximately daily intervals, and scored 15 hours later by the hemadsorption test to determine the fraction of HAD⁻ cells.

viruses, respectively. The data show that the HAD⁻ state is maintained in all of the cells for about 1 to 2 days and then is lost at a relatively rapid and constant rate, independent of cell density. In 3 of the 4 experiments illustrated, after 1 to 2 days, the half-life of the return to an NDV-susceptible state was about 10-15 hours. In cells kept for several days beyond the test period noted in Figures 3 and 4 there was a reversal of the loss of the HAD⁻ state and a gradual return to a confluent monolayer of 100% NDV-refractory cells.

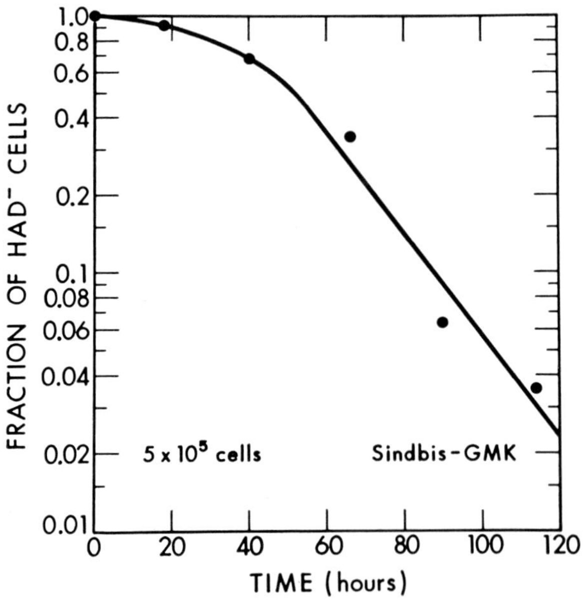

Figure 4. Loss of intrinsic interference following subculture of Sindbis virus-infected cells. Monolayers of GMK cells were infected with Sindbis virus to induce 100% NDV-refractory cells and then treated as described in Figure 3.

Requirement for Active Virus. Stock preparations of Sindbis virus were inactivated with ultraviolet light to a survival level of 5×10^{-5} and compared with active virus in their capacity to induce an NDV-refractory state. The results of our first experiment are presented as the bottom curve in Figure 2, and those of two other experiments in Figure 5. Clearly, the capacity to induce intrinsic interference is lost following inactivation of plaque forming capacity with ultraviolet light. These same preparations of UV-inactivated Sindbis virus are still capable of inducing high levels of interferon production.

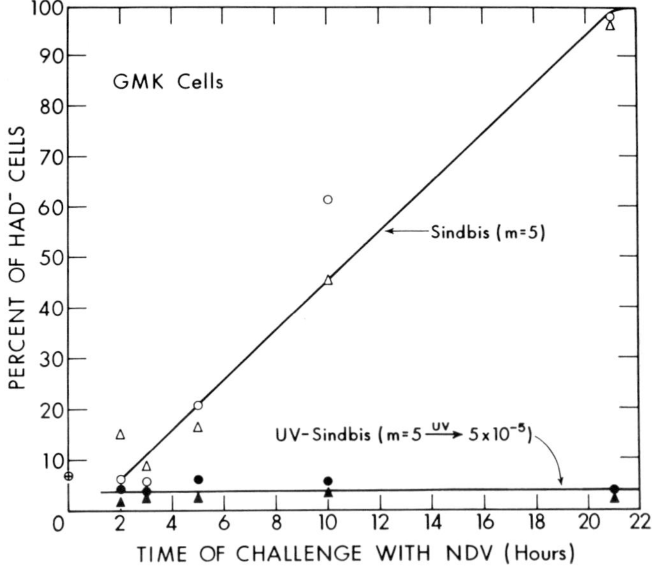

Figure 5. Loss of intrinsic interference inducing capacity of Sindbis virus following UV-irradiation. Monolayers of GMK cells were infected with active Sindbis virus (m = 5 PFP), or the same preparation irradiated with ultraviolet light to a level of 5 x 10⁻⁵ PFP survivors. The rate of development of an NDV refractory state in the cells was determined in the usual manner.

Insensitivity to Actinomycin D. A slight modification of the intrinsic interference test was introduced to determine whether Sindbis virus could induce an NDV-refractory state in cells precluded from synthesizing cellular mRNA. GMK cell monolayers were treated with actinomycin D (30 μg/3 ml medium/plate) for 20 minutes prior to infection with a high multiplicity of Sindbis virus. This treatment had no discernible effect on the hemadsorption-producing capacity of NDV, but reduced H[3]-uridine incorporation by 95% (19). At various intervals after infection with Sindbis virus the cultures were challenged with NDV and the fraction of HAD⁻ cells determined in the usual manner. The results presented in Figure 6 show that Sindbis virus can

induce intrinsic interference in cells treated with actinomycin D
and no longer able to synthesize cellular mRNA. Actually, the
rate of induction consistently is greater in actinomycin D-treated
cells than in untreated cultures.

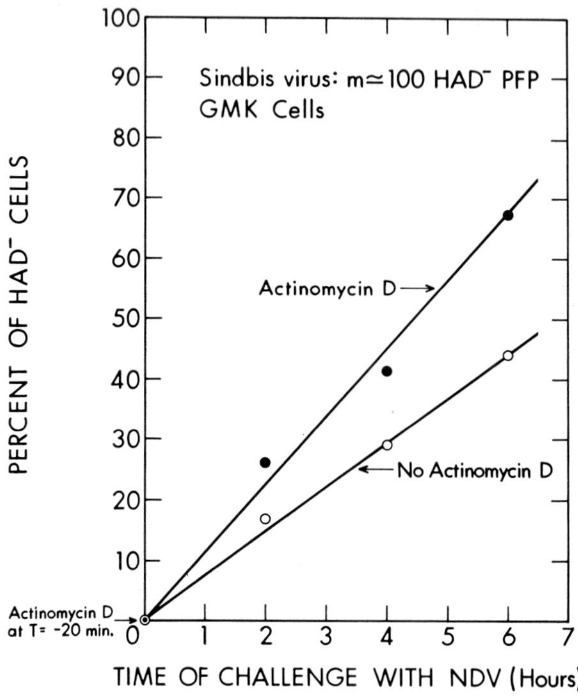

Figure 6. Induction of intrinsic interference in the presence of Actinomycin
D. Monolayers of GMK cells were treated for 20 minutes with actinomycin D
(10 μg/ml), infected with Sindbis virus (m = 100 PFP), and challenged with
NDV after 2, 4, and 6 hours. Fifteen hours later the fraction of HAD⁻ cells
was determined as usual.

Requirement for Protein Synthesis. The effect of inhibition
of protein synthesis on the induction by Sindbis virus of an NDV-
refractory state in GMK cells was examined as follows. Cells were
treated with puromycin (50 μg/ml) for 45 minutes, infected with
a high multiplicity (m=20 PFP) of Sindbis virus, and the virus-
cell complexes were incubated in the presence of puromycin
(*with puromycin*). The drug was washed out, the infected cells
were incubated for various intervals *without puromycin,* chal-

lenged with NDV, and the fraction of HAD⁻ cells was determined as in the standard intrinsic interference test. The results illustrated in Figure 7 show that the percentage of virus-cell complexes which become refractory to NDV (HAD⁻) after a sojourn with and without puromycin is essentially equivalent to that expected if the inductive process were functioning only during the time of incubation *without puromycin.* The rate of induction of the NDV-refractory state in control preparations of Sindbis virus-cell complexes never exposed to puromycin is shown as the dashed line on the right hand side of Figure 7. The interval during which cells were incubated *with* and *without* puromycin is indicated on the left and right, respectively, of the ordinate, with the solid circles representing the percentage of HAD⁻ cells at the end of the 6-hour test period. From the data presented in Figure 7 we

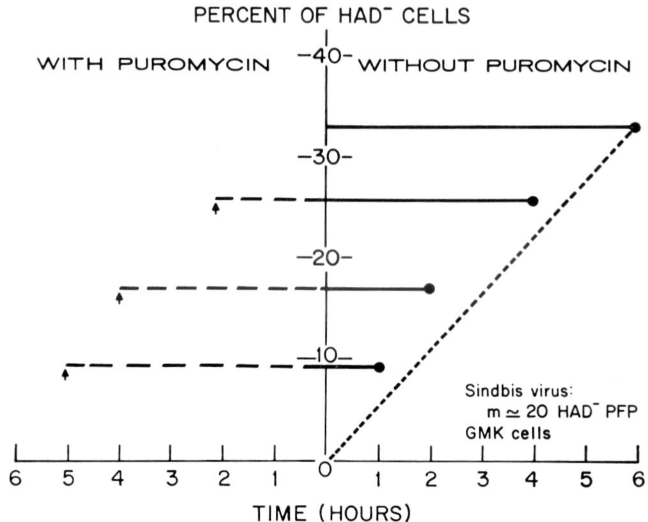

Figure 7. Inhibition of intrinsic interference induction in the presence of puromycin. Monolayers of GMK cells infected with Sindbis virus (m = 20 PFP) were incubated for 0, 2, 4, and 5 hours in puromycin (50 μg/ml), washed, and incubated for 6, 4, 2, and 1 hour (s), respectively, in the absence of the drug. At the end of the 6 hour total incubation period all preparations were challenged with NDV and the fraction of HAD- cells (solid circles) determined 15 hours later by the hemadsorption test. The dotted line on the right represents the rate of development of the NDV-refractory state in the complete absence of puromycin.

conclude that Sindbis virus requires protein synthesis to induce an NDV-refractory state in GMK cells.

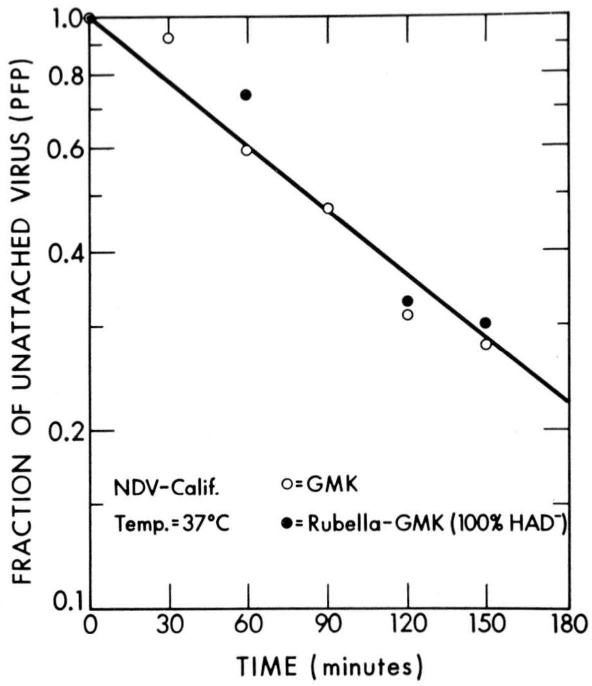

Figure 8. Attachment kinetics of Newcastle disease virus on GMK monolayers containing 100% NDV-refractory cells (100% HAD⁻). GMK cell monolayers infected with rubella virus and incubated to produce 100% NDV-refractory cells were exposed to NDV contained in 0.3 ml attachment solution. At various intervals the supernatant fluids were assayed on chick embryo monolayers for unattached NDV-PFP.

NDV Attachment, Infective Center Assay and Eclipse. NDV binding to cell monolayers was carried out as described previously (15). The kinetics of NDV attachment to monolayers containing 100 per cent NDV refractory GMK cells induced by rubella virus is shown in Figure 8, along with virus adsorption to uninfected control cells. No significant difference in the rates of NDV adsorption are noted. Similar results were obtained using monolayers of FHM cells induced to the 100% HAD⁻ state with Sindbis virus. In this latter instance, it was necessary to prevent plaque formation of residual Sindbis virus by plating NDV in the presence of Sindbis-antiserum on aged CEC monolayers (5).

TABLE III
INFECTIVE CENTER PLAQUING OF RUBELLA VIRUS-INFECTED
GMK CELLS CHALLENGED WITH NDV

State of Cells	Per cent of Maximum NDV Plaque Count on CEC Monolayers	
	Infective Centers: *Intact*	*Infective Centers:* *3X Freeze-Thaw*
HAD^{+a}	100	8.2
HAD^{-b}	5.9	6.4

[a]Control monolayers of GMK cells. All cells susceptible to NDV, i.e., score as HAD$^+$ when challenged.
[b]GMK cell monolayers infected with rubella virus and incubated for 3 days. All cells rendered refractory to NDV, i.e., score as HAD$^-$ when challenged.

The results in Table III present the plaquing efficiency of NDV as infective centers from GMK monolayers made 100% refractory to NDV (HAD$^-$) by infection with rubella virus, compared with that from uninfected control cells (HAD$^+$). Although NDV susceptible and refractory cells adsorb NDV equally well, in the latter case there is a 94% loss in plaquing efficiency of the intact infective centers. When infective centers are freeze-thawed 3 times the control cells show the usual 90% loss in plaque count, representing viral eclipse and loss of viral replicative capacity of the cells. The NDV-refractory cells (HAD$^-$) show no further change in plaquing efficiency from the low level observed with the intact virus-cell complexes, indicating that a constant small per cent of noneclipsed and nonneutralized input virus is present in sensitive and refractory cells alike.

Viruses Inducing Intrinsic Interference

The first step used to screen viruses for their capacity to induce intrinsic interference involves a simple test of their ability to form HAD$^-$ plaques (18). To date, no virus which has displayed this particular trait has failed to meet the more stringent criteria used as the final determinant of intrinsic interference-inducing capacity. These criteria are: 1) the viruses ability to induce an NDV-refractory state in the presence of actinomycin D, and 2) the susceptibility of the NDV-refractory cells to a broad spectrum of viruses. Table I lists the viruses which have met these criteria, the exception being the agent (s) responsible for the NDV-refractory state of the V5 mouse cell line carrying Rauscher murine leukemia virus. In this latter case, the HAD⁻ state is maintained in the presence of actinomycin D, but it has not been possible to initiate the NDV-refractory condition in another line of cells, and hence test inducibility in the drugs' presence. Also included in Table I is a virulent strain of type 1 poliovirus that causes a rapid and extensive cytopathic effect (CPE) so as to preclude testing for an NDV-refractory state. However, in the presence of guanidine cytopathic effects are delayed long enough so the cell may be challenged with NDV and tested for hemadsorption. These conditions permit partial expression of the input poliovirus genome but not its replication (3). The following experiment was done in conjunction with Dr. D. Summers. HeLa cells from spinner culture were suspended in medium containing actinomycin D (5 μg/ml) and guanidine hydrochloride (50 μg/ml) and infected with poliovirus at an adsorbed m = 200 PFP. Control cells were treated similarly but were not exposed to poliovirus. Poliovirus-cell complexes were incubated for $1\frac{1}{2}$ hours at 37°C (a period sufficient to strip most host cell mRNA from polysomes), chilled to 4°C and NDV attached for an adsorbed m = 20 PFP. The cells were washed, plated in Attachment Solution containing guanidine and incubated at 37°C. After 2 hours, the cells were firmly attached and could be treated with NDV-antiserum to neutralize nonengulfed NDV. Periodically, samples were tested for HAD⁺ and HAD⁻ cells. The results presented in Table IV show that 99% of the actinomycin D, guanidine-treated control cells, not infected with poliovirus, became HAD⁺ when challanged with NDV, attesting to their susceptibility. In contrast, infection with poliovirus rendered all of the cells refractory to NDV and these were scored as HAD⁻.

Viruses Insensitive to Intrinsic Interference

Table II lists the challenge viruses that can infect cells made refractory to NDV by the interference inducing viruses shown in Table I. All challenge viruses were used at high multiplicities, i.e., m=10 infectious particles. Cytopathic effects on NDV-refractory cells resulting from the action of these viruses were indistinguishable in their time of onset and severity from control cells challenged at the same time. In the case of the noncytopathic, hemadsorbing simian virus, infectivity was scored by the hemadsorption-positive state it conferred on cells. Thus, rubella virus-infected GMK cells that were completely HAD⁻ upon challenge

TABLE IV
NDV-REFRACTORY STATE INDUCED BY CYTOPATHIC POLIOVIRUS
IN HELA CELLS

Cell Treatment	Time of NDV Challenge (Hours after Poliovirus Attachment)	Number of Cells		Fraction of HAD⁻ cells
		HAD⁺	HAD⁻	
Actinomycin D	4[a]	426	75	0.18
+				
guanidine	6	510	5	0.01
	17	505	3	0.01
Actinomycin D	4	0	450	1.0
+				
guanidine	6	0	525	1.0
+				
poliovirus[b]	17	0	750	1.0

[a] At 2½ hours, i.e., ½ hour post NDV-antiserum treatment, there were no HAD⁺ cells.
[b] 1n = 200 PFP

with NDV, became 100 per cent HAD⁺ when exposed to the simian virus. This produced a cell that was simultaneously expressing the characteristics of the two infecting viruses — rubella virus-induced NDV-refractoriness and simian virus-induced hemadsorption positivity. The 4 strains of NDV thus far tested (Table II) are all excluded by intrinsic interference.

DISCUSSION

Data presented here extend our earlier study of rubella virus-induced interference to Newcastle disease virus (18) and demonstrate that several unrelated viruses can induce a similar pattern of interference. Our experiments show that the interference inducing capacity of these viruses is lost upon irradiation with ultraviolet light, does not take place in the presence of puromycin, but is expressed in the presence of actinomycin D. In concert, these results point out the necessity of protein synthesis, coded for by a functional viral genome, in the induction of the interference. As documented previously (18), the NDV-refractory state is an intrinsic property of the individual cell that has been infected, in contrast to cells manifesting interferon-mediated interference. The latter type of interference is characterized by cellular resistance to a broad spectrum of viruses and may be induced in uninfected cells by the addition of an extrinsic factor, interferon (12, 22). The NDV-refractory state described here is not mediated by the action of interferon since interference can develop in cells blocked by actinomycin D from synthesizing the translation inhibitory protein responsible for the action of interferon (20). Cells refractory to all multiplicities of NDV demonstrate their usual susceptibility to a whole spectrum of other viruses when challenged at high multiplicity. The list of these challenge viruses has been extended to include vesicular stomatitis virus, encephalomyocarditis virus, and SV5 (Table II). Whether strains of NDV will continue to constitute the only viruses excluded by this type of interference remains to be established.

Because of these considerations, we term the NDV-refractory state *intrinsic interference* and define it as: a viral genome induced cellular state of resistance to challenge by Newcastle disease virus, coexistent with a state of susceptibility to a broad spectrum of other viruses.

The saturated hemadsorption pattern seen on HAD+ cells at the conclusion of the NDV challenge is significant, and suggests that the NDV-refractory state represents an all-or-none block, since a cell with small amounts of NDV hemagglutinin on its surface, either because of a delay in initiation of hemagglutinin synthesis, or a lower level of synthesis, can be detected as a result of its reduced capacity to bind red blood cells (17).

The linear rate of development of intrinsic interference shows a dependence on multiplicity of the inducing virus at the time of infection. Virus added at a later time does not appear to influence the rate of development of interference (Lipschitz, Marcus and Carver, In preparation). We interpret this to mean that only the initial input strands of the inducing virus may function as mRNA for the protein(s) that presumably act to bring about interference. Consistent with this interpretation is the development of an NDV-refractory state in HeLa cells infected with poliovirus under conditions (presence of guanidine) that prevent replication of viral RNA (3).

When long-standing cultures of NDV-refractory cells are removed from confluent monolayers and set down in fresh medium intrinsic interference disappears, slowly at first and then with a half-time of about 10-15 hours. After the NDV-refractory state is lost in about 99% of the cells, usually over a period of 6-7 days, there is a gradual return to the hemadsorption-negative state until all cells are again refractory to NDV. We are studying this process and at present can only speculate that the half-life of the refractory state expressed after subculture is perhaps dependent upon cell division and some dilution effect, or a break in contact-inhibition regulated events. The small fraction of infective centers found in these cultures could represent the source of virus for reinfection once the cells were again susceptible. We have omitted from consideration the possible gradual loss of the inducing virus from the cell population. Clearly, the situation is complex and requires further investigation.

Although no obvious taxonomic relationship is apparent between the different types of viruses that can induce intrinsic interference, they share two characteristics in common: 1) they are noncytopathic (the special test conditions for virulent poliovirus may be thought of as delaying cytopathicity and providing a temporary noncytopathic state (2)), and 2) they are RNA viruses insensitive to the action of actinomycin D. Only further investigation will reveal the biochemical event(s) common to the interference inducing capacity of these viruses. We have not tested noncytopathic myxoviruses as inducers of intrinsic interference because of possible complications attending receptor destruction by viral neuraminidases (4, 16). However, it is apparent that not all noncytopathic viruses manifest intrinsic interference. Thus, cells overtly infected with, or carrying, parainfluenza viruses (9, 14), including SV5 (P. Choppin, personal communication) and hog cholera virus (13) appear to be as sensitive, or more so, to NDV infection. Cells transformed by tumor viruses do not appear to be refractory to NDV (H. Hanafusa, H. Temin, personal communication unpublished observations).

Although NDV attaches to cells manifesting intrinsic interference, and enters an eclipse phase in an apparently normal manner, the ultimate fate of the NDV genome is not known, nor is the mechanism controlling its expression understood. We are continuing this aspect of our study. Our approach to this problem is modeled after experiments which have shown that viral RNA is not translated when complexed with ribosomes from interferon-treated cells (20). This inhibition is brought about by the presence of a trypsin-sensitive translation inhibitory protein on the ribosomes of the interferon-treated cell. Experiments are underway to look for a similar, but more specific translation inhibitory protein, in this case presumably coded for by the genome of the inducing virus, that acts to block translation of only NDV-RNA.

The NDV-refractory state induced by intrinsic interference viruses represents a unique example of heterotypic interference. However, in the biochemical sense it is not without precedent that taxonomically unrelated viruses initiate similar biochemical events; for example, the induction of interferon production, or the turn-off of host cell macromolecular synthesis. It would not be profitable to compare other types of viral interference with intrinsic interference until more is known about its molecular basis. But, perhaps mention of one example is appropriate since it raises the intriguing question of whether intrinsic interference inducing viruses can interfere with each other. In this connection, Allen and Lockart (1) have published an abstract in which they report on interference of a cytopathic strain of Western equine encephalomyelitis virus by a slightly cytopathic strain, in the presence of actinomycin D. Other examples of non-interferon mediated interference have been reported (6, 11, 21).

ACKNOWLEDGMENTS

This investigation was supported by grants AI-03619-06 VR and GM-12646-02A1 from the National Institutes of Health. P.I.M. is a Research Career Development Awardee of the National Institute of Allergy and Infectious Diseases (5-K3-GM-15,461-06). D.H.C. is a Kennedy Scholar. The contributions of Miss Harriet Lipschitz (Ph.D. candidate) and Dr. Alfred Gilbert in performing certain experiments are gratefully acknowledged as are their valuable suggestions made during the course of these studies. Miss Helene Weiss provided excellent technical assistance.

LITERATURE CITED

1. ALLEN, P. T., and LOCKART, R. Z.: An example of viral interference not mediated by interferon. *Bacteriol Proceed. (Amer. Soc. Microbiol.)* p. 98, 1965.
2. BABLANIAN, R., EGGERS, H. J., and TAMM, I.: Studies of the mechanism of poliovirus-induced cell damage. I. *Virology, 26*:100-113, 1965.
3. BALTIMORE, D., EGGERS, H. J., FRANKLIN, R. M., and TAMM, I.: Poliovirus-induced RNA polymerase and the effects of virus-specific inhibitors on its production. *Proc. Natl. Acad. Sci., 49*:843-849, 1963.
4. BALUDA, M.: Loss of viral receptors in homologous interference by ultraviolet-irradiated Newcastle disease virus. *Virology, 7*:315-327, 1959.
5. CARVER, D. H., and MARCUS, P. I.: Enhanced interferon production from chick embryo cells aged *in vitro*. *Virology,* In Press, June, 1967.
6. CORDS, C. E., and HOLLAND, J. J.: Interference between enteroviruses and conditions effecting its reversal. *Virology, 22*:226-234, 1964.
7. GRAVELL, M., and MALSBERGER, R. G.: A permanent cell line from the fathead minnow (*Pimephales promelas*). *Ann. New York Acad. Sci., 126*:555-565, 1965.
8. HAM, R. G.: An improved nutrient solution for diploid chinese hamster and human cell lines. *Exper. Cell Res., 29*:515-526, 1963.
9. HERMODSSON, S.: Inhibition of interferon by an infection with parainfluenza virus type 3. *Virology, 20*:333-343, 1963.
10. HOTCHIN, J. E., DEIBEL, R., and BENSON, L. M.: Location of noncytopathic myxovirus plaques by hemadsorption. *Virology, 10*:275-280, 1960.
11. HUANG, A. S., and WAGNER, R. R.: Defective T particles of vesicular stomatitis virus. II. *Virology, 30*:173-181, 1966.
12. ISAACS, A.: Interferon. *Advances Virus Res., 10*:1-38, 1963.
13. KUMAGAI, T., SHIMIZU, T., IKEDA, S., and MATSUMOTO, M.: A new *in vitro* method (END) for detection and measurement of hog cholera virus and its antibody by means of effect of HC virus on NDV in swine tissue culture. I. *J. Immunol., 87*:245-256, 1961.
14. MAENO, K., YOSHII, S., NAGATA, I., and MATSUMOTO, T.: Growth of Newcastle disease virus in a HVJ carrier culture of HeLa cells. *Virology, 29*:255-263, 1966.
15. MARCUS, P. I.: Host-cell interaction of animal viruses. II. *Virology, 9*:546-563, 1959.
16. MARCUS, P. I.: Single cell techniques in tracing virus-host interactions. *Bacteriol. Rev., 23*:232-249, 1959.
17. MARCUS, P. I.: Dynamics of surface modification in myxovirus-infected cells. *Cold Spring Harbor Symp. Quant. Biol., 27*:351-356, 1962.
18. MARCUS, P. I., and CARVER, D. H.: Hemadsorption-negative plaque test: New assay for rubella virus revealing a unique interference. *Science, 149*:983-986, 1965.
19. MARCUS, P. I., and ROBBINS, E.: Viral inhibition in the metaphase-arrest cell. *Proc. Natl. Acad. Sci., 50*:1156-1164, 1963.

20. Marcus, P. I., and Salb, J. M.: Molecular basis of interferon action: Inhibition of viral RNA translation. *Virology, 30:*502-516, 1966.
21. Roizman, B.: Abortive infection of canine cells by herpes simplex virus. III. *Virology, 27:*113-117, 1965.
22. Wagner, R. R.: Interferon. *Amer. J. Med., 38:*726-737, 1965.

PERSISTENT VIRAL INFECTION IN CELL CULTURES[1]

DUARD L. WALKER, M.D.

$\mathbf{M}$Y PURPOSE IS TO CONSIDER persistent viral infections of cells in culture and to point to some features of the cell-virus relationships in such cultures that may be of interest. I want first to review briefly the characteristics of the types of persistent infections that so far have been described and then to dwell at greater length on one type that is particularly interesting because of the cell-to-daughter-cell transmission through mitosis that is seen in it and because of the type of response to superinfection found in such cultures.

MAJOR TYPES OF PERSISTENTLY INFECTED CELL CULTURES

In the time that has elapsed since the use of cell cultures for cultivating animal viruses became widespread, many observations have been reported of infections in which there is persistent multiplication of a virus while the cell culture continues to survive and to grow. The number and variety of these (Walker, 1964) and the fragmentary information that is available about many of them has tended to make this area of virology rather confusing. I believe, however, that if for each culture the answers to the following questions can be obtained, a general understanding of the important processes and, particularly, an understanding of the mechanisms of cell protection in persistently infected cultures becomes possible. The questions are: a) What fraction of the cell population is infected? b) Do infected cells divide and grow into infected clones? c) Must antibody or other anti-viral factors be supplied in the culture medium to maintain equilibrium? d) Can the cultures be freed of virus (cured) by serial cultivation in a

[1]Studies by the author were supported in part by research grant AI-06334 from the National Institute of Allergy and Infectious Diseases.

medium containing antiviral antibody, or can virus-free clones readily be obtained by cloning under antibody? Are cured cultures or virus-free clones resistant to the carried virus? e) Is the culture resistant to superinfection by the carried virus, and is it resistant to infection by other viruses? f) Can interfering factors be demonstrated in the culture?

TABLE I
CHARACTERISTICS OF PERSISTENT INFECTIONS IN CULTURES OF
GENETICALLY RESISTANT CELLS

1. Only a small fraction of the cell population is infected.
    The virus cycle is a standard one in infected cells and the cells are destroyed. Infected cells do not divide or grow into infected clones.
2. Most cells are uninfected and are protected because they are genetically relatively resistant.
    The culture is relatively resistant to superinfection with homologous virus and the resistance is retained in cured cultures and in uninfected clones.
3. The culture can be cured by antiviral serum in the medium. Clones of uninfected cells are easily obtained by cloning cells in an antibody-containing medium.
4. Antibody or other antiviral factors need not be supplied in the medium. Interfering factors are not essential to cultural stability.

It sometimes appears that more than one mechanism of cell protection is operating in a culture but, nonetheless, if one looks at those cultures where the major mechanism can be discerned, four distinct groups stand out.

The first type of persistently infected culture is one in which most cells are genetically quite resistant to the carried virus and the infection is perpetuated in a minority of cells in the population. The general characteristics of this type are listed in Table I. Examples of infections of this kind are those of Coxsackie A9 virus in HeLa cells described by Takemoto and Habel (1958), and poliovirus in HeLa cells described by Pácsa (1961).

The second type includes cultures in which the cells are genetically susceptible to the carried virus, but the transfer of virus from cell to cell is limited by anti-viral factors in the medium. The characteristics of this group are listed in Table II. The best examples of this type of infection are found in the cultures of herpes simplex virus in various cell lines described by Wheeler and Canby (1959) and many other investigators.

TABLE II

CHARACTERISTICS OF PERSISTENT INFECTIONS IN CULTURES OF
GENETICALLY SUSCEPTIBLE CELLS PROTECTED BY ANTIVIRAL
FACTORS IN THE MEDIUM

1. Only a small fraction of the cell population is infected.
   The virus cycle is a standard one in infected cells and the cells are destroyed. Infected cells do not divide or grow into clones.
2. Most cells are uninfected and are protected by antiviral factors in the medium.
   If antiviral factors are removed from the medium, the culture is not resistant to superinfection by the carried virus, and it is not resistant to unrelated viruses.
   Clones of uninfected cells are easily obtained by cloning cells in an antibody-containing medium.
3. Antibody or other antiviral factors must be supplied in the culture medium. Interfering factors are not essential to cultural stability.

The third type is that in which the cells are genetically susceptible, but most cells are uninfected and are made resistant to infection by interfering factors produced within the culture. The general features are shown in Table III. Good examples of cultures of this kind are found in the cultures of Newcastle disease virus in L cells studied extensively by Henle and coworkers (Henle, 1963) and the cultures of Western equine encephalitis virus in L cells studied by Lockart (1960).

The fourth category includes a group of cultures quite different from those just described and is a type of viral infection only

TABLE III

CHARACTERISTICS OF PERSISTENT INFECTIONS IN CULTURES OF
GENETICALLY SUSCEPTIBLE CELLS PROTECTED BY
INTERFERENCE AND INTERFERON

1. Only a small fraction of the cell population is infected.
   The virus cycle in infected cells tends to be an abortive one that yields incomplete virus and interferon in addition to some infectious virus. Infected cells do not divide or grow into colonies.
2. Most cells are uninfected and are protected by a resistance induced by incomplete virus or interferon.
   The culture shows increased resistance to superinfection by the carried virus and to infection by other viruses. The increased resistance is lost after the culture is cured and is not found in uninfected clones.
3. The culture can be cured by use of antiviral serum in the medium. Clones of uninfected cells are obtained by cloning in an antibody-containing medium.
4. Antibody or other antiviral factors need not be supplied in the medium.

recently recognized and studied. The characteristics of these infections are listed briefly in Table IV and I want to expand on these characteristics in the time remaining to me.

TABLE IV

CHARACTERISTICS OF PERSISTENT, NONCYTOCIDAL INFECTIONS
TRANSMITTED THROUGH CELL DIVISION

1. All, or a large fraction, of the cells are infected.
2. Infected cells divide and grow into infected clones. Most clones of cells obtained from the culture are infected.
3. The culture is resistant to superinfection by the infecting virus but is not resistant to unrelated viruses.
4. Antibody or other antiviral factors need not be supplied in the culture medium.
5. The culture is not cured by addition of antiserum to the medium.

PERSISTENT, NONCYTOCIDAL INFECTIONS TRANSMITTED THROUGH CELL DIVISION

Quite a number of viruses, particularly ones of the paramyxovirus group, have been shown to establish in cell cultures a cell-virus relationship in which there is a cytoplasmic infection of all or a very large fraction of the cells without evident cytopathic effect and in which transmission of the infection is from cell to daughter cell through mitosis. Table V lists some virus cell systems that appear to be of this type. Data are not complete for all of them, but there is enough information to make it quite likely that they constitute a group of similar infections. I shall later mention other virus-cell systems that might be added to this list.

The mumps virus system that Hinze and I have studied (Walker and Hinze, 1962a, 1962b) is illustrative of most of the

TABLE V

PERSISTENT, NONCYTOCIDAL INFECTIONS TRANSMITTED THROUGH CELL DIVISION

Virus	*Cell*	*Description*
Mumps	Human conjunctiva	Walker and Hinze (1962a, 1962b)
Measles	HeLa	Rustigian (1962)
HA2	HeLa	Ishida *et al.* (1964)
(Parainfluenza 1)		Miyamoto *et al.* (1965)
Sendai	HeLa	Maeno (1963)
		Maeno *et al.* (1965)
HA1		
(Parainfluenza 3)	Human conjunctiva	Cole and Hetrick (1965)
SV5	Primary monkey kidney	Choppin (1964)

characteristics of infections of this type. Mumps virus in human conjunctiva cells easily establishes an infection that in growing cultures does not cause apparent damage to the cells. There is no requirement for antibody in the system and the cultures usually grow at a rate similar to that of uninfected conjunctiva cells for an apparently unlimited period of time. We have maintained infected cultures for as many as 150 subcultures.

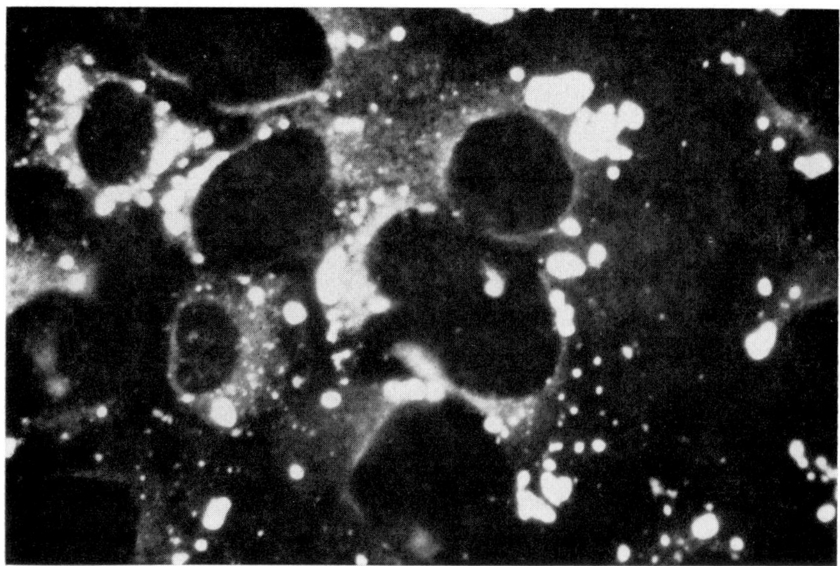

Figure 1. A persistent infection of mumps virus in human conjunctiva cells (C-M cultures). The cells were fixed in acetone and stained with fluorescein-conjugated anti-mumps virus serum. Mumps virus antigen appears as brightly stained, discrete foci in the cytoplasm. Magnification-X400.

Fluorescent antibody can be used to demonstrate that most of the cells in such persistently infected cultures (we have called them C-M cultures) contain cytoplasmic foci of viral antigen (Fig. 1). The antigen appears as discrete, sharply circumscribed masses. Electronmicroscopic study of sections through infected cells shows (Fig. 2) these cytoplasmic masses to be dense collections of strands that look like the helical internal component of the virus (unpublished studies of G. ZuRhein and D. Walker).

Isolation of single cells from the cultures and growth of them into clones under antibody has indicated (Table VI) that usually 98% or more of cells are infected. Development of infected clones from single infected cells also provides evidence that infected cells can divide and can do so repeatedly.

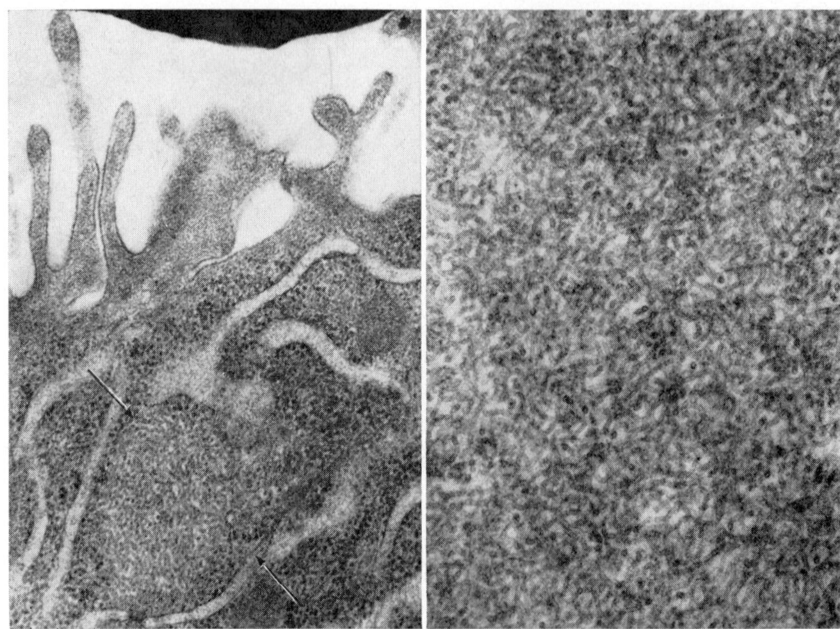

Figure 2. Electronmicrographs of sections of human conjunctiva cells with persistent mumps virus infection. *Left*: A collection of strands of viral internal component (*between arrows*) in the cell cytoplasm beneath the cell surface. A portion of an erythrocyte is seen at the top of the photograph adhering to surface villi. X 52,000. *Right*: Detail of a cytoplasmic collection of strands of internal component. Some strands are seen in cross section. X 98,000.

In spite of the fact that almost all cells are infected, in actively growing cultures only about 1-5% of cells have sufficient virus at their surface to cause erythrocytes to adsorb to the cell. It appears likely, therefore, that in actively growing cultures only 1-5% of cells are actively excreting virus. This situation can be changed, however, by manipulating the environment of the cells (Walker *et al.*, 1966). If the monolayer is allowed to become complete, or the incubation temperature is reduced, or the serum in the

TABLE VI
ISOLATION OF CLONES FROM C-M CULTURES

C-M culture	Cloning Efficiency (Per Cent)	No. of Clones Examined	No. of Clones Containing Antigen
Culture 1	60	51	46
Culture 2	43	76	76
Culture 3	100	89	89
Culture 4	46	53	52
Culture 5	40	52	52

medium is reduced, and particularly if all of these are done at once, there is then a marked increase in the number of cells to which erythrocytes absorb and, also, a marked increase in the amount of virus in the medium (Table VII). Since the manipulations mentioned are ones that slow or stop cell multiplication, we have felt that the simplest explanation of this observation is that even though multiplication of the cells is slowed, multiplication of virus can continue and virus accumulates until it reaches levels that result in the appearance of virus at the cell surface and in virus release by most of the cells. This would suggest that multiplication of virus and cell are quite independent processes in this type of infection just as they are in the usual cytocidal viral infection. Although there is not time here to present the pertinent data (see Walker *et al.*, 1966), this view is supported by our finding that if viral multiplication is slowed by raising the incubation to 38°C and antibody is added to the medium to protect uninfected cells that appear, then in the course of many subcultures there is an accumulation of uninfected cells that become virus-free as the virus is diluted out during cell division. These observations do not exclude, however, the possibility that there is some form of repression on viral multiplication in multiplying cells that is at least partially relieved when cell division is inhibited.

Of particular interest are observations regarding resistance to superinfection in persistently infected cultures of this type. The available information from 5 culture systems is summarized in Table VIII. Where it has been tested, it has been found that addition of more of the carried virus (homologous virus) or of a more virulent line of the homologous virus does not result in cytopathic effect or increased virus production. Thus, the cultures appear to be resistant to superinfection with homologous virus. Maeno (1963) has found with the HeLa-Sendai system that the infected cells adsorb homologous challenge virus as well as uninfected cells, so the resistance seems to be at a point after adsorption. Some degree of resistance has sometimes been found to viruses closely related to the carried virus as, for instance, in the resistance to mumps and Sendai viruses in the HeLa cell-HA2 system of Ishida *et al.*, (1964) and the small degree of resistance to influenza A and Sendai viruses found in the conjunctiva cell-mumps system of Walker and Hinze (1962). But the cultures usually have been found to lack resistance to heterologous viruses when tested by any of several methods.

TABLE VII
EFFECT OF INHIBITION OF CELL MULTIPLICATION
ON VIRUS EXCRETION BY CELLS IN C-M CULTURES

Cultural Conditions	Initial Cell Count (Cells/Vial)	Final Cell Count[a] (Cells/Vial)	Virus in Medium[a] (PFU/Cell)	HAd+Cells[b]/ No. Counted	Per Cent HA+Cells
Growth medium at 37°C	72,000	384,000	0.0034	82/1408	5.8
Low-serum medium at 35°C	644,000	605,000	0.27	986/1773	55.6

[a]After 72 hr under experimental conditions.
[b]Hemadsorption-positive cells. From Walker, D., Chang, P., Northrop, R., and Hinze, H.: *J. Bacteriol.*, 92:983-989, 1966.

Maeno *et al.* (1966) have recently reported that Newcastle disease virus (NDV) produces larger plaques and more plaques in their Sendia-infected HeLa cell cultures than it does in control HeLa cells, and that after infection with a low multiplicity of NDV, the carrier cultures yield more NDV than do control HeLa cultures. They find a similar enhancement of Sindbis virus multiplication. They present evidence that indicates that their persistently infected cultures do not produce interferon upon stimulation with NDV and are less susceptible to the virus-blocking action of interferon than are normal HeLa cells, and they conclude that this refractoriness explains the enhanced activity of NDV in their carrier cultures. Similar refractoriness to interferon production or action has not yet been demonstrated for the other

TABLE VIII

RESISTANCE OF PERSISTENTLY INFECTED CULTURES
TO SUPERINFECTION WITH HOMOLOGOUS AND HETEROLOGOUS VIRUSES

| Virus | Persistent Infection | Superinfection | | Reference |
	Cells	Resistance	No Resistance	
Mumps	Human conjunctiva	Mumps Influenza A (small) Sendai (small)	Newcastle disease Vaccinia Vesicular stomatitis	Walker and Hinze (1962a)
Sendai	HeLa	Sendai	Newcastle disease Poliovirus 1 Sindbis	Maeno *et al.* (1965, 1966)
HA1 (Parainfluenza 3)	Human conjunctiva	ECHO 6 Poliovirus 2	Vesicular stomatitis Guaroa Herpes simplex	Cole and Hetrick (1965)
HA2 (Parainfluenza 1)	HeLa	Sendai Mumps	Poliovirus Adenovirus Vaccinia	Ishida *et al.* (1964)
SV5	Primary monkey kidney		Coxsackie A9 Coxsackie B4 ECHO 7 ECHO 12 Influenza A2 Poliovirus 2 Vaccinia Vesicular stomatitis	Choppin (1964)

cellvirus systems in Table VIII, but where interferon has been tested for directly it has not been found and the general lack of resistance to heterologous viruses certainly argues that interferon is not operative in these infections.

There is much to be done before we can have a clear understanding of the cell-virus relationship in these cultures. Most of the work done on them so far just outlines the problems. It appears, however, that here are some noncytocidal infections that do not inhibit cell multiplication, nor other cell functions that we know about, and the infections appear not to stimulate interferon production. Yet, in infected cells, there is a resistance to homologous virus and some resistance to closely related viruses. The resistance to ECHO 6 and polio-2 viruses in the conjunctiva cell-HA1 virus system studied by Cole and Hetrick is not explainable at this time, but serves to emphasize the incompleteness of our information about this kind of infection.

There is not time here to discuss them in detail, but it should be mentioned that there are other viruses that sometimes can establish infections similar in several respects to those I have just discussed. Although rubella virus usually produces a noncytocidal infection it seems frequently to cause release of interfering factors into the medium (Maassab and Veronelli, 1966). However, Rawls and Melnick (1966) have found that in cultures of human cells derived from the tissues of infants infected *in utero* practically all the cells carry rubella virus but continue to divide and do not produce detectable interferon. Their cultures are resistant to some heterologous viruses but not to others; a state rather like that just noted in the conjunctiva cell-HA1 system of Cole and Hetrick. Some of the arboviruses seem to establish in some cells infections in which there is a high proportion of infected cells in the culture, and, apparently, continued multiplication of infected cells (Maguire and Miles, 1965; Mayer, 1962). These, however, usually have interferon in the medium to complicate their analysis. It appears, also, that rabies virus and lymphocytic choriomeningitis virus can establish persistent, noncytocidal infections in cells in which there is some enhancement of the rabies infection by the lymphocytic choriomeningitis virus (Fernandes *et al.*, 1964; Wiktor *et al.*, 1966). And, finally, although the viruses so far mentioned are all cytoplasmic, enveloped, RNA viruses that are released by cell membrane budding, there is the rather peculiar example described by Hare and Morgan (1964) of a persistent, noncytocidal infection of polyoma virus in L cells that appears to be mainly a cytoplasmic infection that does not cause interferon production nor resistance to vaccinia virus although there is a variable degree of resistance in the culture to encephalo-myocarditis virus and considerable resistance to superinfection with polyoma virus.

I am sure that the studies on enhancement of viral growth in acute infections involving two viruses that are being reported by Hermodsson (1963), by Frothingham (1965), and by Cantell (Valle and Cantell, 1965) will be helpful in understanding some of the phenomena observed in these persistent, noncytocidal infections.

REFERENCES

1. CHOPPIN, P. W.: Multiplication of a myxovirus (SV5) with minimal cytopathic effects and without interference. *Virology, 23:*224-233, 1964.

2. COLE, F. E., JR., and HETRICK, F. M.: Persistent infection of human conjunctiva cell cultures by myxovirus parainfluenza 3. *Can. J. Microbiol., 11:*513-521, 1965.

3. FERNANDES, M. V., WIKTOR, T. J., and KOPROWSKI, H.: Endosymbiotic relationship between animal viruses and host cells. *J. Exper. Med., 120:*1099-1116, 1964.

4. FROTHINGHAM, T. E.: Further observations on cell cultures infected concurrently with mumps and Sindbis viruses. *J. Immunol., 94:*521-529, 1965.

5. HARE, J. D., and MORGAN, H. R.: Polyoma virus and L cell relationship| II. A curable carrier system not dependent on interferon. *J. Nat. Cancer Inst., 33:* 765-775, 1964.

6. HENLE, W.: Interference and interferon in persistent viral infection of cell cultures. *J. Immunol., 91:*145-150, 1963.

7. HERMODSSON, S.: Inhibition of interferon by an infection with parainfluenza virus type 3 (PIV-3). *Virology, 20:*333-343, 1963.

8. ISHIDA, N., HOMMA, M., OSATA, T., HINUMA, Y., and MIYAMOTO, T.: Persistent infection in HeLa cells with hemadsorption virus type 2. *Virology, 24:* 670-672, 1964.

9. LOCKART, R. Z.: The production and maintenance of a virus carrier state: The possible role of homologous interference. *Virology, 10:*198-210, 1960.

10. MAASSAB, H. F., and VERONELLI, J. A.: Characteristics of serially propagated monkey kidney cell cultures with persistent rubella infection. *J. Bacteriol., 91:*436-441, 1966.

11. MAENO, K.: Studies on the persistent infection in HeLa cell cultures with myxovirus parainfluenzae 1 (HVJ). *Virus (Osaka), 13:*16-23, 1963.

12. MAENO, K., YOSHII, S., NOGATA, I., and MATSUMOTO, T.: Persistent infection of

cell cultures with myxovirus parainfluenzae 1 (HVJ). *Symp. Cell Chem. Kanazawa, Japan, 15:*131-150, 1965.

13. MAENO, K., YOSHII, S., NOGATA, I., and MATSUMOTO, T.: Growth of Newcastle disease virus in a HVJ carrier culture of HeLa cells. *Virology, 29:*255-263, 1966.

14. MAGUIRE, T., and MILES, J. A. R.: The arbovirus carrier state in tissue cultures. *Arch. Ges. Virusforsch., 15:*457-474, 1965.

15. MAYOR, V.: Interaction of mammalian cells with tick-borne encephalitis virus. II. Persisting infection of cells. *Acta Virol., 6:*317-326, 1962.

16. MIYAMOTO, T., HINUMA, Y., and ISHIDA, N.: Intracellular transfer of hemadsorption type 2 virus antigen during persistent infection of HeLa cell cultures. *Virology, 27:*28-36, 1965.

17. PÁCSA, S.: Poliovirus-carrying lines of HeLa cells: Their establishment and sensitivity to viruses. *Acta Microbiol., 8:*329-332, 1961.

18. RAWLS, W. E., and MELNICK, J. L.: Rubella virus carrier cultures derived from congenitally infected infants. *J. Exper. Med., 123:*795-816, 1966.

19. RUSTIGIAN, R.: A carrier state in HeLa cells with measles virus (Edmonston strain) apparently associated with noninfectious virus. *Virology, 16:*101-104, 1962.

20. TAKEMOTO, K. K., and HABEL, K.: Virus-cell relationships in a carrier culture of HeLa cells and Coxsackie A9. *Virology, 7:*28-44, 1959.

21. VALLE, M., and CANTELL, K.: The ability of Sendai virus to overcome cellular resistance to vesicular stomatitis virus. *Ann. Med. Exper. Fenn., 43:*57-60, 1965.

22. WALKER, D. L.: The viral carrier state in animal cell cultures. *Progr. Med. Virol., 6:*111-148, 1964.

23. WALKER, D. L., CHANG, P., NORTHROP, R. L., and HINZE, H. C.: Persistent, non-cytocidal viral infection: Nonsynchrony of viral and cellular multiplication. *J. Bacteriol., 92:*983-989, 1966.

24. WALKER, D. L., and HINZE, H. C.: A carrier state of mumps virus in human conjunctiva cells. I. General characteristics. *J. Exper. Med., 116:*739-750, 1962a.

25. WALKER, D. L., and HINZE, H. C.: A carrier state of mumps virus in human conjunctiva cells. II. Observations on intercellular transfer of virus and virus release. *J. Exper. Med., 116:*751-758, 1962b.

26. WIKTOR, T. J., KAPLAN, M. M., and KOPROWSKI, H.: Rabies and lymphocytic choriomeningitis virus (LCMV). *Ann Med. Exper. Fenn., 44:*290-296, 1966.

27. WHEELER, C. E., and CANBY, C. M.: Influence of specific antibody on herpes simplex infections in tissue culture. *Arch. Derm., 79:*86-95, 1959.

BEHAVIOUR OF THE AMNIOTIC MEMBRANE DURING VIRAL INFECTION IN VITRO

C. CHANY, I. GRESSER, E. FALCOFF and F. FOURNIER

In the animal host, dissemination of virus within a given tissue is often limited, and diffuse cellular destruction is seldom observed. The liberation of interferon and other antiviral substances by infected cells may account in part for some instances of viral inhibition, although it seems unlikely that such mechanism are solely responsible for the localization of viral lesions. Human amniotic cells cultivated *in vitro* are fully susceptible to poliovirus.

In the intact membrane, however, previous investigators were not able to demonstrate polioviral multiplication. Shortly after trypsinization, the cells are also resistant to poliovirus. Holland attributed this lack of sensitivity of the cells to the lack of receptors in the organs — thus the receptor would appear only in tissue culture (1). Our study confirms most of Holland's observations, proposing, however, an alternative explanation to this phenomenon (2, 3).

Preliminary studies have shown that a considerable quantity of neutralizing antibody is so deeply absorbed into the tissues, that they cannot be washed out (2). This might account partially for the resistance of the membrane to poliovirus. When the membrane is kept for 3 days in tissue culture medium at room temperature, absorbed antibodies are slowly liberated in the supernatant. The membrane can then be infected with poliovirus but the behaviour of the cell in the membrane is strikingly different when compared with cells *in vitro*. Whatever the initial multiplicity of infection may be, poliovirus is eliminated in the supernatant for long periods (Fig. 1).

However, in this system at a given time, only a small proportion of the cells can be infected (Table I). Only a small amount of interferon is liberated in the supernatant.

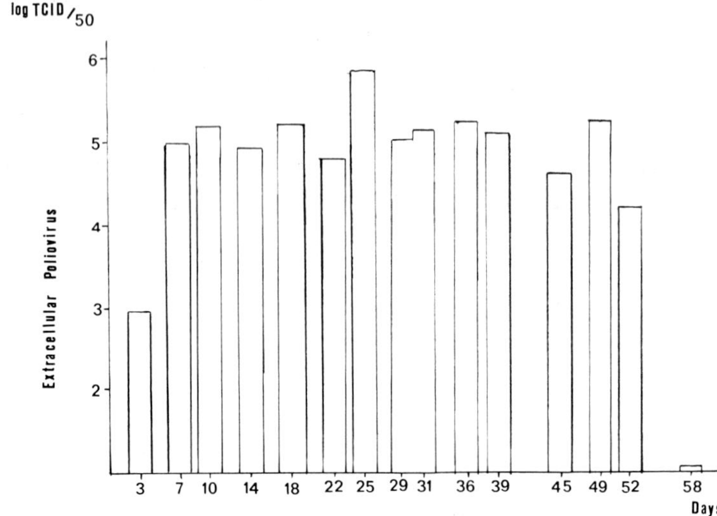

Figure 1. Extracellular polio virus. Each column represents the titer of polio virus in the supernatant at a given day.

TABLE 1

PROPORTION OF AMNION CELLS INFECTED WITH POLIOVIRUS

Membrane	Day after Infection	Total No. of Cells (millions)	Infected Cells
			%
I	14	2.39	.1
	20	1.00	2.8
II	14	3.34	2.0
	16	0.80	1.0
	21	0.60	0.8

Similarly, when cells are infected after trypsinization, these cells are also resistant to the poliovirus. The cells become fully susceptible only 3-4 days after the culture was started (Fig. 2). This strikingly different behaviour of the amniotic membrane, when compared to freshly trypsinized amniotic cells, can be explained by the presence of a mechanical barrier in the organs. This barrier is not destroyed by trypsin, but is liberated slowly from the cells during their maintenance in tissue culture medium. This view is supported by two observations:

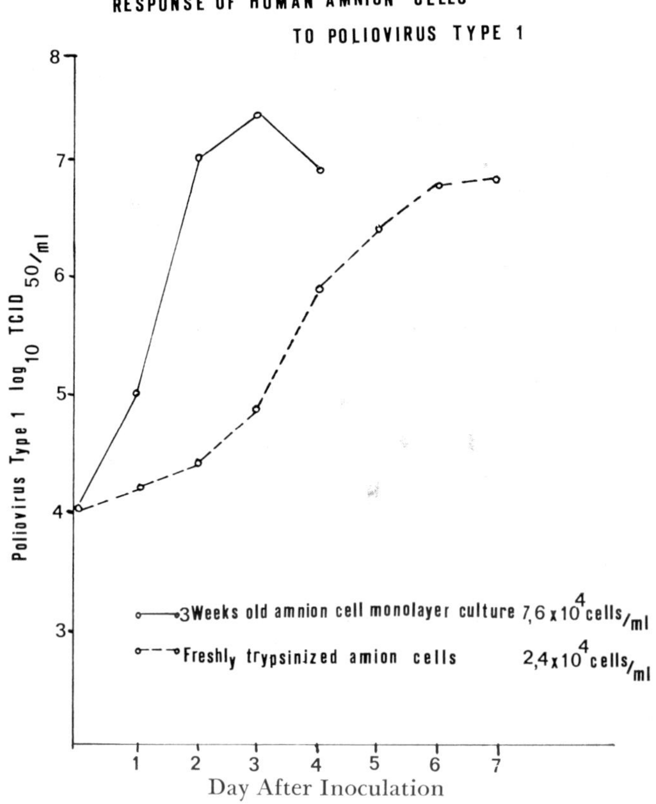

Figure 2. Response of human amnion cells to polio virus type 1. o———o Three weeks old amnion cell monolayer culture 7.6 x 10⁴ cells/ml. o----o Freshly trypsinized amnion cells 2.4 x 10⁴ cells/ml.

1) When amniotic membrane strips are treated with a silver nitrate solution, the presence of a well demarcated argentophil envelope separating the cells can be demonstrated. This technique was described by Ranvier for the study of the peritonium, since silver grains emphasized the intercellular matrix. After trypsinization, using the same technique, a thick envelope was visualized around either single cells or clumps of cells. When the cells were maintained in tissue culture medium, during the ensuing days the amnion cells flattened and spread out on the glass. The silver impregnation method showed then less sharply delineated cell boundaries (Figs. 3 and 4).

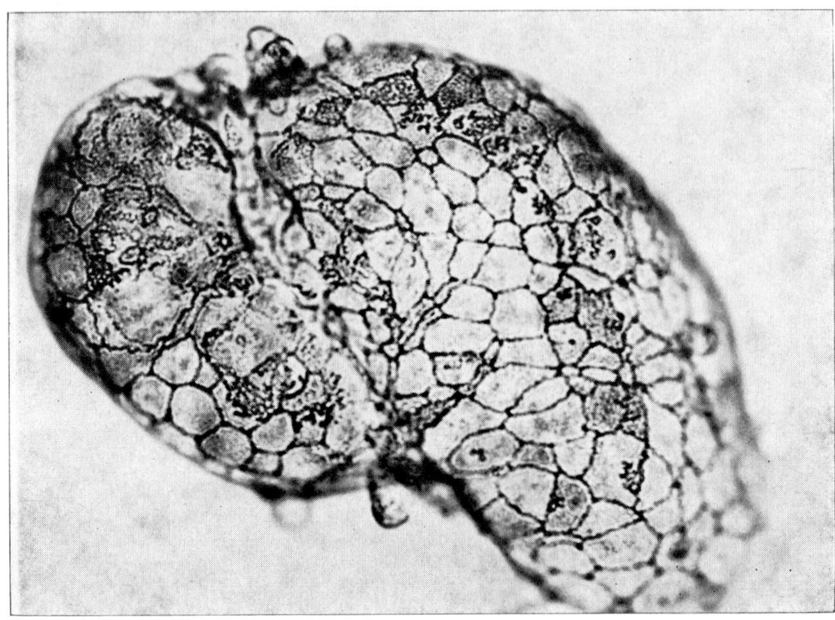

Figure 3. Aggregate of freshly trypsinized amnion cells. Note argentophil matrix between the cells.

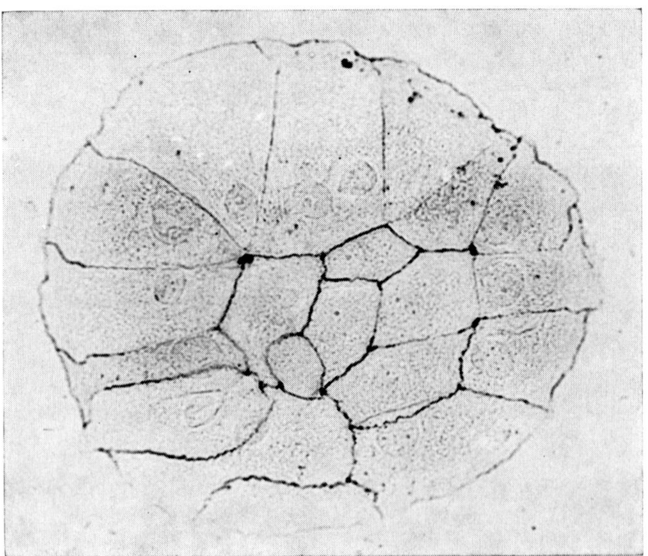

Figure 4. Human amnion cells, 96 hours in culture. Note diminished impregnation by silver (800 x).

2) Since animal cells incorporate viral particles by pinocytosis, it was of interest to compare the phagocytic capacity of the culture of amnion cells, and the same cells in the amniotic membrane. When the intact membrane was exposed to a suspension of carbon particles, these particles formed a film above the cellular layer and did not penetrate into the cytoplasm. Likewise, carbon particles were rarely incorporated in freshly trypsinized cells (Fig. 5).

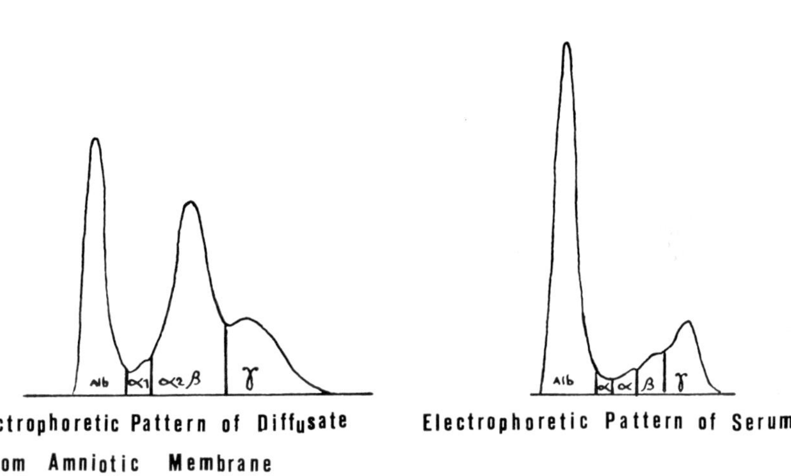

Figure 5. Electrophoretic pattern of diffusate from amniotic membrane (*left side*). Electrophoretic pattern of serum (*right side*).

Cells of monolayer cultures engulfed this material when they were exposed to carbon particles after 5-6 days after trypsinization. When these cells were trypsinized, sedimented by centrifugation, fixed and stained, the histologic sections of the cell block clearly demonstrated the carbon particles within the cytoplasm of the amnion cells. It was postulated, therefore, that this argentophil material represents a mechanical obstacle to the penetration of poliovirions into the cells, blocking the cell receptors. When this material disappears in tissue culture, the cell demonstrates a marked phagocytic activity and becomes susceptible to polioviral infection.

PENETRATION OF MYXOVIRUSES

The mucoprotanic nature of this intercellular matrix was suspected because of the electrophoretic pattern of the tissue culture medium in which the amniotic membranes were maintained. They showed clearly the presence of an increased peak of β globulins which contained in general mucoproteins (Fig. 5). Since myxoviruses possess neurominidase activity, it was postulated that they might penetrate into the cell of the amniotic membrane without previous preparation. However, as shown before, these viruses produce only incomplete cycles in amniotic tissue culture cells, liberating in the supernatant considerable amounts of interferon.

When an amniotic membrane was infected at high multiplicities with parainfluenza I (Sendai) virus, within 24 hours the cells produced a considerable amount of interferon. However, when studied by gel filtration in a Sephadex G 100 or G 200 column, in presence of reference proteins, its molecular weight was found to be 160,000. In spite of this heavy molecular weight, this interferon behaves in a similar manner to the white cell interferon (molecular weight 25,000). It is pH² stable; it is destroyed by trypsin; and its intercellular action is blocked by actinomycin (Table II).

TABLE 2
SOME BASAL PHYSICO-CHEMICAL AND BIOLOGICAL PROPERTIES
OF HAM INTERFERON

	Effect
Trypsin 250γ/ml	destruction
Ribonuclease 100γ/ml	destruction
Desoxyribonuclease 100γ/ml	no destruction
pH 2 - pH 10	no destruction
Conservation	—20°C
Cell specificity	yes
Molecular weight	160.000
Actinomycin	inhibits its action

CONCLUSIONS

The cells in the amniotic membrane, and perhaps in other tissues, are protected by the presence of an intercellular matrix which acts as a barrier to viral dissemination. Duran-Reynals already underlined the importance of the ground substance of the mesenchyme, stating that *in vitro* viral infection is more than "conflicts between infectious agents and cells." His comments seem pertinent to the problem here investigated (4).

It can be postulated, therefore, that if viruses have access to the cell receptors, they multiply in the cells in a similar fashion, regardless of whether the cells are located in the organ or isolated *in vitro*. They produce both, virus and interferon.

REFERENCES

1. HOLLAND, J. J.: *Virology, 15:*312-326, 1961.
2. GRESSER, I., CHANY, C., and ENDERS, J. F.: *J. Bacteriology, 89:*470, 1965.
3. CHANY, C., GRESSER, I., VENDRELY, C., and ROBBE-FOSSAT, F.: *Proc. Soc. Exper. Biol. & Med., 123:*960, 1966.
4. DURAN-REYNALS, F.: *La Sem. des Hôpit. de Paris, 25:*1, 1952.

DUAL INFECTION OF MICE WITH NEWCASTLE DISEASE VIRUS AND MOUSE CYTOMEGALOVIRUS*

JUNE E. OSBORN, M.D.

Dual viral infection occurs not only in cell culture systems but also in the intact animal. I would like to describe some experiments in which it was found that mice undergoing acute murine cytomegalovirus (MCMV) infection became susceptible to Newcastle disease virus (NDV) with multiplication of both viruses in the spleen.

Extensive studies of acute and chronic MCMV infection in the laboratory of Donald N. Medearis, Jr., (1, 2, 3) raised the question of the role of interferon in pathogenesis of MCMV disease. In association with Dr. Medearis, I undertook to explore this problem, and found no evidence that interferon played a direct role. In fact, using NDV as a stimulant of circulating mouse interferon, it was found that mice undergoing acute cytomegalovirus infection were markedly inhibited in their capacity to respond to intravenous NDV with interferon (4).

Once this phenomenon of interferon inhibition was recognized, a question arose as to the fate of NDV in such mice. NDV multiplies poorly if at all in mouse tissue under ordinary circumstances, and striking interferon response (5) associated with rapid elimination of virus is the usual result of NDV inoculation of the normal animal. The failure of interferon response in our MCMV-infected mice suggested that those animals might respond differently to NDV challenge in other respects as well.

Figure 1 shows the format of experiments designed to determine the fate of infectious NDV in control animals and in mice injected on the fifth day of their MCMV infection. After a dose of $10^{8.2}$ plaque-forming units (pfu) of NDV administered intra-

*Aided by grants HD-01536 and G. R. S. G. FR-05507 from the National Institute of Child Health and Human Development.

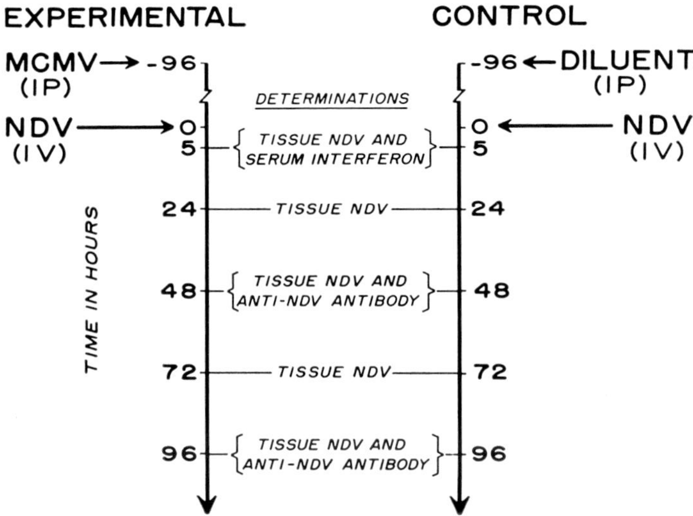

Figure 1. Plan of experiments to study dual infection with MCMV and NDV in mice.

venously, representatives of each group were sacrificed at various intervals and their blood and viscera tested for NDV by plaque assay in primary chick embryo cell cultures. In addition, studies of interferon response and neutralizing antibody appearance were conducted on sera from appropriate groups of the same animals.

As indicated in Figure 2, the spleen was found to sustain striking multiplication of NDV in the first 72 hours following inoculation. Control animals eliminated NDV with the usual rapidity, and viremia — indicated in the figure by an asterisk — was seen in only one group of controls beyond 5 hours. In contrast, the titer of NDV in spleen of MCMV-infected animals failed to drop at 24 hours and showed a definite increase at 48 and 72 hours. Low grade ND-viremia was a frequent occurrence at 24 and 48 hours and in one group was found to occur as late as 72 hours. The occurrence of viremia correlated with a high NDV titer in the spleen. The multiplication of NDV terminated spontaneously and the virus was eliminated rapidly before 96 hours.

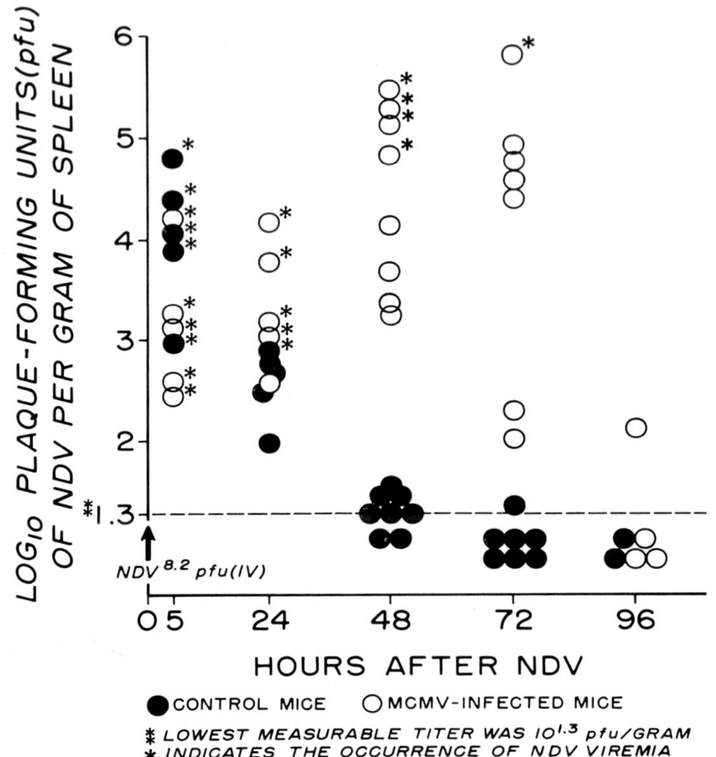

Figure 2. Titer of infectious NDV in mouse spleen at intervals after intravenous injection of 10^8.2 pfu of NDV. Each point represents result of testing spleens of 2-5 mice.

Figure 3 shows the same data in terms of mean spleen titer at each interval, and in addition, mean NDV concentration in lung and liver are shown. While the spleen is clearly the primary site of involvement, there is a definite delay in elimination of NDV from the lungs of MCMV-infected mice. The liver is apparently the most efficient of the three organs tested in eliminating or inactivating NDV in both groups; the small amount of NDV occasionally demonstrated in the livers of dually infected mice can be correlated with occurrence and titer of viremia in the same animals.

One factor which could contribute at least to the abnormal persistence if not to the multiplication of NDV in these mice would be abnormal clearance (6) . Figure 4 shows data concerning disappearance of NDV from the circulation of infected and control animals in the first five hours after administration. The lines connect mean titers at each interval, and it is notable that in all instances the mean NDV titer in the blood of cytomegalovirus-infected animals is slightly higher than controls. The difference at five hours is significant at the 0.05 level. Thus it appears that clearance is somewhat impaired in these dually infected animals.

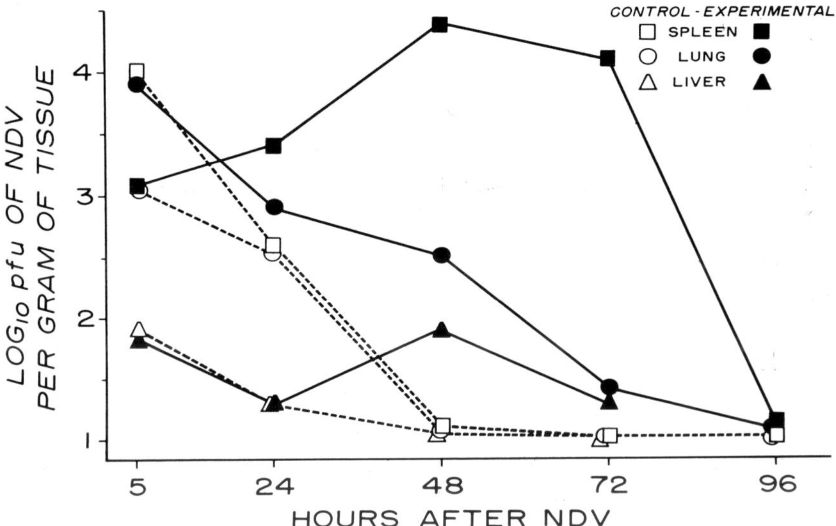

Figure 3. Mean titer of NDV in spleen, lung and liver after intravenous injection of $10^{8.2}$ pfu NDV: MCMV-infected (closed symbols) and control mice *(open symbols)* .

The possible role of neutralizing antibody response to NDV at such an early stage of infection is difficult to assess; but it was found that, as with interferon, antibody synthesis was markedly suppressed during acute MCMV infection. Figure 5 shows representative data concerning this phenomenon. Groups of control animals had good levels of neutralizing antibody 10 days after NDV, as measured by plaque reduction neutralization of NDV assayed in chick embryo cell culture. In contrast, MCMV-infected animals had barely detectable levels of neutralizing antibody at 10 days.

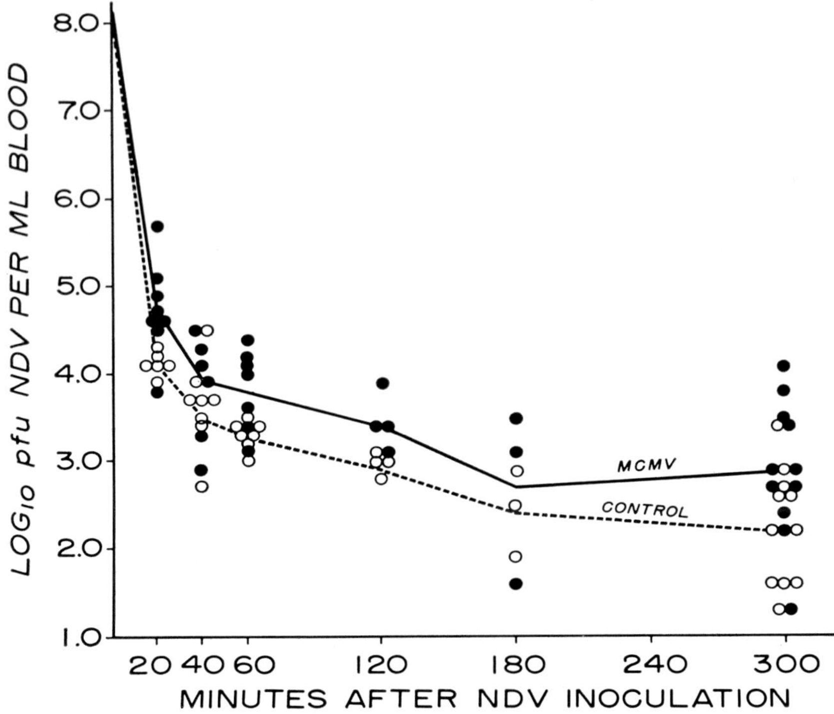

Figure 4. Clearance of NDV after intravenous injection of $10^{8.2}$ pfu NDV in MCMV-infected (closed symbols) and control (*open symbol*) mice.

As mentioned earlier, interferon response to intravenous NDV is markedly inhibited in acutely MCMV-infected mice. Figure 6 shows the plan of experiments in which an attempt was made to suppress NDV multiplication in these animals by administration of exogenous interferon. Interferon titer is expressed in terms of plaque reduction of encephalomyocarditis (EMC) virus in secondary mouse embryo cell culture, one unit being the reciprocal of that dilution of serum which causes a 50 per cent reduction in plaque count when challenged with 60-120 pfu of EMC. MCMV-infected animals were given intraperitoneal injections of either normal mouse serum or 320 units of mouse serum interferon one hour before NDV and twice that amount two hours after. Then, spleens were harvested at 48 hours and assayed for NDV.

	SOURCE OF SERUM	
MCMV-infected Mice 10 Days after NDV (IV)	*Control Mice 10 Days after NDV (IV)*	*Normal Uninoc- ulated Mice*
5*, 5, 10, 10	>40, >40, >40 80, 80, 80	<5, <5, <5

Figure 5. Suppression of NDV neutralizing antibody response in MCMV-infected mice.

*Each figure is result of testing pool of serum collected from mice and is the reciprocal of highest serum dilution which reduced control NDV plaque count by more than 50%.

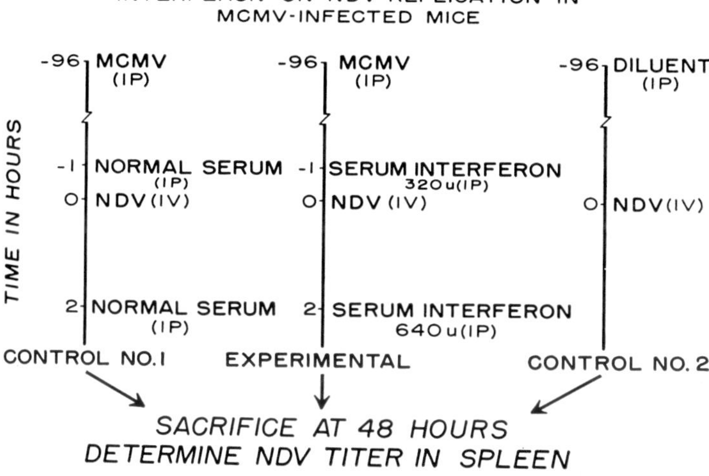

Figure 6. Plan of experiments to determine effect of exogenous interferon administration on course of NDV infection in MCMV-infected mice.

Figure 7 shows the results of these experiments. If the datum 1.3 is omitted from Group 1 as an aberrant datum, the mean reduction in NDV titer in interferon-treated mice is significant at the 0.01 level.

In considering this dual-infection model, it is important to know that, during the first week of MCMV infection, the spleen is the site of most vigorous cytomegalovirus multiplication, with the reticulo-endothelial organs as a group being primary loci of acute involvement (7). The salivary gland, the site of chronic infection in the mouse, becomes infected only secondarily.

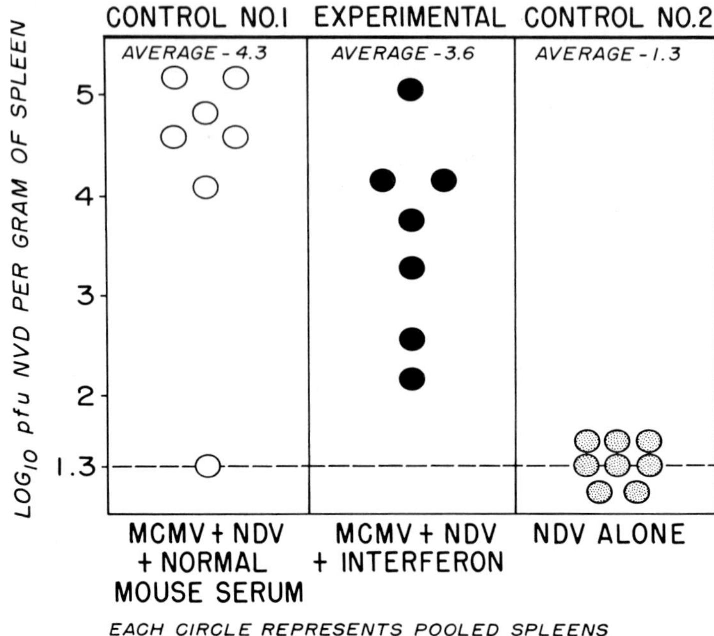

Figure 7. Effect of administering exogenous interferon on titer of NDV in mouse spleens 48 hours after NDV, $10^{8.2}$ pfu IV.

Thus, in the dually infected animals described here, both viruses are undergoing their most efficient multiplication in the same organ simultaneously. The factors which terminate NDV multiplication in this situation are not at all clear, but this termination coincides with a rapid drop in splenic MCMV titer, and it is possible that common determinants are operative.

The fact that both interferon and antibody synthesis are inhibited in these animals directs attention again to the spleen, thought by many to be important in both kinds of anti-viral response. It is of interest that the only other viruses that have thus far been shown to suppress these two diverse proteins are members of the murine leukeomogenic group of agents: Rauscher and Friend viruses (8, 9, 10).

It is possible that the failure of interferon and antibody response represents a simple cell-deletion phenomenon, but that seems intuitively less likely than a more selective biochemical inhibition. General suppression of protein synthesis seems an untenable explanation, for the involved animals often appear quite healthy even during the peak of cytomegalovirus multiplication.

In this context, it is pertinent to mention another experiment, data from which is shown in Figure 8. Here the question broached was whether endotoxin-induced interferon response was inhibited in MCMV-infected mice. When mice were bled at optimal times, after either endotoxin (11) or NDV, the endotoxin group responded entirely normally in contrast to the marked inhibition noted after NDV. Since endotoxin-induced interferon response necessitates *de novo* synthesis, this finding may lend indirect support to the idea that requisite synthetic mechanisms are inoperative in MCMV-infected animals (12) while preformed interferon is normally available. It is entirely possible, however, that this observation may be quite differently explained. For instance, it could simply represent the fact that two separate cell populations are involved.

INTERFERON TITER IN SERUM

Group	2 Hrs after IV Endotoxin, 250 ug E. coli 0113	5 Hrs after IV NDV, $10^{8.2}$ pfu
MCMV-infected[a]	80[b,c], 160, 160, 320	<5, 5, 5, 20
Control	80, 320, 320, 320	80, 160, 160, 160

a) Mice injected with interferon stimulus on Day 4 or 5 of MCMV infection.
b) Interferon titer expressed as the reciprocal of the highest serum dilution causing a 50% reduction of 60-120 pfu EMC in MECC.
c) Each figure represents pooled sera from 5-10 animals.

Figure 8. Comparison of circulating interferon response of MCMV-infected mice to endotoxin or NDV IV.

Efforts are currently underway to further elucidate mechanisms of this dual infection model. It is hoped that fluorescent-antibody straining of dually infected spleens will help to determine site and cell types involved in multiplication of each virus. Further, it has been found that in addition to NDV-neutralizing antibody, sheep erythrocyte hemagglutinin response is suppressed during acute MCMV infection and this model is being pursued.

Finally, efforts are underway to establish an *in vitro* model for study of MCMV suppression of interferon.

ACKNOWLEDGMENT

Acknowledgment is made to *proceedings of the Society for Experimental Biology and Medicine* for permission to use many of the illustrations and a part of the text which appeared in Vol. 124, page 347, 1967, of that journal.

REFERENCES

1. MANNINI, A., and MEDEARIS, D. N., JR.: *Amer. J. Hyg., 73:*329, 1961.
2. MEDEARIS, D. N., JR.: *Ibid., 80:*103, 1964.
3. MEDEARIS, D. N., JR.: *Ibid., 80:*113, 1964.
4. OSBORN, J. E., and MEDEARIS, D. N., JR.: *Proc. Soc. Exper. Biol. & Med., 121:* 819, 1966; *124:*347, 1967.
5. BARON, S., and BUCKLER, C. E.: *Science, 141:*1061, 1963.
6. WAGNER, H. N., JR., IIO, M., and HORNICK, R. B.: *J. Clin. Invest., 42:*990, 1963.
7. RUEBNER, B. H., HIRANO, T., SLUSSER, R., OSBORN, J., and MEDEARIS, D. N., JR.: *Amer. J. Path., 48:*971, 1966.
8. WHEELOCK, E. F.: *Proc. Nat'l. Acad. Sci. (U. S. A.), 55:*774, 1966.
9. SALAMAN, M. H., and WEDDERBURN, N.: *Immunology, 10:*445, 1966.
10. SIEGEL, B. V., and MORTON, J. I.: *Immunology, 10:*559, 1966.
11. STINEBRING, W. R., and YOUNGNER, J. S.: *Nature, 204:*712, 1964.
12. HO, M., and KONO, Y.: *Proc. Nat'l. Acad. Sci. (U. S. A.), 53:*220, 1964.

ENHANCEMENT AND INACTIVATION OF REOVIRUS INFECTIVITY BY PROTEOLYTIC ENZYMES

REX S. SPENDLOVE, Ph.D. and MARY E. McCLAIN, Ph.D.

WE SHOULD LIKE TO REPORT BRIEFLY on some reovirus studies conducted in Dr. Lennette's laboratory at the California State Department of Public Health. Early work was done in collaboration with Dr. F. L. Schaffer; the electron microscopy was done by Dr. Lyndon Oshiro.

The infectivity of most preparations of reovirus of the three serologic types was found to be enhanced by treatment with chymotrypsin[1]. Enzyme treatment did not produce a simultaneous increase in hemagglutinating activity indicating that the enzyme was not breaking up aggregates of virus. Sedimentation experiments showed that the property of enhanceability was closely associated with the virus particles themselves, i.e., pelleted virus resuspended in fresh medium retained its enhanceability.

While most preparations were enhanced by treatment with proteolytic enzymes, an occasional preparation was inactivated by the enzymes. Studies with plaque purified virus showed that the prototype strains of reovirus types 1 and 3 were heterogeneous and contained some virus that was inactivated by proteolytic enzymes. Kinetic studies of inactivation of enzyme sensitive virus showed that the infectivity of these viruses was enhanced before they were inactivated (Fig. 1).

In this experiment, virus was incubated at 37°C with one of the 4 concentrations of chymotrypsin as shown. The control included virus incubated without enzyme. Samples taken at the time intervals indicated were diluted and assayed for infectivity. The infectivity of the untreated control remained constant at approximately 2.2×10^7 infectious units/ml throughout the experiment. There was an inverse relationship between the concentra-

---

[1]SPENDLOVE, R. S., and SCHAFFER, F. L.: Enzymatic enhancement of infectivity of reovirus. *J. Bacteriol.*, *89*:597-602, 1965.

tion of enzyme used, and the time of enhancement and subsequent inactivation of infectivity.

It was later found that the enzyme sensitive virus could be stabilized by the addition of ethylene-diamine-tetra-acetic acid (EDTA), a chelating agent, to the virus or by dialyzing the virus against distilled water before enzyme treatment. Figure 2 shows the protective effect of EDTA on virus treated with chymotryp-

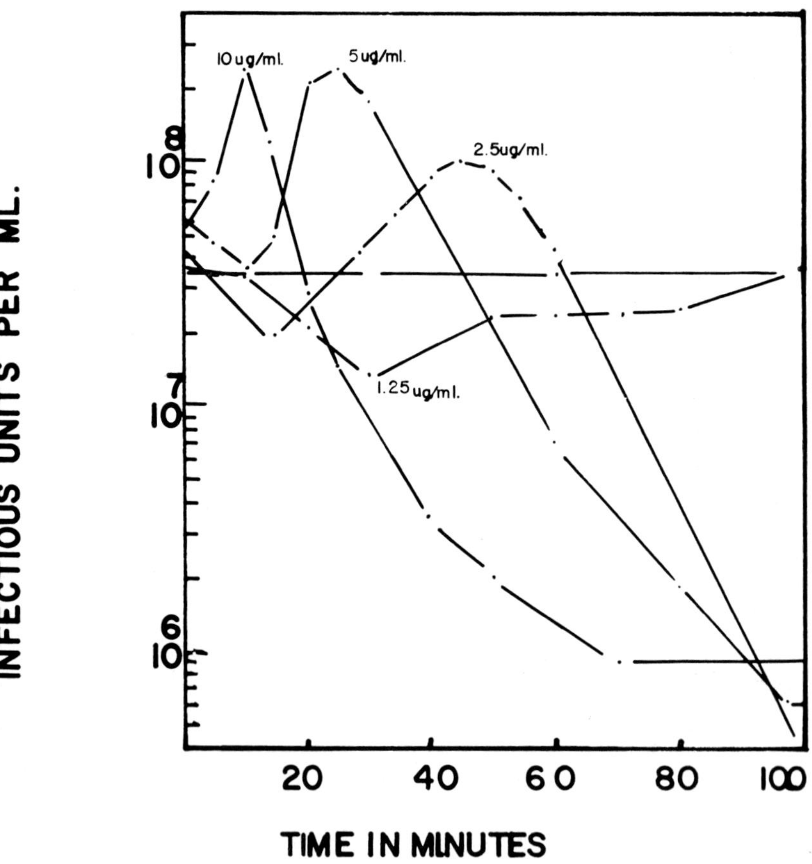

Figure 1. Kinetics of enhancement and inactivation of infectivity of a chymotrypsin sensitive strain of reovirus.

sin. The infectivity of virus treated with enzyme in the presence of
EDTA was enhanced, but was not subsequently inactivated as
was the preparation exposed to enzyme in the absence of EDTA.
The infectivity of virus incubated with EDTA, or with enzyme
diluent (phosphate buffered saline) did not vary during the
experiment.

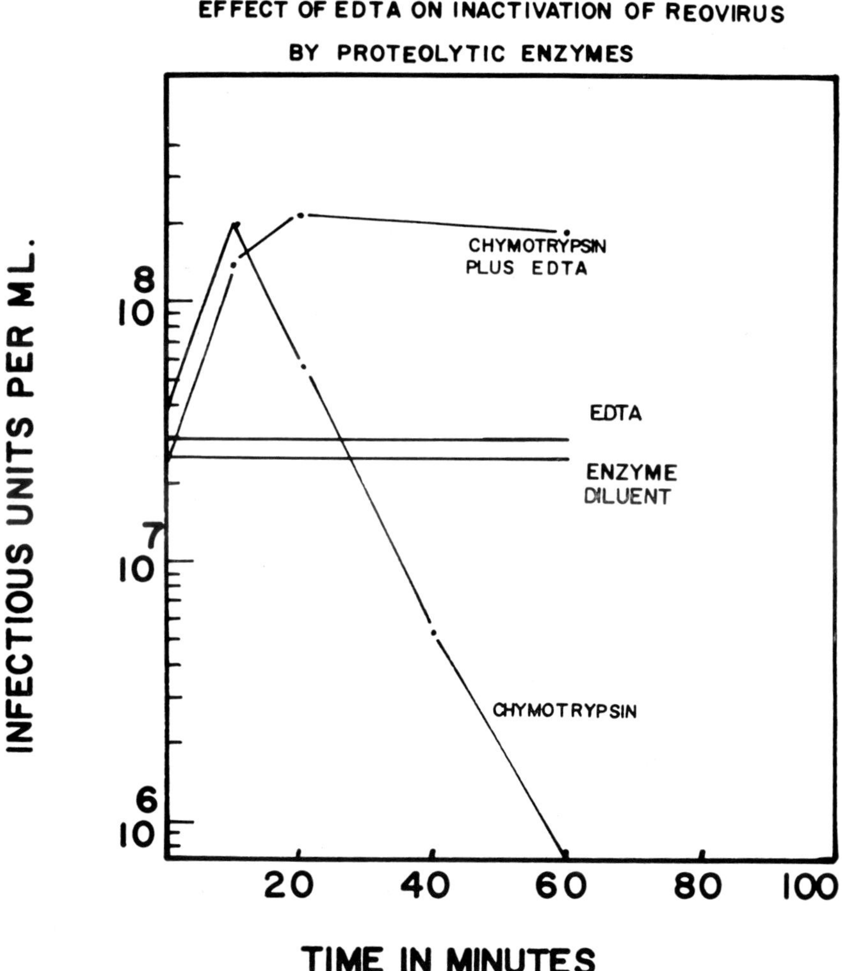

Figure 2. Effect of EDTA on inactivation of reovirus by proteolytic enzymes.

Enzyme sensitive virus stabilized by dialysis against distilled water and then treated with chymotrypsin is inactivated by calcium ions. The reason for this inactivation is currently under investigation.

The above experiments did not permit us to determine whether the substrate upon which the proteolytic enzymes act is a bound protein, i.e., an inhibitor, or whether some portion of the virus coat itself is being removed. Consequently, electron microscopic studies were undertaken.

Reoviruses have been shown by several investigators to have an outer capsid approximately 75 mu in diameter, and an inner protein layer of approximately 45 mu diameter which covers the viral genome. Electron micrographs of enzyme treated reovirus showed the outer capsid to be removed leaving the infectious (activated) virus with a diameter of approximately 50 mu.

In summarizing what is known of the enhancement of reovirus infectivity by proteolytic enzymes, it has been found that:

1) The infectivity of all 16 strains of reovirus tested can be enhanced by exposure to proteolytic enzymes.

2) The prototype strains of reovirus 1 and 3 are heterogeneous and contain some particles which are inactivated by proteolytic enzymes when exposed in the presence of Ca^{++}, and other particles which are resistant to inactivation. Enzyme sensitive virus has not been found in the type 2 prototype strain.

3) The degree of enhancement of most reovirus preparations is from 1 to 5 logs. This shows that the great majority of particles in reovirus preparations are coated with a protease sensitive substance which renders the particles noninfectious under the usual methods of assay. This substance appears to be the outer viral capsid.

4) The non-infectious virus is able to attach to and penetrate the host cell, but in most cases the cellular enzymes are not capable of removing the outer capsid. Consequently, in these cases the viral RNA is not uncoated and the infection is abortive.

OBSERVATIONS ON INTERFERENCE INDUCED BY ULTRAVIOLET-EXPOSED REOVIRUS PARTICLES IN MONOLAYERS OF BSC-1 CELLS

MARY E. McCLAIN

WHEN INFECTIVITY OF reovirus particles is reduced to survival levels of $10^{-2} - 10^{-3}$ by exposure to ultraviolet light (UV) the phenomenon of multiplicity reactivation (MR) can be observed in FL human amnion cells both by immunofluorescent cell count (ICC) and by plaque assay[1]. Similar evidence of MR is seen in ICC titrations of residual infectivity carried out in BSC-1, L929 and RK13 cells. However, plaque assay of UV reovirus in these same cells results in a variety of other effects ranging from MR in L929 cells, cytotoxicity in RK_{13} and interference in BSC-1 cells (McClain, M. E. manuscript in preparation). In the context of this symposium, I would like to focus attention on the interference induced by UV-irradiated type 1 reovirus (Lang) in BSC-1 cells, and describe some exploratory experiments designed to reveal the nature of this interference.

Tables I and II summarize representative results obtained when UV-exposed type 1 reovirus is assayed in BSC-1 cells by ICC and plaque tests. The fact that MR is seen in these cells under the conditions of immunofluorescent assay (essentially single cycle growth under fluid medium) while pronounced interference is observed in lower assay dilutions under conditions of plaque development (multiple cycles of growth with limited diffusion under agar) suggested involvement of interferon or similar substance in this effect. This possibility was strengthened by the demonstration that yields of infectious particles/fluorescent cell were comparable whether infection was initiated by single un-

The work on which this article is based was supported by Grant AI-01475 from the National Institute of Allergy and Infectious Diseases, National Institutes of Health, United States Public Health Service, Department of Health, Education, and Welfare.

131

irradiated particles, single UV-exposed particles or several UV particles under conditions of MR.

Experiments to explore the interference effect were of two types; those designed to demonstrate an interferon-like effect or substance in reovirus-infected cell suspensions, and those which examined the effect of agents commonly used to suppress interferon production.

Response to Vesicular Stomatitis Virus (VSV). BSC-1 cells inoculated with 4 x 10^6 (originally infective) Lang UV-exposed particles 12 hours previously were completely resistant to challenge with 100 pfu of VSV. Viability of these cells, evaluated by morphology and neutral red uptake, did not differ in any obvious way from uninoculated controls at 72 hours.

TABLE I

MR OF TYPE 1 REOVIRUS IN FOUR CELL LINES

(ICC ASSAY)

UV Dose (Min)	Virus Dilution (log 10)	Fluorescent Cells/0.02 ml				EOI/FL		
		FL	BS-C-1	L929	RK13	BS-C-1	L929	RK13
0	4.5	133	120	82	235	0.9	0.6	1.8
	2.0	930	467	93	374	0.5	0.1	0.4
10	2.5	72	36	9	42	0.5	0.1	0.6
	3.0	5	3	0	6	0.6	$<$0.2	1.0

Number of cells per coverslip culture were 4.5, 4.7, 6.1 and 5.2 x 10^5 for FL, BS-C-1, L929 and RK13 respectively.

Maximum infectivity in control suspension — 7.5 x 10^6/0.02 ml.

Efficiency of infection (EOI) in indicated cell compared with that in FL cells.

TABLE II

MULTIPLICITY REACTIVATION AND PLAQUE INHIBITION

IN ASSAYS OF UV EXPOSED TYPE 1 REOVIRUS

Virus Strain	UV Dose (Min)	Virus Dilution (log 10)	Plaque Counts[a]		
			BSC	L929	RK13
		5.0	$>$250, partial confluence	315	200
	0	5.0	120	100	105
		6.0	39	29	32
		1.5	Pronounced inhibition. No countable plaques	1500–2000	Cell death
Lang	10	2.0	Inhibition with late development of 300 "plaque" areas	240	Cell death
		2.5	82-Atypical plaques	31	Cells viable No plaques
		3.0	32-Normal plaques	7	Cells viable No plaques

[a]Average count of four cultures in each group.

Interference with Heterologous Reoviruses. To demonstrate the interfering activity of reovirus suspensions, advantage was taken of differences in plaque size and slower rate of development between a strain of type 2 reovirus (States) and the prototype strains of types 1 and 3. Monolayers of BSC-1 cells were inoculated with 4-40 x 10^5 UV States particles and harvested at 24 and 48 hours. Mixture of such suspensions with unirradiated Lang or Abney viruses induced complete suppression of plaque development when tested undiluted and approximately 50% reduction in plaque count with hundred-fold dilution. This interference was evident when tested both in BSC-1 and L929 cell monolayers, indicating lack of host cell specificity. Furthermore, the interference effect was completely removed from these suspensions by filtration through 0.45 mu millipore filter and was neutralized by mixture with reovirus immune serum. Treatment with low concentrations of chymotrypsin *increased* interfering activity.

In summary, results of the above experiments led to a conclusion that interferon possessing the usual properties probably was not involved in the reovirus interference described. In contrast, however, experiments involving effects of interferon-suppressing agents on reovirus interference yielded results consistent with a positive interpretation.

Effect of Actinomycin on Reovirus Multiplication. Actinomycin $(0.1\gamma/\text{ml})$ incorporated in the agar layer of BSC-1 monolayers resulted in cell death before reovirus plaque development occurred, precluding direct test on the interference effect associated with UV particles. Indirect evidence of interferon activity was obtained from effects of actinomycin in single cycle growth of reoviruses in BSC-1 and L929 cells.

Under the experimental conditions used, actinomycin at concentrates of $0.25\gamma/\text{ml}$ per ml more results in significant reduction of infectious yields when added 2 hours after infection. With actinomycin concentrations of $0.1\gamma/\text{ml}$ added in the period of 2—6 hours after infection, the yield of virus at 20 hours was increased 3-5 fold over that of controls. Details of the enhancing effect varied with the host cell, virus type and whether infecting particles had been exposed to chymotrypsin prior to inoculation. Since an interferon-like substance has not yet been identified in control suspensions, the increased yield cannot be ascribed to suppression of interferon formation, although the results are consistent with such an interpretation.

Effect of Ultraviolet Irradiation of Host Cells. De Maeyer and De Maeyer[2] reported that exposure of rat embryo cells to ultraviolet light (UV) increased plaque size of Sindbis virus, an effect attributed to suppression of interferon formation. Irradiation of BSC-1 monolayers with 450 erg/mm^2 UV prior to inoculation with UV exposed type 1 reovirus particles suppressed the interference effect under examination. Although similar interference had not been observed in L929 cells, exposure of these cells to UV prior to inoculation revealed a striking effect on reovirus plaque development. This effect was particularly evident with reovirus strains which plate with low efficiency on normal cells, and with inocula not treated with chymotrypsin before assay. In irradiated cells, appearance of plaques was accelerated by several days, size was increased by as much as ten times and total numbers were 2-10 times greater than in normal cells. With chymotrypsin treated suspensions, the enhancing effect on size and rate of development was observed in UV cells, but final counts were comparable with those in unirradiated cells. Similar, less striking, effects were seen in BSC-1 cells.

The experimental results with actinomycin and UV irradiation of host cells are consistent with the view that interferon may be involved in reovirus interference or produced during reovirus replication, particularly in L929 cells. However, both agents exert a variety of effects in biological systems in addition to those on interferon formation. Failure to demonstrate that the interfering activity, clearly demonstrable in reovirus suspensions, exhibits the properties of classical interferons precludes an unequivocal statement at this time about the existence of a reovirus induced interferon.

ADDENDUM*

Recent experiments have established that the interference effects induced by UV-reovirus suspensions can be separated into two components. One is associated with reovirus particles and is

---

*Made subsequent to the Symposium.

characterized by neutralization with homologous antisera and by the absence of host cell specificity. A second component possesses properties of typical interferon including failure to sediment at centrifugal speeds which remove virus particles, resistance to neuralization by reovirus immune serum and activity only in homologous host cells.

REFERENCES

1. McClain, M. E., and Spendlove, R. S.: Multiplicity reactivation of reovirus particles after exposure to ultraviolent light. *J. Bact., 92:*1422-1429, 1966.
2. De Maeyer-Guignard, J., and De Maeyer, E.: Inhibition of interferon synthesis and stimulation of virus plaque development in mammalian cell cultures after ultraviolet irradiation. *Nature, 205:*985-987, 1965.

INTERFERENCE — DISCUSSION

DR. CHANY: This work was done in association with Dr. Verser and Dr. and Mrs. Falcoff in Paris. In the animal host, the simulation of the virus in a given tissue is often limited and diffuse cellular destruction is seldom observed. The liberation of interferon or other antiviral substances by infected cells may account in part, in some instances, for minor viral inhibition. However, it seems unlikely that such mechanisms are solely responsible for the localization of viral lesions in the hosts. Human amniotic cells cultivated *in vitro* are fully susceptible to poliovirus. In the intact amniotic membrane, however, previous investigators were not able to demonstrate polioviral multiplications.

DR. SMORODINTSEV: I would like to discuss not only the basic fundamental problems of interference presented in the preceding pages, but also the prospects of practical use of this most available form of inhibition of viral activity. We have had some experience in the role of interchanging interference induced in human beings in the control of influenza by respiratory use of various live vaccines during the period of our outbreaks, usually lasting four to six weeks. At the present time, we are still waiting for an urgent situation for testing interference by oral application of some safe biological compound such as Statolon. The inoculation of live vaccine preparations appears to be the most realistic method for substantial increase of local defense of the respiratory tract against natural influenza viruses, especially during the time when even successful immunization against previously known variants may be hopeless in the face of new antigenic species. The interference does provoke and can be measured also against superinfection by control of the live influenza virus vaccine introduced one to five days later. The evidence is that this mechanism suppresses very well the production and immunogenic response in secondary infection during the period tested. It is more reasonable to start with inoculation of most specific of the influenza vaccine of Types A2 and B and to repeat this procedure twice during the first ten days. This way, we have the advantage not only to materialize the antigenic interference but to stimulate the development of specific immunity for the later period of outbreak. After that, another new live vaccine should be inoculated such as live polio vaccine which is very well existant and provokes interferon very well. In the respiratory tract, measles or mumps combined live vaccine are also able to induce interference. Rhinoviruses can also be prepared for this purpose and there we have unlimited choice of suitable agents. Trials in Leningrad and Moscow with the efficiency of this kind of emergency prevention have already shown favorable results similar on each statistical rate with the routine vaccinations.

DR. MARCUS: Results obtained by Dr. Carver and me agree in all major stances where we have further information. We have found that the effect of aging, that is, the loss in our case of plaquing efficiency of Sindbis virus on cell aging, is essentially obliterated if the chick embryo cultures are incubated at 31 °C. So, aging phenomenon must require some events at 37 °C. Unlike Dr. Lockart, we apparently can slow the aging process if we give daily medium changes to the cells. Now, this is perhaps not as mysterious as it sounds, when I mention the third point, namely, that we have found a striking difference in the effect of aging by the use of different media. In fact, it became quite clear when we began working with Sindbis virus for the first time, not too many years ago. There were two kinds of people in the field — those who consistently got high titers and those workers who obtained varying titers. After speaking to Drs. Levine and Simpson, it appears the particular medium which they used obviated the aging phenomenon. In other words, the loss in plaquing efficiency is much less when their medium is used. I think that the medium that we use probably produces effective aging. Our final conclusion is somewhat stronger than Dr. Lockart's in that, in every case, we have been able to find an increased production of interferon of the aged cell, and we feel that it is primarily the fact that the aged cell for reasons unknown, can produce large amounts of interferon upon stimulation at the time of plaquing.

DR. MAHDY: We have some experiments on the interference of rubella, Sindbis and active monkey kidney. I may add that Sindbis virus in the active monkey kidney cultures is able to produce slow evidence of interferon and this production of interferon in the monkey kidney is inhibited with Actinomycin. Sindbis virus produced in the active monkey kidney inactivates the kidney by heat. Testing for any interference capacity in this preparation is negative. This may show that what you have described as interference involved by the viral genome inactivated by UV could be demonstrated by the method which I have used and which perhaps utilizes inactivation by heat. We have been carrying out some experiments with WEE in aged chick cultures and have observed that chick embryo cell cultures cultivated at $37°C$ for a period of 72 hours, then transferred to temperatures of $29°C$ and kept up to 15 days, become resistant to infection either by Sindbis virus or vesicular stomatitis. This resistance could be partially overcome by treatment of cultures with Actinomycin D prior to infection with the virus. Furthermore, the resistance is more manifested against Sindbis virus than against vesicular stomatitis.

DR. REGELSON: For many years, we have been working with synthetic polyantimes as mitotic inhibitors, inhibitors of tumor growth. We have recently become very concerned with them as they affect host resistance. It appears that these substances induce interferon and pretreatment of mice inhibits the Friend leukemia virus. As a result, we have explored a number of compounds, all of which, in an action similar to the synthetic polyantimes, stimulate the reticuloendothelial system. Statolon does the same. Pertussis vaccine, for example, aborts the protective action of these compounds possibly by interfering with RNAse activity. What I want to specifically refer to is the effect of these polyantimes on the cell surface. They have a striking aggregate effect on cell surface producing hydrophilic gels when added in sufficient quantity to washed tumor cells. In a sense, the phenomenon that one sees is similar to that seen with phytohemagglutinate. However, the effect on the cells is to induce loss of RNA and a protein leak from these cells, although the viability of the cells is not affected unless a tremendous amount of the material is added. Simultaneously, with protein and RNA, we get changes in the nucleus with histones and RNA "leaking" out, the DNA not appearing free in the medium but adhering to the cell surface. One question that I would like to address to the phenomenon of interferon induction relates to uptake of the material the virus or whatever the stimulating agent would be, for example, Statolon by the cell? We know that with stimulation of the reticuloendothelial system we have the open door concept. In other words, has anybody ever examined the possibility of RNA and of protein leak-out of these cells in proportion to the degree of resistance that develops?

DR. FELLOWS: Our microbiology section has been investigating the carrier problem in foot and mouth disease virus. We have suspected this to be the case for a long time as have many others. We were able to pick up a carrier virus in a completely immune animal by obtaining tissue juices and cells from a specific area in the pharynx where much of the virus is residual. The animal was apparently completely protected yet the virus was taken out, treated with hydrocarbon to remove any possible chance for antibody and returned to the tongue of the animal the same day. Infection resulted! It is also remarkable that foot and mouth disease virus produces very little interferon. One strain of low infectivity produces relatively large amounts of interferon, yet such strains are not susceptible to the wrong interferon. An outside indicator must be employed.

DR. LOCKART: I wonder if it is really fair at this stage to extend the idea of helper viruses to helper systems? I am not really quite sure just what helper systems mean. It appears in some of the cases presented that the virus was exerting a blocking effect within the cell that the virus propagated. I am not sure that such a situation is really a helper virus.

At the beginning of our experiments, we found the term "intrinsic interference" helped distinguish it from the term, "extrinsic interference." In our minds, this is synonomous with the action of interferon since one can take the supernatant culture, add an extrinsic factor to a new population of cells and convey the interference. Now, the intrinsic aspect of the interference we were talking about becomes more apparent when one realizes that if there is a challenge of monolayer cells with any of the intrinsic inducing viruses and then a superinfection with NDV, at any given time individual cells in the population will display complete refractoriness to NDV. This refractory nature remains an intrinsic part of the cell.

We can trypsinize these monolayer populations and still score these cells as refractory. It is this striking difference that differentiates the extrinsic interferon type from the intrinsic.

DR. HENLE: In this regard, I wanted to ask Dr. Walker whether his mumps-infected cultures do not reveal the same situation — are the cultures resistant to certain myxoviruses but not to others?

DR. WALKER: Yes, I think the curious thing in Dr. Marcus's cultures is that this interference with a particular virus, with NDV, where with these persistently infected cultures, there may be interference with the resistance to the homologous virus and some related ones, but not to

NDV particularly; although in other aspects, they would look pretty much the same, but there is not this peculiar resistance to NDV.

Dr. Sigel: My comments are addressed to Dr. Chany, Dr. Spendlove and Dr. Lockart. It is conceivable that the phenomena described are, at least in part, due to an immunoglobulin. This need not be an antibody specific for the viruses in question. It could be nonimmune (in the sense of viral specificity) gamma A immunoglobulin. Such a substance would be capable of serving as a mechanical barrier in the case of Dr. Chany's experiments by coating the amnion cells. With regard to Dr. Spendlove's experiments, this immunoglobulin could perhaps attach itself to the capsid and thereby reduce the infectivity of the virus.

The age-mediated phenomenon described by Dr. Lockart may be selective in nature since opposite effects are encountered with other viruses. For example, aged cultures of human amnion are more susceptible to adenovirus than are young cultures.

My last comment concerns the effect of cortisone. Several years ago, Ann Beasley and I demonstrated that cortisone could affect the susceptibility of HeLa cells to poliovirus. Very recently, my colleagues, Drs. J. D. Connor and A. Marti, showed that cortisone can bring about an escape from contact inhibition in amnion cells which have formed a complete monolayer. Such cells are normally restrained from further active proliferation in the monolayer. The hormone releases them from the restraint with the result of additional cell growth.

Dr. Henle: There is one form of cellular resistance that we haven't discussed at all. I refer to Dr. Crowell's work.

Dr. Crowell: A few years ago, we described a system whereby the Coxsackie viruses which were Coxsackie B3 and B5 viruses, separately, were able to maintain a chronic infection of HeLa cells, and I think the most striking feature of this culture was that they were antibody dependent and that a large amount of virus was produced in the cultures and that the chronically infected cells were specifically resistant to superinfection by Coxsackie group B viruses and were susceptible to infection-vaccinia, herpes, adeno and the polio viruses as well as Coxsackie A viruses. There was a very specific kind of interference that apparently was not interferon-mediated and which turned out to be interpreted from experimental results as a result of an antibody virus complex residing at the cell surface, to render the cells saturated really in this regard to specific receptors. These receptors for Coxsackie B viruses appear to be closely related but probably not identical. They are certainly very different from the poliovirus receptors of HeLa cells, the ECHO viruses and the group A Coxsackie virus receptors. I think that it is an interesting thing that cells are perfectly normal; no change in morphology and in some cases, no change in virus susceptibility. That perhaps summarizes the earlier work that has been published. Subsequently, we have been interested in characterizing receptors on HeLa cells to the enteroviruses and Dr. Chany's discussion here renewing the interest that Dr. Holland originally described in terms of the amnion insusceptibility and susceptibility of amnion cells when they are grown in the tissue culture. I would like to ask Dr. Chany if he has actually measured virus attachments to his amnion preparations?

DR. CHANY: We have measured the virus attachments, and we found that there was none when the amniotic membrane was taken fresh, but we also found that at this time the membrane contained a considerable amount of antibody, poliovirus in this case, and this had to be washed off. After washing off the live cells there was some attachment, which was difficult to measure because there was only one or two total population of the virusus and this was difficult to evaluate by the usual technique. We noted that the amnion cells do not take up poliovirus but after elimination of the antibody it is impossible to measure.

DR. CROWELL: This confuses me just a little bit, because if you had antibody would you not get an apparent attachment even if it were cell bound antibody to the virus?

DR. CHANY: It is not cell bound, it is tissue bound. But you can't wash it off. You use $1\frac{1}{2}$ liters of saline and there is no measurable antibody activity.

DR. CROWELL: In regard to measuring virus attachment to cells, a mysterious development occurred. We were studying the comparative attachment rates of Coxsackie group B viruses, B1, B2, B3, B4, B5 and B6. It is easy to work with B1, B3 and B5 and you have seen most of the work in the literature relating to these viruses as far as basic studies are concerned. When we finally got around to checking B2, B4, and B6, the only data available were Choppin's of two years ago showing that B4 attaches very slowly to cells. With HeLa cells B2, B4, and B6, we could demonstrate no significant attachment by ordinary techniques, that is, looking for an unattached virus in 5 or 10 million cells per ml. The HeLa cells were exquisitely sensitive. They produced virus to 10^{10} titers and made very large plaques. Yet, here we could not demonstrate virus attachment. Later, it was found necessary to boost the cell concentration somewhere around 150 million per ml to get significant virus attachment. Perhaps this reflects the amount of receptors per cell as being important in considering the susceptibility and the availability of specific receptor sites which must be on a cell surface to allow this initial attachment. If a particular cell has a very few receptors, it may or may not control its over-all susceptibility since it takes only one receptor per cell for a virus particle. In view of these findings, we should perhaps re-evaluate some interesting studies which relate to the Coxsackie A viruses in infant mouse tissues. You are aware that the Coxsackie A viruses kill infant mice with paralysis and virus multiplies in muscle tissue to high titers, but when one attempts to culture these tissues in tissue culture systems, no Coxsackie A viruses can be demonstrated. Dr. Cain and I published results which showed that we could demonstrate no attachment of these viruses to the tissues in tissue culture or in preparations. We concluded that the receptors were presumably present initially *in vivo* but in taking the tissues out of the animal, the receptors were destroyed or inactivated. Now, finding the B2, B4, and B6 viruses poor attachers on HeLa cells, even though the cells are fully susceptible, we really should re-evaluate our complete interpretation that receptors are not present in tissue culture. It is just the opposite of what Dr. Chany and Dr. Holland found in the monkey kidney showing the formation of receptors and a tissue culture system.

Dr. Ho: The interference that is needed by the virus genome, I believe, is very similar to what has been called infection interference. Now, Henderson and Taylor, some years ago, described the interference of arbovirus with chick and guinea virus and they described two fractions of this virus or the products of this virus that mediated interference. There was the inhibitor which is identical to Dr. Marcus's extrinsic interference or interferon mediated interference, designated as an infection interference. I think the term infection interference has some merit in that it points to the possible application. It is well known the vaccines that are infectious and alive may have some effect beyond, for example, either their ability to use antibodies or interferon. There is some infection by itself. Hence, I would just like to relate intrinsic interference to this phenomenon. The one thing that is very interesting about this intrinsic interference is as Dr. Marcus pointed out, that it interferes with only one virus. I wonder if it is actually true because in the African green monkey system it is precisely this system that is used to demonstrate the presence of rubella virus by interference. This finding is related to the intrinsic interference.

In regard to susceptibility of cells with age, I think it is clear we cannot generalize. Dr. Sigel, I believe has mentioned relative to something that decreases susceptibility with age. Clearly, there are cell virus systems in which the susceptibility to virus increases with age. I think that one of the first to show this was Frothingham who described, in 1958, the increased susceptibility. In fact, one of the essential conditions of the replication of one type of RMC of poliovirus in human amnion cells is that it be aged and that it had to be beyond a certain age. This is true to a lesser extent also of Sindbis virus.

DR. MARCUS: Dr. Walker has mentioned receptor destruction instances in which influenza or parainfluenza have been used in carrier systems, and it is quite clear that in many instances these cultures are susceptible to NDV. This does not seem to be the class of viruses which induce intrinsic interference as we defined it. With respect to a comment on the rubella and the green monkey kidney system; quite clearly we are aware of the fact that this was the first system in which rubella interference was demonstrated and we were struck with the point that 25% of the cells are totally refractory to NDV. If we take that culture and test it in the standard rubella virus manner, that is with a 100 or a 1,000 plaque forming doses of ECHO 11 without medium change, that entire culture in a sense will be refractory. So, it will show the classic rubella interference. However, that same monolayer of cells washed and then infected in multiples of ten or more with that same virus, ECHO 11, is completely susceptible and shows CPE in the same manner as controlled cultures. The experimental conditions are different here.

DR. KAPLAN: I just want to make a comment about the aged and the young cells. You are dealing here with a strikingly different situation in that the cells are in different physiological states. One must understand what changes occur physiologically before you can again understand what the virus is going to do to the cell, that is why the virus will grow or not grow. For example, we have found that rabbit kidney cells in log phase have practically no phosphatases while other cells are loaded with phosphatases. Here is a case where perhaps you can make nucleic acids necessary for the RNA viruses at a later state.

DR. MAHDY: In regard to rubella, we have some similar experiments, where after about 7 or 8 days of infection of the rubella virus of the African green monkey kidney cells, it was possible to harvest the fluids containing the infected rubella virus, replace them with fresh media and harvest again 24 hours later and keep doing this for about 11 or 12 days getting a titer of about 10:40 interferon doses or more.

In relation to amnion absorption and the role of antibody, I think a good model to attach the viruses to the primary amnion fragments of tissue would be to use a virus for which antibodies are not commonly encountered in the human being. I think Newcastle disease virus or vesicular stomatitis virus would offer a reasonable model rather than poliovirus which has a very high infection rate either naturally or through vaccination.

DR. MARCUS: With respect to the constant yields of rubella virus over a period of time with medium change, I am very happy to hear about that since we did in fact publish this back in August a year ago, and this is precisely what we found. These same cultures are totally refractory to NDV and totally susceptible to this other list of viruses. One further comment to the effect of Actinomycin D on green monkey kidney cells not inhibiting the production of interferon; in our experiments we have used high enough concentrations of Actinomycin D on the primary green monkey cells to be assured by Uradine Corporation that there was no new cellular message formation and no interferon production under these conditions.

DR. GORDON: Perhaps Actinomycin D brings ribosomes that are otherwise engaged in a physiologically active cell back into the monosome pool. Perhaps in aged culture cells having occupied themselves in other matters, treatment with Actinomycin D might very well afford an opportunity for virus polysomes to be produced at much of an increased rate rather promptly. Conceivably, this notion of Dr. Kaplan's like everybody else's, that physiological changes are involved is correct. It is a common experience that treating the cell with Actinomycin D at the time of infection with RNA virus often results in greatly increased output of virus 10 or a 100-fold.

DR. WALKER: I refrained from talking about the effects of Actinomycin

D this morning but people have used these persistently infected cultures too often. One of the things wrong with what has been said is that in these crowded aged cultures the nucleus is turned off, and the ribosomes are free to take up any activity that may come along. In these persistently infected cultures, we find that as the cells do become crowded and are no longer growing, there is inhibition and then they do begin to produce more virus. It seemed to us that what you are suggesting that the nucleus is inhibited and the ribosomes are not kept supplied with messenger from the nucleus, that they do become available then in a competitive fashion, they become available to the virus and this has been the simple competition in the cell in which the nucleus can usually provide far more message than can the virus. When the nucleus was "turned off" with Actinomycin, will this result in increased production on these persistently infected cultures? In fact, it does. There is an increased production. The only thing wrong with this is that presumably then the competition was for the ribosome for production of viral protein, probably, and that if the Actinomycin results in greater production of viral protein we ought to be able to block this with Puramycin, but that doesn't work. Puramycin, in fact, adds to the effect of the Actinomycin. There is a further increment in the virus produced from the persistently infected cells. I did not get into this this morning because I don't understand this completely but with the future information on the effects of Actinomycin D and Puramycin, maybe this will become evident as to what is going on.

Dr. Levine: Perhaps this point on aging is a little bit belabored. It has been pointed out that one could get many different effects. I think that we should perhaps point out that much of this work on aging has been done in primary cultures. The point that Dr. Marcus made about medium, may be very important. I think that in primary cultures one not only deals with the question of physiology of the cell but also the possibility of selecting different cells at different times depending upon what you are doing and depending upon what you call aging which is simply letting the culture sit for a certain amount of time, in which case you may get overgrowth in other types of cells. We have had experiments where we have gotten the opposite effects of Dr. Marcus. In this case, we saw that there was another kind of cell that was taking over in the population.

THE INTERFERON SYSTEM AT THE CELLULAR LEVEL

SAMUEL BARON, CHARLES E. BUCKLER,
HILTON B. LEVY, KEN WONG

THERE ARE SEVERAL recent reviews which have considered the interferon system at the cellular level (1, 2). The available evidence suggests that the interferon system is composed of two different proteins — the first, interferon itself, does not inactivate virus directly but can stimulate cells to produce the second, an intracellular, antiviral protein which seems to be the antiviral component of the interferon system. Production of interferon is a function of the animal cell which has been stimulated by virus or certain other substances. The time of production and the total amount of interferon produced is influenced by many factors. To exert an antiviral action, interferon must react with cells and thereby apparently induce the production of the intracellular antiviral protein. Available evidence favors the view that an undetectably small fraction of mouse or chick interferon must be irreversibly bound to cells during induction of the antiviral protein. Mouse and chicken cells which have reacted with interferon for more than seven hours develop a fairly constant level of antiviral activity which probably results from equal rates of production and decay of the antiviral substance. The degree of inhibition of virus multiplication is usually dependent upon the concentration of interferon applied to cells in culture. It seems well established that interferon does not induce resistance to virus in cells of unrelated animal species. In certain interferon-treated cell systems a nongenetically determined fraction of virus populations retains ability to multiply. Inhibition of viral oncogenesis by the interferon system can result from intracellular inhibition of transformation and also from inhibition of virus growth, thereby making less virus available for transformation.

This chapter will consider several of the more recently recognized aspects of the interferon system. They are: (a) How much uptake of interferon by cells is required to induce antiviral activity? (b) What is the nature of the interaction of interferon with the cell leading to antiviral activity? and (c) What amount of viral interference in cell culture is attributable to the interferon system?

HOW MUCH UPTAKE OF INTERFERON BY CELLS IS REQUIRED TO INDUCE ANTIVIRAL ACTIVITY?

Several laboratories have contributed information on the disappearance or lack of disappearance of interferon from culture fluids during the induction of antiviral activity. One group of laboratories (see references in 3, 4-6) observes substantial losses of interferon from culture fluids. Consistent with this loss is their finding that induction of antiviral activity is dependent on the total amount (volume) as well as the concentration of interferon applied (6, 7). However, loss from the medium could be due to either specific or nonspecific causes.

In other laboratories (see references in 3, 8), no detectable disappearance of interferon from culture fluids could be observed

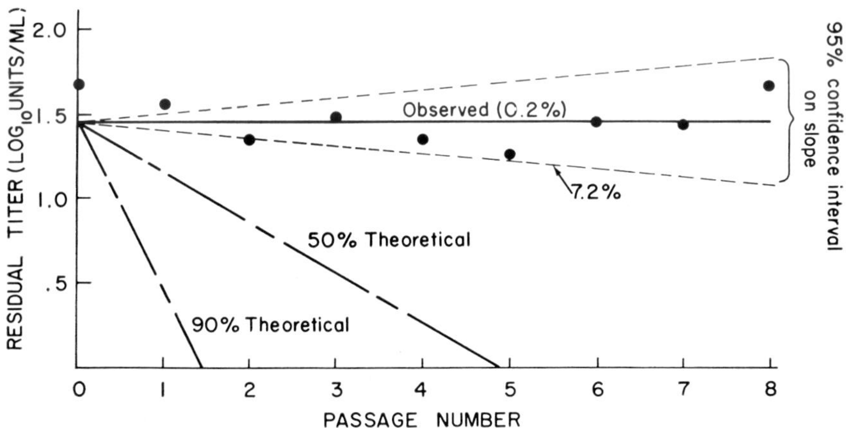

Figure 1. Residual titer of a preparation of chicken interferon after eight, serial, three-hour incubations on chick embryo cell cultures. The statistical procedure was applied to assays of the serially transferred preparations, the slope of the fitted line being expressed as percentage consumption per transfer. The confidence interval on this slope was calculated from the estimated residual titers.

during the development of antiviral activity (Fig. 1). Consistent with undetectable loss in these laboratories is the dependence of the induction of antiviral activity only on the concentration and not on total amount of interferon applied (9, 10) (Table I). Also consistent is the dependence of the continuation of antiviral activity on the continued presence of the original concentration of extracellular interferon (9) (Fig. 2).

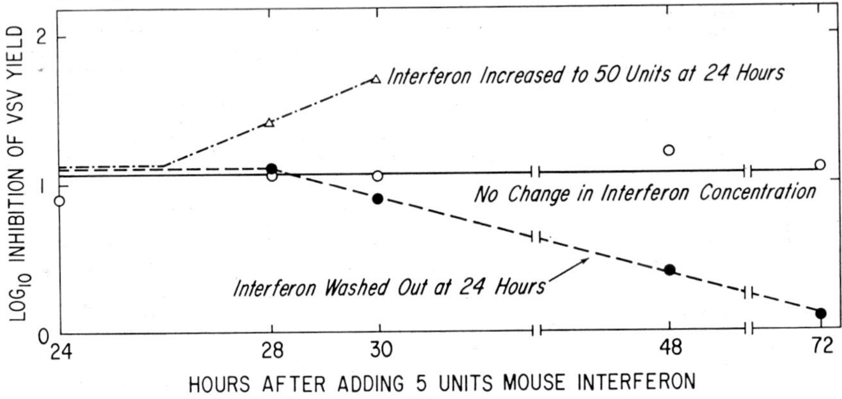

Figure 2. Effect of changing the concentration of interferon on the preexisting level of resistance to vesicular stomatitis virus by mouse embryo cultures pretreated with interferon.

It seems reasonable to assume that all the experimental observations are correct. Since, under certain experimental conditions, antiviral activity can develop in the absence of detectable loss of interferon it necessarily follows that disappearance of measurable quantities of interferon is not a prerequisite for development of antiviral activity. This conclusion would be invalid only if antiviral activity is induced by different mechanisms under the differing experimental conditions.

If we are willing to accept this reasoning, then two questions arise: (a) What is the cause of the proposed nonspecific disappearance of interferon in tissue culture and *in vivo* under certain experimental conditions, and (b) By what mechanism can interferon induce an antiviral effect and yet not be measurably depleted under other conditions? The cause of the proposed nonspecific loss of interferon (question a) is probably unrelated

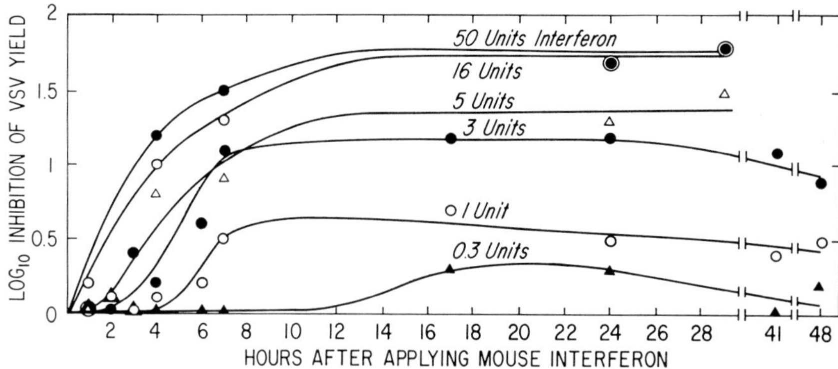

Figure 4. Time-course of development of resistance to multiplication of vesicular stomatitis virus, by cultures of mouse embryo cells treated with varying amounts of interferon.

Time Course of Development of Intracellular Antiviral Activity. The time-course of development of antiviral activity by cultured mouse embryo cells exposed to varying amounts of interferon is shown in Figure 4. The level of antiviral activity is expressed as $\log_{10}$ inhibition of VSV yield. It may be seen that the time of onset of antiviral activity is directly correlated with concentration of interferon applied. The rate of increase of antiviral activity at each dose level began to diminish at seven hours. These findings are similar to those reported for mouse L cells exposed to interferon (14). After seven hours, the level of antiviral activity at each dose level remained relatively constant for the duration of the experiment, and the degree of this steady state antiviral activity was proportionate to the undiminished concentration of interferon in the culture fluid (Fig. 4). The apparent lack of increased inhibition of VSV by 50 units/ml of interferon, as compared with 16 units/ml, is actually due to unremoved virus from the large virus inoculum. With more careful washing to remove inoculated virus, increased final inhibition of VSV was observed through 100 units/ml of interferon and is indicated by the dashed line in Figure 3.

These findings demonstrate that application of interferon to mouse embryo cell cultures is followed in several hours by an increasing degree of intracellular antiviral activity which approaches maximum shortly after seven hours. Subsequently there is no increase in the level of intracellular antiviral activity despite an undiminished extracellular concentration of interferon.

Effect of Changing Concentration of Interferon on the Degree of Antiviral Activity. Two possible reasons for the failure of cells to continue to increase resistance to virus after seven hours are: a) cells no longer respond to interferon stimulation, and b) the intracellular antiviral substance decays at a rate which balances its rate of continuing synthesis. To examine these possibilities, the effect of changing the concentration of interferon on the level of pre-established antiviral activity was determined.

Mouse embryo cultures were treated with 5 units/ml of interferon for 24 hours so that their antiviral activity had reached a steady level (Fig. 2). At 24 hours, the 5 units/ml of interferon was replaced by a 10-fold higher concentration of interferon (50 units/ml) and the level of antiviral activity was determined four and six hours later. It may be seen that the degree of antiviral activity increased at four and six hours. The degree of activity at six hours was equal to that of control cultures which had not been pretreated with interferon but were exposed to 50 units/ml of interferon for six hours. The results indicate that cells at a steady level of antiviral activity have not developed a refractoriness to the action of interferon.

In the same experiment, another group of cultures were pretreated with 5 units of interferon for 24 hours and then washed five times with Earle's balanced salt solution to remove interferon and refed Eagle's medium containing 2% fetal bovine serum (maintenance medium). Figure 2 shows that following removal of extracellular interferon, there was no significant loss of antiviral activity during the first seven hours of continued incubation at 37°C in the absence of interferon. Thereafter, the level of antiviral activity decreased but at a slower rate than the initial rise of antiviral activity, as shown in Figure 4. This rate of decline is comparable with previous reports of decline of antiviral activity following removal of interferon (14-16).

The eventual decline of steady state antiviral activity after removal of interferon indicates that the continued presence of extracellular interferon is necessary for maintenance of the hypothesized intracellular antiviral substance. The intracellular level of antiviral substance could possibly be maintained at a constant level, in the presence of a constant concentration of interferon, as a result of an equilibrium between continuing production of antiviral substance and its equally rapid rate of decay. According to this interpretation, removal of interferon could result in decreased production of the antiviral protein by decreasing the induction of its messenger RNA (17). The initial seven-hour delay in loss of the previously established antiviral activity and its relatively slow decline after removal of interferon (Fig. 2) could be accounted for by the time required to stop induction of messenger RNA (mRNA) and by the time required for decay of the previously induced mRNA. This hypothesis can be tested by determining the rate of decline of established antiviral activity in the presence of inhibitors of synthesis of cellular proteins or RNA.

Effect of Inhibition by Fluorophenylalanine (FPA) of Synthesis of Functional Cellular Proteins on the Degree of Antiviral Activity. If interferon acts to continually induce formation of an antiviral protein during the steady state of antiviral activity, then application of FPA during the time of constant antiviral activity should result in the production of nonfunctioning antiviral protein. If, in addition, there is continuing decay of the preexisting antiviral protein, then the equilibrium between production and decay of the antiviral substance will be altered so that there would be a new loss of antiviral substance.

The results of a representative experiment are shown in Figure 5. Mouse embryo cells were treated with 10 units/ml of interferon for 24 hours at 37°C. The cultures were then washed three times and one group was refed with maintenance medium containing the same concentration of interferon. To another group of cultures was added 60 µg/ml FPA in phenylalanine free Eagle's medium. At intervals, after adding FPA, cultures were washed three times, fed maintenance medium containing 48 µg/ml phenylalanine and challenged with a high multiplicity of VSV to determine yield of VSV at 24 hours. Cultures treated with maintenance medium alone or phenylalanine-free medium plus FPA served to determine control virus yields. FPA treated virus controls produced as much virus as untreated controls when FPA was reversed with Eagle's medium three hours after beginning FPA treatment, produced 0.5 $\log_{10}$ less virus when reversed after seven hours and produced about 1.0 $\log_{10}$ less virus when reversed after 24 hours. Antiviral activity of interferon and FPA treated cultures were compared to that of the appropriate control cultures.

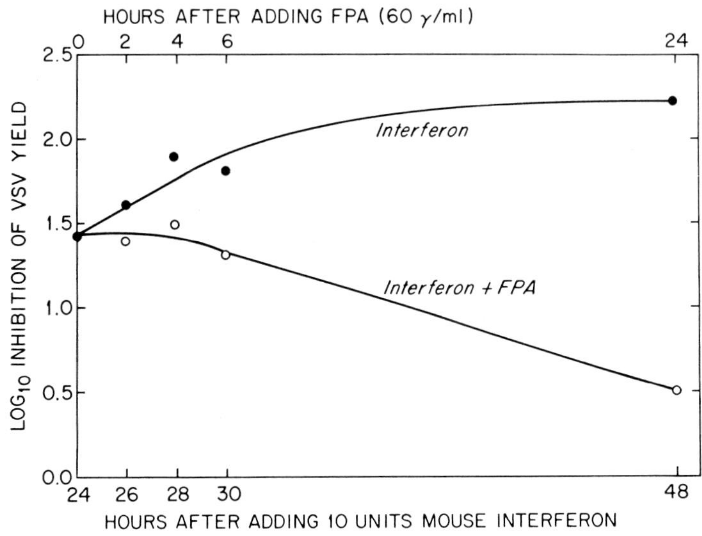

Figure 5. Effect of fluorophenylalanine on the preexisting level of resistance to vesicular stomatitis virus by mouse embryo cultures pretreated with interferon.

Studies of incorporation of ^{14}C-phenylalanine showed 61% inhibition of specific activity at four hours, 61% inhibition at six hours, and 74% inhibition at 24 hours in cultures treated with 60 μg/ml. FPA. At the same time, there was no inhibition and sometimes enhancement of incorporation of ^{3}H uridine. As may be seen in Figure 5, addition of FPA resulted in a decrease of antiviral activity despite the continued presence of interferon. The rise of antiviral activity in interferon treated cultures between 24 and 28 hours was frequently associated with feeding fresh medium. These results strongly support the interpretation that the steady state of antiviral activity is due to an equilibrium between production and decay of an unstable antiviral protein.

Additional experiments were done to compare the loss of antiviral activity in the presence and absence of interferon and/or FPA. Cultures were pretreated with interferon, washed to remove extracellular interferon and then refed either interferon-free medium or fed phenylalanine-free and interferon-free medium containing 30 µg/ml FPA. As shown in Figure 6, the loss of antiviral activity in cultures with interferon removed and treated with FPA was more rapid than in cultures with interferon removed. These findings indicate that removal of extracellular interferon is followed by a slowly decreasing production of antiviral protein, which in turn suggests that removal of interferon may act primarily to decrease production of messenger RNA encoded for the antiviral protein. This interpretation could be tested by comparing the effect on antiviral activity of inhibition of cellular RNA synthesis with the effect of FPA.

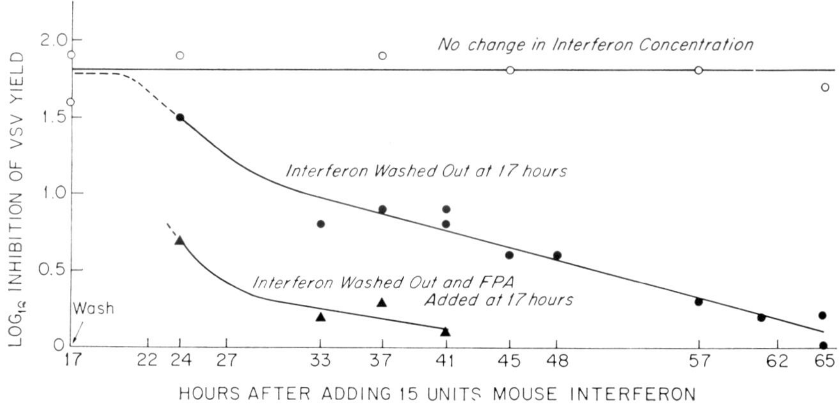

Figure 6. Effect of fluorophenylalanine on the loss of resistance to vesicular stomatitis virus by mouse embryo cultures pretreated with interferon and then washed free of interferon.

Effect of Inhibition of Cellular RNA Synthesis on the Degree of Antiviral Activity.

If, during the steady state of antiviral activity, interferon acts to continuously induce mRNA for an antiviral protein, then actinomycin D should prevent formation of new mRNA (18) and preexisting mRNA would continue to decay. As a result, the production of antiviral protein would diminish in much the same way as the postulated effect of removal of extracellular interferon. Since actinomycin D would permit some continued synthesis of antiviral protein from preexisting mRNA, then the rate of decline of antiviral activity of cells exposed to interferon and treated with actinomycin D should be slower than the rate of decline of cells exposed to interferon and treated with FPA which would act to inhibit formation of any new, functional, antiviral protein.

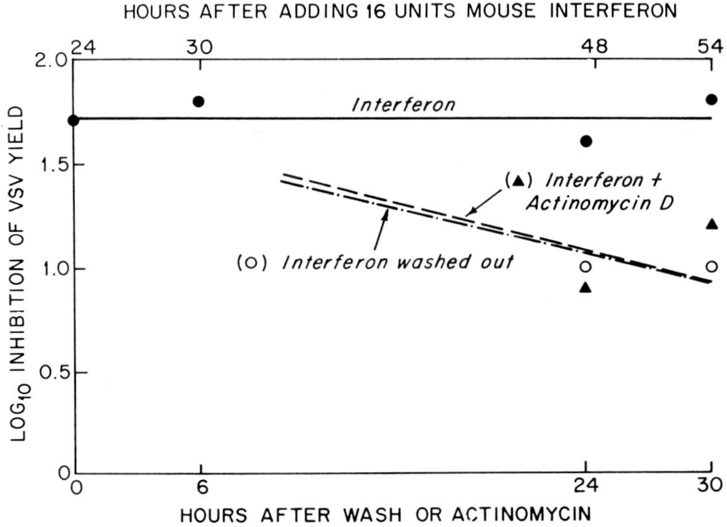

Figure 7. Effect of actinomycin D on the preexisting level of resistance to vesicular stomatitis virus by mouse embryo cultures pretreated with interferon.

The results of a representative experiment are shown in Figure 7. Cell cultures were pretreated with interferon for 24 hours and then divided into three groups. One group was left unchanged. To another group, actinomycin D was added to make a final concentration of 0.1 μg/ml. The third group of cultures was washed to remove interferon and refed interferon-free maintenance medium. At the intervals shown in Figure 7, inhibition of yield of VSV was compared to appropriate virus controls which were either treated or not treated with actinomycin D. Virus controls treated with actinomycin D yielded as much virus when challenged after two hours treatment as untreated virus controls, 1.0 $\log_{10}$ less virus after four and six hours, 1.4 $\log_{10}$ less virus after 17 hours, and 1.8 $\log_{10}$ less virus after 24 hours. VSV yields in interferon and actinomycin D treated cultures were compared to that of appropriate control cultures.

Biochemical controls, using the incorporation of labeled uridine or phenylalanine showed that by 8 hours and 24 hours after treatment with actinomycin D, RNA synthesis was inhibited more than 95%. Protein synthesis was inhibited 46% eight hours after actinomycin D, and protein synthesis was inhibited 85% 24 hours after actinomycin D. These results are consistent with the current viewpoint that actinomycin D acts primarily to inhibit the synthesis of DNA primed RNA, including mRNA, and as a consequence of decay of preexisting mRNA, a secondary inhibition of protein synthesis occurs (18). The assumption is made that the decay of ribosomes during this period would not be sufficient to account for the observed decrease of protein synthesis. Therefore, a decrease in antiviral protein following treatment of resistant cells with actinomycin D would probably reflect a decreased production due to decreased availability of its mRNA.

It may be seen in Figure 7 that actinomycin D decreased the antiviral activity of interferon treated cells, but the rate of decrease was less than that following treatment with FPA (Figs. 5 and 6). This finding supports the interpretation that continued formation of mRNA is necessary to maintain constant production of antiviral protein. Figure 7 also shows that the decline of antiviral activity following removal of interferon occurs at about the same rate as the decline following addition of actinomycin D. This result is consistent with the interpretation that removal of interferon results in inhibition of production of mRNA. The apparent lack of change of antiviral activity between 24 and 30 hours in cultures washed free of interferon, or treated with actinomycin D, demonstrates the relatively slow loss of antiviral activity following removal of interferon as was shown in Figure 7.

Taken together, the findings indicate that application of interferon to cells in culture results in the development of antiviral activity which begins at one to four hours, reaches a peak at about seven hours, and remains relatively constant thereafter. The time

of onset of antiviral activity and its final level are dependent upon
the concentration of interferon applied to the cells. The constant
level of antiviral activity requires the continued presence of the
same concentration of interferon, synthesis of functional protein
and synthesis of RNA.

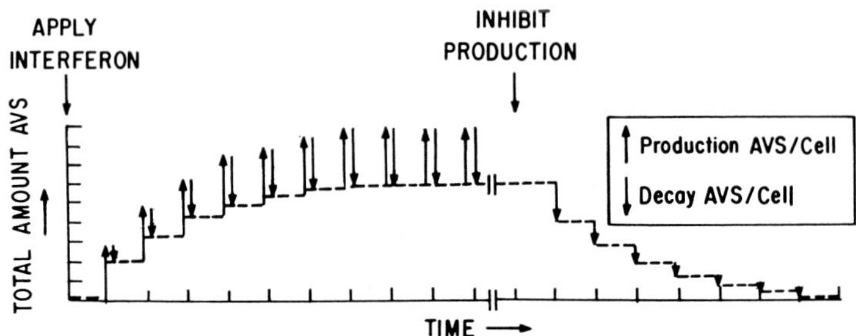

Figure 8. Illustration of working hypothesis of the development of a constant level of antiviral activity in cells treated with interferon. AVS = Antiviral Substance.

A working hypothesis to explain these results is presented below and is illustrated in Figure 8. As originally proposed by Taylor (17), interferon may react with cells to induce formation of a mRNA that codes for an antiviral protein or polypeptide which is subsequently produced. The initial rise in antiviral activity in cells exposed to interferon for one to four hours may be due to the first production of the antiviral protein. Increasing antiviral activity up to about seven hours can be explained by continued production and accumulation of the antiviral substance. As the antiviral protein accumulates, more becomes available for decay although its rate of production probably remains constant. When the amount which decays per unit time has increased to balance the amount produced per unit time (after 7 hours), then a constant level of antiviral substance and antiviral activity will be maintained in cells. Increasing production of antiviral protein by increasing the concentration of interferon would result in a higher equilibrium level of antiviral activity. Decreasing production of antiviral protein by decreasing the concentration of interferon or inhibiting protein or RNA synthesis would result in a lower equilibrium level of antiviral activity.

TABLE I

RELATIONSHIP OF CONCENTRATION OF INTERFERON
AND TOTAL AMOUNT OF INTERFERON TO THE DEGREE OF ANTIVIRAL ACTIVITY

Amount Interferon Applied (ml)	*Log$_{10}$ Dilution of Interferon*	*Hours Between Interferon & VSV*	*Log$_{10}$ Inhibition of Yield of VSV*
0.1	—3.5	4	0.4
0.3	”	”	0.5
1.0	”	”	0.6
0.1	—4.0	”	0.3
0.3	”	”	0.2
1.0	”	”	0.2
0.1	—4.0	7	0.4
0.3	”	”	0.4
1.0	”	”	0.4
0.1	—4.5	”	0.2
0.3	”	”	0.0
1.0	”	”	0.1

BIOLOGICAL SIGNIFICANCE OF THE INTERFERON PROTEIN

If we are willing to assume that interferon reacts with cells to induce an antiviral substance, then we must accept a message-carrying function for this protein. (Message-carrying function is used in the hormonal sense rather than in the genetic code sense.) Since the other message-carrying polypeptides and proteins within the body are generally classified as hormones (19), it is tempting to compare the interferon proteins with hormones. An important property of hormones is their effect on cells distant to the site of their production. Similarly, interferon may be distributed throughout the body by the bloodstream and exert its effect at distant sites. Hormones are produced by specialized cells within the body, but the DNA information for their production is thought to be carried by all body cells (20), though expressed only by the specialized cells. The virus-infected cell which produces interferon may or may not be considered to be analogous to a specialized cell. The response to a hormone is generally greater than the input amount of hormone. Similarly, reaction of cells with interferon gives rise to an antiviral effect with undetectable consumption of interferon. Polypeptide hormones range in molecular weight from 1200 to 45,000. The molecular weight of most forms of interferon is about 30,000, but it may be as high as 160,000. Several hormones are thought to act by inducing the formation of messenger RNA and specific polypeptide (21, 22).

Interferon is also thought to act by inducing formation of messenger RNA and specific polypeptide. Certain hormones may act without being detectably bound within cells (22, 23). Similarly, available evidence is consistent with the concept that interferon protein may not be measurably bound within cells. Table II compares some properties of interferon with those of thyroid stimulating hormone (TSH) and growth hormone (GH). Although this comparison cannot establish the hormonal nature of interferon, it is consistent with the possibility.

TABLE II
PROPERTIES OF INTERFERON THYROID STIMULATING HORMONE (TSH)
AND GROWTH HORMONE (GH) (32)

Property	Interferon	TSH*	GH
Chemical messenger	+	+	+
Prod. by specialized cells	0	+	+
Acts on distant cells	+	+	+
Acts on specialized cells	0	+	0
Production influenced by product	+	+	
Amplification	+	+	+
Action prevented by inhibitors of protein and RNA synthesis	+	0	+
Negligible consumption	+	+	+
Animal Species Specificity	+	0	+
Polypeptide	+	+	+

*See reference 32.

WHAT AMOUNT OF VIRAL INTERFERENCE IS ATTRIBUTABLE TO THE INTERFERON SYSTEM?

Viral interference may occur by inhibition of attachment to cells or by intracellular inhibition of challenge virus (24, 25). Intracellular interference may be divided into two categories by the manifestation of inhibition of a narrow range or of a broad range of challenge viruses (24-28). The intracellular interference induced by interferon may be characterized by: (a) inhibition of multiplication of a broad range of viruses (29); (b) quantitatively different inhibition of different viruses (29), and (c) requirement for host cell RNA and protein synthesis for establishment of antiviral activity (17). Using these criteria, intracellular interference induced by viruses and other stimuli may be characterized to help determine whether the inhibitory effect is related to the interferon system. These criteria were generally fulfilled for the interference between heat inactivated chikungunya virus and infectious chikungunya virus, Semlike forest virus and Newcastle disease virus in chick embryo cell cultures (30).

A comparison of the "interference" induced in African green monkey kidney cells by several viral and non-viral stimuli is shown in Table III. As may be seen, the resistance to viral multiplication induced by Sindbis virus, rubella virus and statolon,* extended to a broad range of viruses and inhibited the interferon-sensitive viruses to a greater degree than the relatively interferon-resistant poliovirus. On only one occasion did the interference against poliovirus approach that against the other viruses (Table III, rubella virus, 24 hours). This aberration was not confirmed in other experiments.

TABLE III
INDUCTION OF INTERFERON AND RESISTANCE TO VIRUSES IN GMK
CELL CULTURES TREATED WITH VARIOUS INDUCERS

Inducer	Hours	Interferon Yeild Units/ml	Log_{10} Inhibition of Challenge Virus Sindbis	Vesicular Stomatitis	Semliki Forest	Polio
Interferon	12	<1	1.1	0.9	1.1	0.6
3 units/ml	24	<1	1.2	0.6	0.9	0.1
Sindbis virus*	7	<2	—	3.8	3.9	0.5
	8	6	—	>3.8	—	—
Rubella virus†	8	<1	0.3	1.4	0.8	0.1
		10	3.0	3.9	3.0	2.8
Statolon						
20 μg/ml	48	<1	0.6	0.4	0.4	0.0
100 μg/ml	48	10	—	4.3	—	—

*More than 20 pfu/cell.
†10 InD 50/cell.

It is of interest that resistance attributable to the interferon system often occurred in the absence of detectable interferon production. This observation suggests that either undetectable amounts of intracellular interferon are capable of inducing the proposed antiviral protein of the interferon system (17) or virus and statolon can induce the hypothesized antiviral protein independently of the interferon protein.

Another property of the resistance induced by these stimuli is its prevention by prior treatment with actinomcycin D, indicating a requirement for cellular RNA synthesis (Table IV). The dose of

*Obtained through the courtesy of Dr. W. Kleinschmidt, Ely Lilly Co.

actinomycin D used (32 μg/ml) in the experiment shown in Table IV, although large by some criteria, results in only near maximum inhibition of RNA synthesis in African green monkey kidney cells (31). When the dose was increased to 100 μg/ml, the small degree of residual resistance shown in Table IV was virtually abolished.

TABLE IV

EFFECT OF ACTINOMYCIN D ON INDUCTION OF INTERFERON AND
RESISTANCE IN GMK CELL CULTURES TREATED WITH VARIOUS INDUCERS

		Interferon Yield (Units/ml)		*Log_{10} Inhibition of:* VSV		Polio I	
Inducer	*Hours*	*None*	*Act. D**	*None*	*Act. D*	*None*	*Act. D*
Interferon	12	<1	<1	1.4	0.7	—	—
Sindbis Virus	8	8	<1	3.7	2.4	1.1	0.3
Rubella Virus	20	14	2	—	—	2.0	0.4
	24	—	—	4.7	0.7	3.2	0.5
Statolon	24	3	<1	1.6	0.3	—	—

*30 μg/ml actinomycin D treatment of cultures at time of application of inducer.

Taken together, the findings suggest that the major part of interference induced in African green monkey cell cultures by Sindbis virus and rubella virus is due to the proposed antiviral protein of the interferon system. To help determine what fraction of the many instances of viral interference is mediated by the interferon system, it will be necessary to characterize many such interference phenomena.

SUMMARY AND CONCLUSIONS

Several aspects of the induction of interferon and its antiviral effect in tissue culture have been considered. The evidence concerning the initial interaction of exogenous interferon with cells indicates that under certain experimental conditions only an undetectably small fraction is bound to cells at any one time. Studies of the continuing reaction of interferon with cells suggests that the antiviral activity reaches a constant level which is dependent on the concentration of interferon in the extracellular fluid. Since maintenance of the constant level of antiviral activity also requires cellular RNA and protein synthesis it is suggested that the constant level results from an equilibrium between production and decay of the proposed antiviral component of the interferon system (17). Characterization of antiviral activity induced in African green monkey cells by several viruses indicates that these examples of interference may be attributed to the interferon system.

REFERENCES

1. FINTER, N. B., Editor: *The Interferons.* Amsterdam, North Holland Publishing Co., 1966.
2. BARON, S., and LEVY, H. B.: Interferon. *Ann. Rev. Micro., 20:*291, 1966.
3. BUCKLER, C. E., BARON, S., and LEVY, H. B.: Interferon: Lack of detectable uptake by cells. *Science, 152:*80, 1966.
4. WAGNER, R. R.: Personal communication.
5. BURKE, D. C.: Personal communication.
6. LOCKART, R. Z.: Personal communication.
7. KE, Y., ARMSTRONG, J., and HO, M.: Personal communication.
8. YOUNGNER, J. S., TAUBE, S. E., and STINEBRING, W. R.: Personal communication.
9. BARON, S., BUCKLER, C. E., LEVY, H. B., and FRIEDMAN, R. M.: Submitted for publication.
10. YOUNGNER, J. S., and HALLUM, J.: Unpublished results.
11. BURKE, C. D., and BUCHAN, A.: Interferon production in chick embryo cells. I. Production by ultraviolet-inactivated virus. *Virology, 26:*28, 1965.
12. MERIGAN, T. C., WINGET, C. A., and DIXON, C. B.: Purification and characterization of vertebrate interferons. *J. Molec. Biol., 13:*679, 1965.
13. PASTAIN, I., ROTH, J., and MACCHIN, V.: Binding of hormone to tissue: The first step in polypeptide hormone action. *Proc. Nat'l. Acad. Sci., U.S.,* In press, 1966.
14. LOCKART, R. C.: Interaction of an interferon with L cells. *J. Bact., 85:*996-1002, 1963.
15. ISAACS, A., and WESTWOOD, M. A.: Duration of protective action of interferon against infection with West Nile virus. *Nature, 184:*1232-1233, 1959.
16. PAUKER, K., and CANTELL, K.: Quantitative studies on viral interference in suspended L cells. V. Persistence of protection in growing cultures. *Virology, 21:*22-29, 1963.
17. TAYLOR, J.: Inhibition of interferon action by actinomycin. *Biochem. Biophys. Res. Comm., 14:*447-451, 1964.
18. REICH, E.: Biochemistry of actinomycins. *Cancer Research, 23:*1428-1441, 1963.
19. JENKIN, P.: *Animal Hormones. A Comparative Survey.* Part I, Kinetic and Metabolic Hormones. New York, Pergamm Press, 1962.
20. McCARTHY, B., and HOYER, B.: Identity of DNA and diversity of messenger RNA molecules in normal mouse tissues. *Proc. Nat'l. Acad. Sci., U.S., 52:*915, 1964.
21. FERGUSON, J., JR.: Hormones and macromolecules. *Ann. Int. Med., 60:*925, 1964.
22. DAVIDSON, E.: Hormones and genes. *Sci. Am., 212:*36, 1965.
23. NEWERLY, K., and BERSON, S. E.: Lack of specificity of insulin - I[131] binding by isolated rat diaphragm. *Proc. Soc. Exper. Biol. & Med., 94:*751, 1957.
24. HENLE, W.: Interference phenomena between animal viruses: A review. *J. Immunol., 64:*203, 1950.
25. ISAACS, A.: Viral interference. *Symp. Soc. Gen. Micro., 9:*102, 1959.
26. RUBIN, H.: Response of cell and organism to infection with avian tumor viruses. *Bact. Rev., 26:*1, 1962.

27. HERMODSSON, S.: Action of parainfluenza virus type III on synthesis of inter-
    feron and multiplication of heterologous viruses. *Acta Path. Microbiol.
    Scandinav., 62:*224, 1964.
28. MARCUS, P. I., and CARVER, D. H.: A new type of viral interference. *Bact. Proc.,*
    p. 123, 1966.
29. ISAACS, I.: Interferon. *Adv. Virus Res., 10:*1, 1963.
30. FRIEDMAN, R. M.: Role of interferon in viral interference. *Nature, 201:*848, 1964.
31. WONG, K., BARON, S., and LEVY, H. B.: Submitted for publication.
32. PASTAN, J.: Biochemistry of the nitrogen-containing hormones. *Ann. Rev.
    Biochem., 35:*369, 1966.

ACTION AND INDUCTION OF INTERFERON

HILTON B. LEVY AND WILLIAM A. CARTER

THE MAJOR research effort on interferon has been on its biological and biomedical aspects. More recently, an increasing interest has developed in this protein as an agent of importance in molecular biology. Before dealing with the recent observations that have generated this interest, it might be desirable to review some of the earlier biochemical studies, first on the induction of interferon synthesis, and secondly on its action.

Is the information needed for interferon synthesis resident in the DNA of the cell, or in the viral genome? The fact that interferon showed specificity for the cell and not for the virus, — that is, that interferon from chick cells was effective on other chick cells and not on mouse cells and vice versa, suggested that interferon synthesis was a function of the cell genome. A number of laboratories independently began experimentation with actinomycin D (ACM) which blocks the synthesis of DNA-directed RNA synthesis, without blocking the synthesis of the RNA of some RNA viruses. Heller (1) was the first to report that ACM treatment of cells infected with an RNA virus inhibits the synthesis of interferon, thus establishing that the host cell genome is the source of the information for interferon synthesis. The expression of this cell function is ordinarily repressed and some event in virus infection leads to depression, with subsequent synthesis of interferon. By the timed addition of ACM relative to the addition of virus, one can determine when the mRNA for this protein is synthesized. In one system which we have examined, the mRNA for interferon is made $1\frac{1}{2}$ to $2\frac{1}{2}$ hours after infection (2) ; in other systems somewhat later (3) .

When interferon is placed in contact with new cells, these cells develop a resistance to virus infection and produce less virus than untreated cells. It was early observed (4) that several hours of exposure to interferon were needed to develop resistance to virus, and that the cells had to be in a state of active metabolism during this time. Taylor (5) showed that treatment of the cells with ACM blocked the development of the antiviral state, indicating probably that new RNA synthesis was needed. Friedman and Sonnabend demonstrated that protein synthesis also was needed (6), as did others (7, 8). On the basis of these observations it was hypothesized that treatment of cells with interferon leads to the synthesis of a new protein which is the actual antiviral protein (A.V.P.). It should be emphasized that the evidence for the A.V.P. while strong, is somewhat indirect. Henceforth, when speaking of the action of interferon we are in fact probably discussing the effects of this second protein which is the active antiviral mole-cule.

TABLE I

EFFECT OF PURIFIED INTERFERON ON INCREASE OF RNA, DNA AND PROTEIN CONTENT OF CE CELLS. DATA EXPRESSED AS μg/PETRI DISH.

	Protein		RNA		DNA	
Time	Int	Control	Int	Control	Int	Control
Zero time	20		4.3		5.5	
15 hr	39	35	7.1	5.5	7.5	7.3
18 hr	49	44	6.9	6.9	7.2	7.3
39 hr	76	87	10.3	12.3	8.2	9.1
42 hr	102	118	13.5	14.0	9.8	9.4

Some of the early reports on the biochemical mechanism of action of interferon implicated alterations in major biosynthetic processes of normal, uninfected cells. To test this possibility, we studied the effect of interferon on the development of chick cells in tissue culture using purified interferon (9). Cells were grown in the continued presence of high levels of interferon and the amount of RNA, DNA and protein determined at several times during their growth. As seen in Table I, the presence of inter-feron had no effect on production of these macromolecules by the growing cells. In another part of the study, the rates of synthesis of RNA and protein were determined at several times during the growth of the cells by exposing the cells to a short pulse of either ^{3}H uridine or ^{14}C valine and measuring the incorporation of these compounds into RNA or protein. As shown in Table II, the rates of synthesis of RNA and protein were unaffected by the presence of interferon. It should be emphasized however, that even though, by these criteria, there are no detectable effects of interferon, some more subtle change must be taking place in uninfected cells upon treatment with interferon, since they become resistant to virus.

TABLE II

EFFECT OF PURIFIED INTERFERON ON INCORPORATION OF H³
URIDINE AND C¹⁴ VALINE INTO CE CELLS

| | *Spec. Act. RNA* | | *Spec. Act. RNA* | |
| | *cpm/mg X 10⁻³* | | *cpm/mg X 10⁻³* | |
Hrs.	*Interferon*	*Control*	*Interferon*	*Control*
15	122	142	28	29
18	526	555	190	237
39	70	54	38	40
42	231	189	252	197

Those samples labelled 18 and 42 hours had isotope exposure beginning at 15 and 39 hours, respectively, but had an additional 3 hours exposure in the presence of 1000 fold excess of unlabelled uridine and 1.0 ml more medium.

The step or steps in virus replication that are sensitive to the action of interferon was shown early to be after eclipse, that is, when the infecting viral nucleic acid had been made available to initiate virus replication (10). Several groups (10, 11, 12, 13) have demonstrated that interferon inhibits the synthesis of new

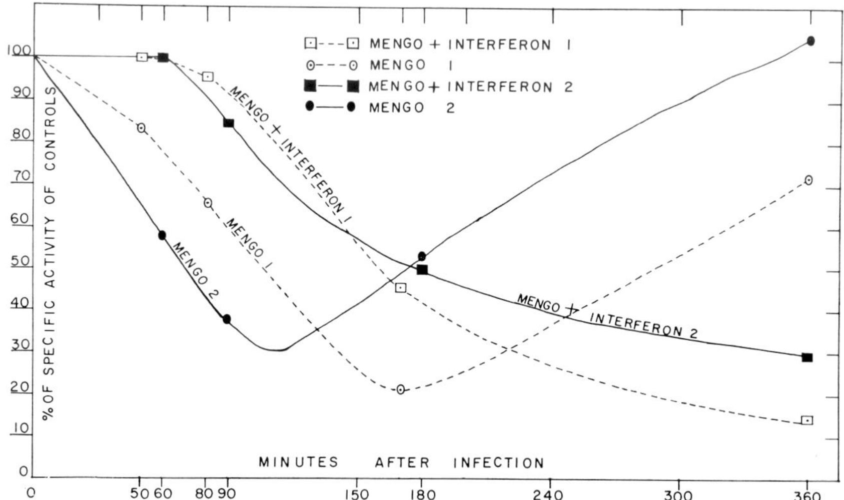

Figure 1. Effect of interferon on inhibition of cell RNA synthesis induced by Mengo virus. RNA synthesis was measured by incorporation of ³H uridine into total RNA.

virus nucleic acid, both as determined by infectivity and by uptake of labeled precursors. Double-stranded as well as single-stranded viral RNA synthesis is also inhibited, (13, 14, 15). Thus the action of interferon is on a step between the release of the infecting nucleic acid, and the production of new nucleic acid. There are several reports that interferon affects an early event in virus replication. One such early manifestation for example, is the inhibition of the induction of thymidine kinase in chick embryo cells that ordnarily follows infection with vaccinia virus (16). An even earlier effect of interferon is revealed by the following: If one infects L cells in suspended culture with Mengo virus there is rapidly developed an inhibition of the rate of cell RNA synthesis, as first shown by Franklin and Baltimore (17). In fact, it can be seen that within 45 minutes post infection the rate of cell RNA synthesis is clearly inhibited (Fig. 1). The inhibition increases until, at 3-4 hours after infection a secondary rise in RNA synthesis occurs, associated with the production of new viral RNA. When these data are compared with those obtained from cells treated with interferon, two differences are noted. First, the later rise in RNA synthesis is not seen in interferon treated cells, showing that viral RNA is not being made. Secondly the early cut-off of cell RNA synthesis, ordinarily detectable by 45′ post infection is delayed about an hour (12).

These actions of interferon occur so early during virus replication as to suggest that the interferon system was affecting the original input virus or some direct manifestation of the input virus. We set out to test this idea by preparing purified Mengo virus containing radioactive nucleic acid and studying the effect of interferon exposure of cells upon the intracellular fate of this virus. The method of preparation of the radioactive virus has been reported (18). Its purity is revealed in Figure 2, which compares the radioactivity and infectivity of fractions obtained from a C_8Cl equilibrium sedimentation of the purified virus. L cells, with and without interferon treatment, were infected with this virus in the presence of actinomycin. Forty-five minutes after infection, the ribosomes were extracted from these cells and analyzed by sucrose gradient sedimentation to see if we could locate the radioactive input viral RNA.

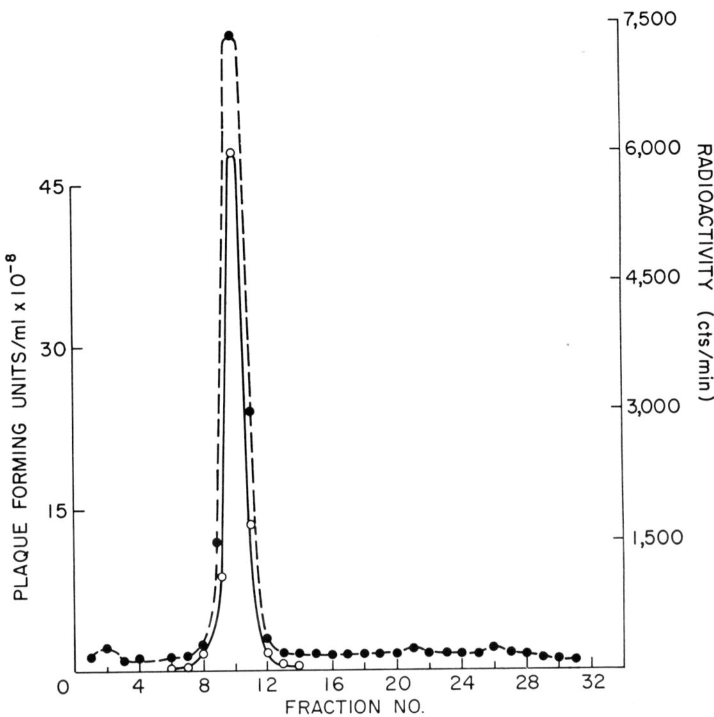

Figure 2. CsCl density gradient analysis of purified ³H uridine labelled Mengo virus. The radioactivity is expressed as cpm/5 λ of each fraction. ● -- ● radioactivity. ○————○ plaque forming units.

Figure 3 shows the results of one such experiment. Since these cells were pretreated with ACM, as well as infected, the amounts of polysomes are greatly diminished. Two radioactive peaks are seen, one at 150 S and one at 50 S, both developing from the radioactivity of the input virus. The 150 S peak represents the free uneclipsed virus in the post mitochondrial supernatant fluid, which was used as the source of polysomes, for the following reasons: first, purified virus sediments at 150 S in a similar gradient; secondly, infectivity determinations on the sucrose gradient fractions reveal that there is a peak of infectivity corresponding with the 150 S radioactive peak. The amount of infectious virus is sufficient to account for the amount of radioactivity. The radioactive peak at 50 S on the other hand cannot be accounted for by infectious virus, nor is it free viral RNA, which sediments at 37 S. However, phenol extraction of this 50 S material releases 37 S viral (Fig. 4). This 50 S peak of radioactivity probably represents the association of viral RNA with a cellular material. Joklik and Becker (19) and Henshaw *et al.* (20) have recently indicated that mRNA binds to a 40 S ribosome subunit prior to assembly into a polysome. The 50 S material seen in the present experiments may also represent an association of Mengo virus RNA with a ribosomal subunit, prior to the incorporation of this RNA into a viral polysome.

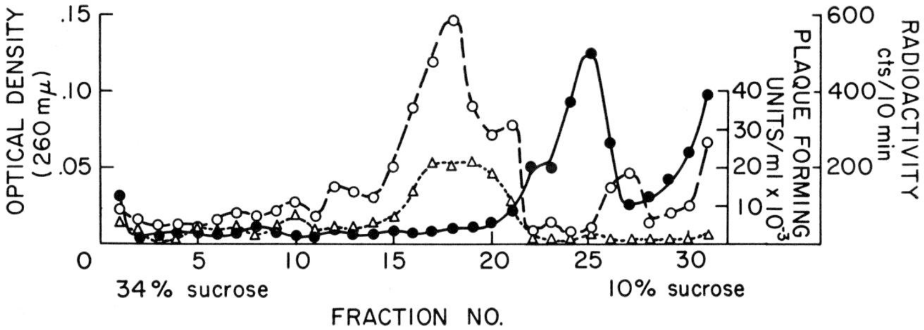

Figure 3. Polysome profile of L cells infected for 45′ with radioactive Mengo virus.

L cells were infected with radioactive Mengo virus at a multiplicity of 2. After 45′ at 37° C, polysomes were prepared, and resolved on a 10-34% sucrose gradient in RSB buffer by centrifugation at 25,000 RPM for 2½ hours in an SW 25 rotor. Each fraction was analyzed for optical density, infectivity and radioactivity. ●————● optical density, △ - - - - △ infectivity, ○————○ radioactivity.

Interferon blocks the incorporation of the viral RNA into this 50 S structure. That is, the 50 S peak of radioactivity either does not appear in ribosome profiles from interferon-treated cells, or is decreased by at least 80%. If the ribosomal subunit-viral RNA is a necessary part of the mechanism of viral polysome assembly, this blockade would inhibit the formation of a functioning viral polysome. In fact, the entry of viral RNA into the polysome region was decreased. To demonstrate this, we used a purified ribosome-polysome complex obtained by precipitating these structures from the post mitochondrial supernatant fluid by 0.07M

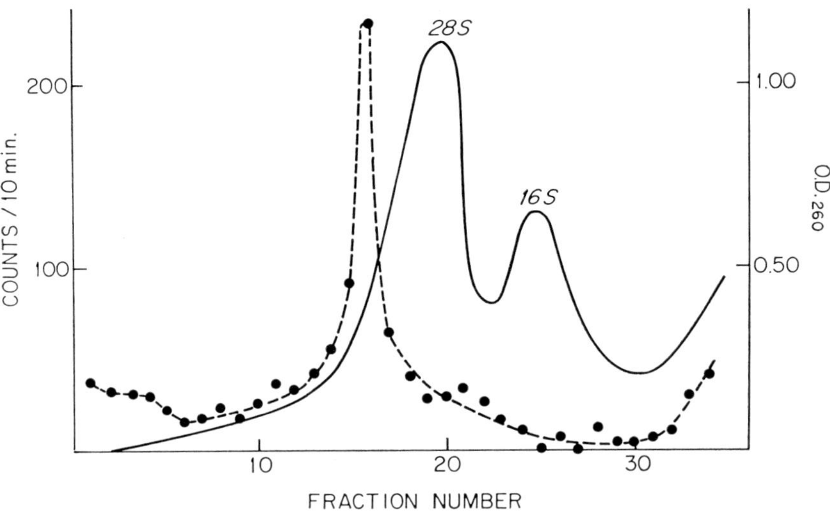

Figure 4. Analysis of the radioactivity of the 50S region of the polysome gradient.

Polysomes were prepared from L cells infected for 45 minutes with ^{3}H labelled Mengo virus, and separated by sucrose gradient density contrifugation. The samples from the 50S region were collected and the RNA extracted at 60°C with phenol-SDS, after the addition of ribosomal RNA as carrier. The RNA was resolved in 5-30% sucrose by centrifugation at 22,000 RPM for 16 hours in an SW 25.1 Spinco rotor.

magnesium, and resuspending the precipitated material at a lower Mg^{++} concentration before examination on sucrose gradients. This procedure, modified from that reported by Warner (21), presented two advantages for us: First, the large quantity of soluble and low molecular weight material found in the upper portion of the gradient was eliminated by Mg^{++} purification of ribosomes, and second only 1-10% of the free radioactive virus was precipitated with the ribosomes, thus enabling us to examine the heavier region of the gradient for polysome-like material. With this technique, interferon-treated or control L cells were exposed to radioactive Mengo virus in the presence of ACM and the infection allowed to proceed 2 to $2\frac{1}{2}$ hours. The polysomes were extracted, precipitated with 0.07 $M\text{-}Mg^{++}$, resuspended in 1 x $10^{-3}M$ Mg^{++}, and analyzed on sucrose gradients (Fig. 5). In the control system, i.e., cells infected with radioactive virus and not exposed to interferon, there are two peaks of radioactivity, one at 240 S, and one at 50 S. There is also a smaller peak at 150 S, representing residual free virus. If the polysome preparation containing the radioactive virus is treated with ribonuclease before applying the sample to the gradient, only the 150 S radioactive peak, the whole virus, persists. Figure 6 compares the data obtained with control infected cells and interferon treated infected cells. The viral RNA has been incorporated into a 50 S component in the control cells, but not in the interferon-treated cells. There is significant entry of viral RNA into the 240 S region in

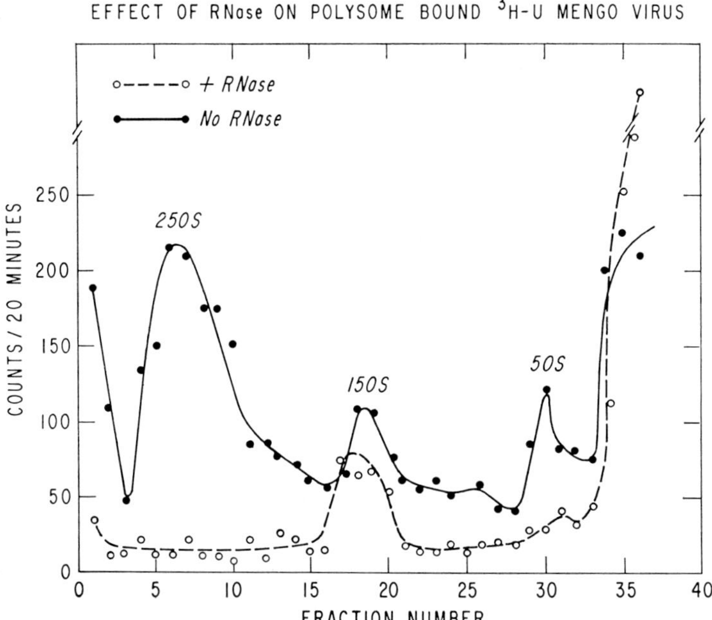

Figure 5. Association of the RNA of ^{3}H labelled Mengo virus with ribosomes of L cells after 2 hours of infection.

L cells were exposed to 2 μg actinomycin D for 2 hours and then injected with ^{3}H labelled Mengo virus for 2 hours in the continued presence of actinomycin ribosomes were prepared and precipitated at 7 X 10^{-2} Mg++. After resuspension in RSB buffer at 10^{-3}M Mg++, they were analyzed in 10-34% sucrose gradients. One part of the ribosomes were treated with 0.5 μg/ml of RNase for 15 minutes at room temperature before applying to the sucrose gradient.

the control cultures, which is inhibited by interferon only 35-40%.

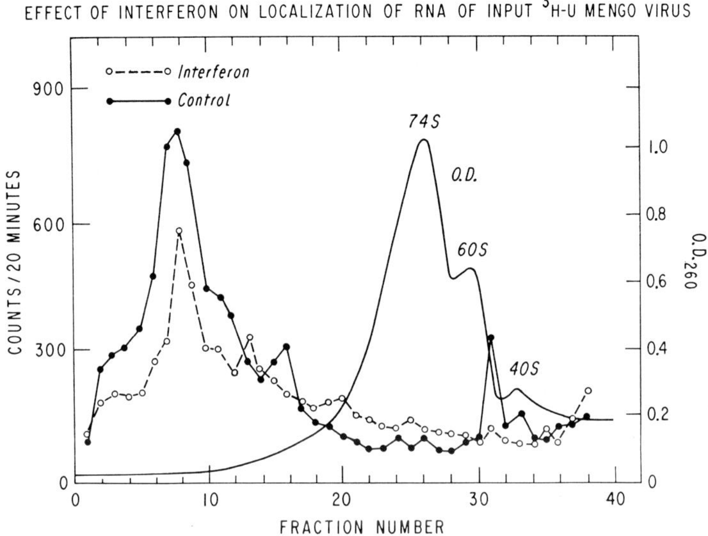

Figure 6. Effect of interferon on association of ^{3}H labelled Mengo virus with L cells.

Details as for Figure 5, except that one culture cell was exposed to 10 units/ml of interferon for 16 hours prior to the beginning of treatment with actinomycin. No RNase was used.

The exact nature of the 240 S material is not known, but it may resemble the 250 S structure found in poliovirus infected cells (22). We thus are observing two presumably related structures which do not appear to be inhibited to the same extent by interferon: i.e., a marked inhibition by interferon of binding of viral RNA into the 50 S material and a much weaker inhibition of binding of the viral RNA into this 240 S material. The 40% decrease of incorporation of input viral RNA is rather small to explain the > 99% inhibition of virus growth by interferon, unless the 240 S structure made in the presence of interferon is non-functional. We have found, as have Baltimore and Girard (22), that this structure is associated with newly synthesized viral RNA. By pulse labelling ACM treated infected cells with ^{3}H uridine and examining the labelled RNA found in the 240 S material we

have found that the particle made in the interferon treated cells
has less than 5% as much new viral RNA as the control preparation; thus, the 240 S structure assembled in interferon exposed
cells is a defective structure.

To recapitulate these last data — the RNA of Mengo virus,
upon infecting L cells, rapidly becomes associated with a ribosomal subunit, the complex sedimenting at 50 S. Subsequently, the
viral RNA is found in a 240 S structure, in which synthesis of
viral RNA can take place.

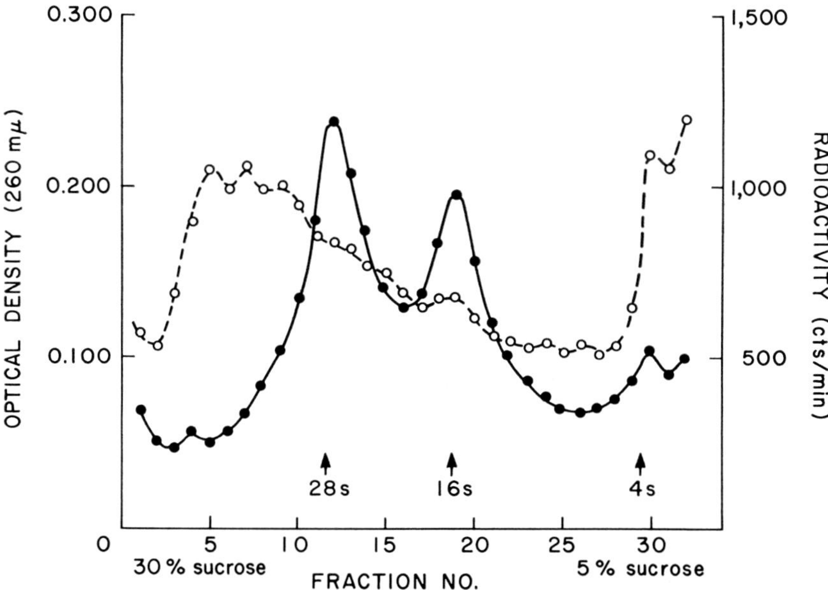

Figure 7. Sucrose gradient sedimentation of rapidly labeled cellular RNA.
The RNA was layered on a linear 5-30% sucrose (w/v) gradient and sedi-
mented at 22,000 rpm for 16 hours at a chamber temperature of 2°C.
Legend: -●-●-●-●-, optical density; -o-o-o-o- radioactivity.

Pretreatment of the cells with interferon leads to marked inhibition of incorporation of the viral RNA into the 50 S structure,
with a lesser inhibition into the 240 S structure. However, the
240 S structure from interferon-treated cells is much less active in
associating newly made viral RNA than the comparable control
structure. If the 240 S particle is the actual site of synthesis of viral

RNA these data raise the interesting possibility that passage of the viral RNA thru a 50 S structure may be a requisite for the formation of a functional 240 S structure.

We have been attempting to study these phenomena in a cell-free system, and some beginnings have been made (23, 24). If the mechanism of interferon action resides in the inability of ribosomes from interferon treated cells to interact effectively with viral RNA's, then certain consequences follow. Ribosomes in a cell free system from both control and interferon treated cells should be able to bind normal cell mRNA since no gross alterations in rates of cell directed protein synthesis are observed. However, ribosomes from interferon treated cells should bind viral RNA less than control ribosomes in a cell free system.

We prepared rapidly labelled L cell RNA. Its sedimentation characteristics are shown in Figure 7. The radioactivity is polydisperse, from 6-45S, with approximately 10% of the radioactivity sedimenting at less than 4s.

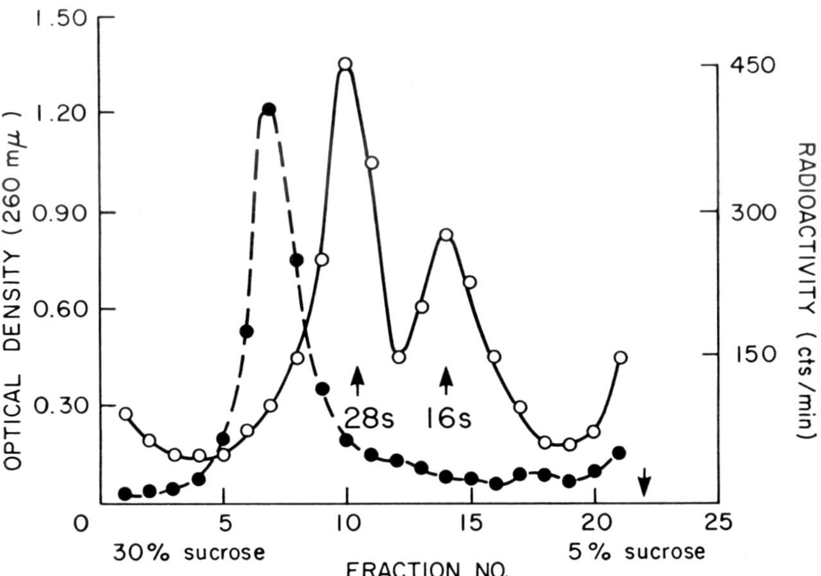

Figure 8. Sedimentation of ^{32}P labeled Mengo RNA. The ^{32}P labeled RNA was sedimented on 5 — 30% (w/v) sucrose gradients for 3 hours at 35,000 rpm in an SW 39 rotor. Legend: -o-o-o-o-, optical density; -●-●-●-●-, radioactivity.

Mengo virus RNA was also prepared and its sedimentation revealed the characteristic 37S peak (Fig. 8). Ribosomes from control L cells and interferon exposed cells were prepared and purified by repeated sedimentation through a layer of 2M sucrose.

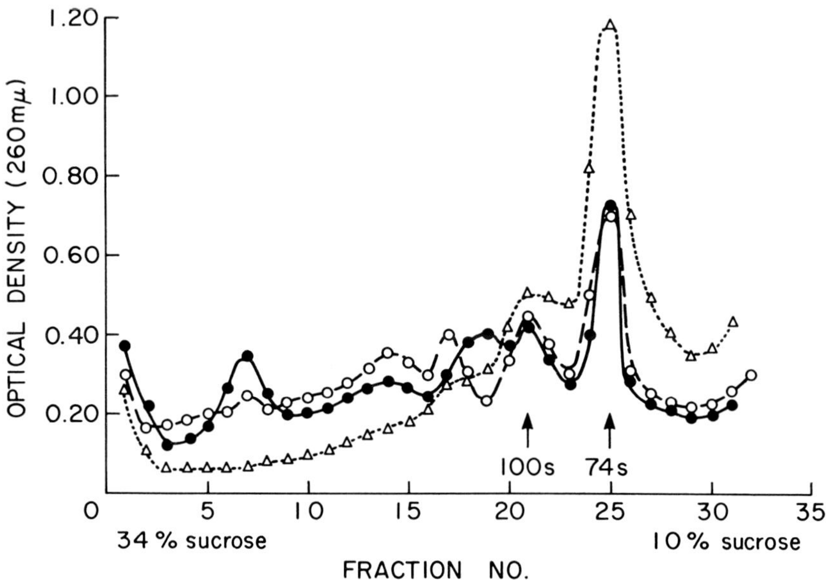

Figure 9. Polysome profiles from normal and interferon exposed cells. Polyribosomes were prepared as described in text from normal, interferon exposed (10 units/ml, 16 hours) or ACM exposed (1 μg/ml, 1½ hours) cells. The polysome pellets were gently resuspended in polysome medium and layered over linear sucrose gradients (10-34%). The samples were sedimented at 25,000 RPM for 3 hours in an SW 25.1 rotor with a chamber temperature of −2°C. Legend: -o-o-o-o-, optical density of normal ribosomes; -●-●-●-●-, optical density of ribosomes from interferon exposed cells; -Δ-Δ-Δ-Δ-, optical density of normal ribosomes exposed to ACM. Fraction number 25 contains the monosome peak (74S) and fraction number 21 the dimer ($\sim$ 100S) peak.

Optical density profiles of polysomes from interferon-treated and control cells are shown in Figure 9. The ribosomes were incubated with the two RNAs under suitable conditions (24) and the reaction mixture examined by sucrose gradient sedimentation for possible interactions.

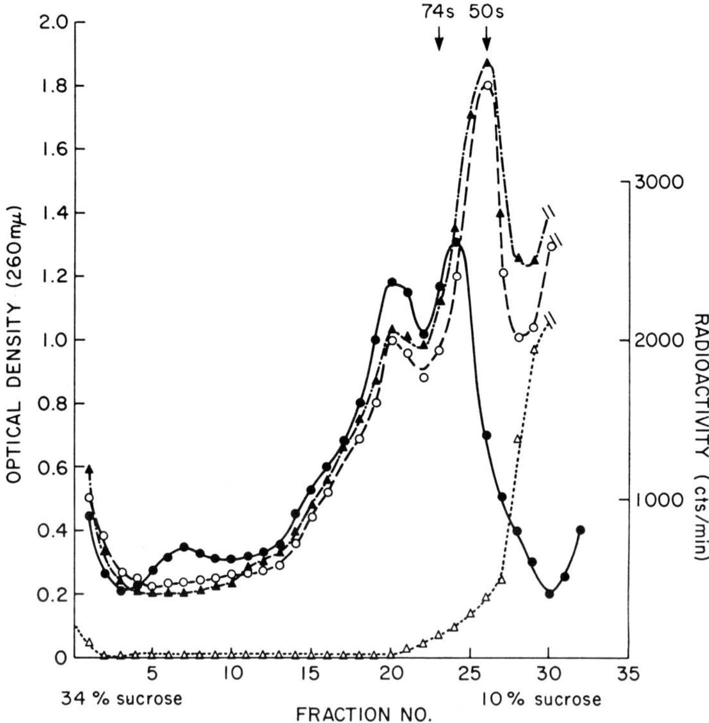

Figure 10. The association of rapidly labeled RNA with ribosomes from normal and interferon exposed cells. Equal amounts of ribosomes were reacted with rapidly labeled L cell RNA for 45 minutes; a third tube contained viral RNA but no ribosomes. The reaction mixtures were layered over linear sucrose gradients (10-34%) and sedimented for 3 hours at 25,000 RPM as described under Figure 9. Legend: -●-●-●-●-, optical density of normal ribosomes; -o-o-o-o-, radioactivity of ^{32}P cell RNA + normal ribosomes; -▲-▲-▲-▲-, radioactivity of ^{32}P cell RNA + ribosomes from interferon exposed cells; -Δ-Δ-Δ-Δ-, radioactivity of ^{32}P cell RNA sedimented in the absence of ribosomes. Fraction number 24 contains the monosome (74S) peak.

Figure 10 demonstrates the altered sedimentation of rapidly labelled cell RNA after incubation with ribosomes from control cells or interferon treated cells. The cell RNA alone, under these conditions of sedimentation, does not move far into this gradient, but, after incubation with either group of ribosomes, sediments into the heavier regions of the gradient. Peaks of association are seen at 100 S and 50 S. There are no detectable differences between the binding of L cell RNA to control ribosomes or those derived from interferon treated cells. The formation of these complexes is blocked by cycloheximide, but not by puromycin (Fig. 11). Figure 12 contrasts the binding of Mengo RNA to these two types of ribosomes. The control ribosomes bind Mengo RNA, with a peak at 50 S but the ribosomes from interferon treated cells shows little or no association. Thus *in vitro* as *in vivo* the ribosomes from interferon treated cells appear to complex poorly with viral RNA, but accept cell RNA normally.

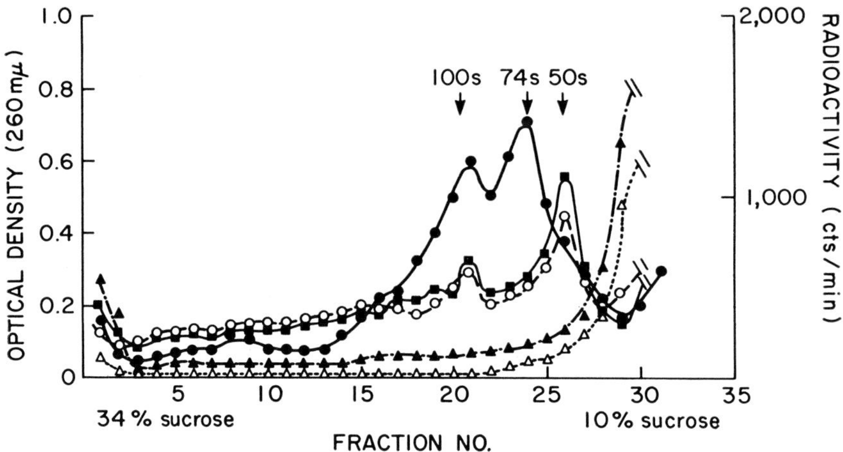

Figure 11. The effect of cycloheximide and puromycin on the rapidly-labeled RNA-ribosome interaction. Reaction mixtures were held at 0-2°C for 45 minutes. When puromycin (10μg/ml) were used, they were added to ribosomal suspensions 5-10 minutes before the addition of the RNA fraction. The 10-34% (w/v) sucrose gradients were centrifuged for 3 hours at 25,000 rpm with a chamber temperature of —2°C. Legends: -●-●-●-●-, optical density; -o-o-o-o-, radioactivity of ribosomes + ^{32}P RNA; -■-■-■-■-, radioactivity of ribosomes + ^{32}P RNA + puromycin; -▲-▲-▲-▲-, radioactivity of ribosomes + ^{32}P RNA + cycloheximide; -Δ-Δ-Δ-Δ-, ^{32}P RNA alone.

Joklik and Merigan (25) have obtained evidence that the mRNA of a DNA virus, vaccinia, is not incorporated into a polysome in interferon treated cells. The formation of a polysome containing viral mRNA is a point of similarity in the replication of DNA and RNA viruses, and thus would form a basis for a single explanation of the action of interferon on both virus types. Differences in the degree of rejection of different viral RNAs by these altered ribosomes could account for differences in sensitivity of these viruses to interferon.

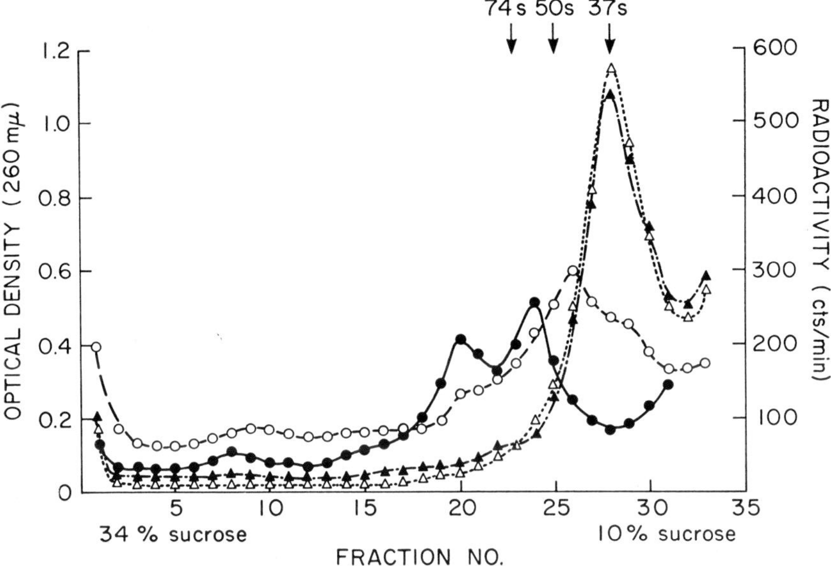

Figure 12. The interaction of viral RNA with ribosomes from normal and interferon exposed cells. Equal amounts of ribosomes from normal and interferon exposed (10 units/ml x 16 hours) cells were reacted with viral RNA at 0-2°C for 45 minutes. Legend: -●-●-●-●, optical density of normal ribosomes; -o-o-o-o-, radioactivity of normal ribosomes + [32]P viral RNA; -▲-▲-▲-▲-, radioactivity of ribosomes from interferon exposed cells + [32]P viral RNA; -Δ-Δ-Δ-Δ-, radioactivity of viral RNA sedimented without ribosomes.

The alteration in ribosomes in interferon treated cells is in accord with the antiviral protein hypothesis discussed earlier. It will be recalled that cells treated with interferon need to synthesize RNA and protein in order to become resistant to viruses. Interferon treatment according to this concept, would lead to the activation of a normally repressed cistron with subsequent production of the anti viral protein. This protein would act, as do a number of cytoplasmic proteins (21), to form a labile attachment to the ribosomes and their subunits, and by controlling the nature or extent of attachment of viral mRNAs would exercise relatively specific control over protein synthesis at the ribosome level. In this regard, interferon would resemble certain polypeptide hormones (26). We are currently attempting to obtain physical evidence for the existance of this protein.

LITERATURE CITED

1. HELLER, E.: *Virology, 21:*652, 1963.
2. LEVY, H. B., AXELROD, D., and BARON, S.: *Proc. Soc. Exper. Biol. & Med., 118:* 384, 1965.
3. WAGNER, R. R.: *Nature, 204:*49, 1964.
4. LINDENMANN, J., BURKE, D. C., and ISAACS, A.: *Brit. J. Exper. Pathol., 38:*551, 1957.
5. TAYLOR, J.: *Biochem. Biophys. Res. Commun., 14:*447, 1964.
6. FRIEDMAN, R. M., and SONNABEND, J. A.: *Nature, 203:*366, 1964.
7. LOCKART, R. Z., JR.: *Biochem. Biophys. Res. Commun., 15:*513, 1964.
8. LEVINE, S.: *Virology, 24:*229, 1964.
9. LEVY, H. B., and MERIGAN, T. C.: *Proc. Soc. Exper. Biol. & Med., 121:*53, 1966.
10. HO, M.: *Proc. Soc. Exper. Biol. & Med., 107:*639, 1961.
11. HO, M.: *Proc. Soc. Exper. Biol. & Med., 112:*511, 1963.
12. LEVY, H. B.: *Virology, 22:*575, 1964.
13. DeSOMER, P., PRINZIE, A., DENYS, P., JR., and SCHONNE, E.: *Virology, 16:*63, 1962.
14. FRIEDMAN, R. M., and SONNABEND, JR.: *Nature, 206:*532, 1965.
15. GORDON, I., CHENAULT, S., STEVENSON, D., and ACTON, J.: *J. Bacteriol., 91:*1230, 1966.
16. GOSH, S. N., and GIFFORD, G. E.: *Virology, 27:*186, 1965.
    OHNO, S., and NOZIMA, T.: Personal communication, 1964.
17. FRANKLIN, R., and BALTIMORE, D.: In *Viruses, Nucleic Acids and Cancer, 310.* Univ. of Texas M.D. Anderson Hosp. & Tumor Inst., Baltimore, Williams & Wilkins, 1964.
18. LEVY, H. B., and CARTER, W. A.: *J. Mol. Biol.,* In press.
19. JOKLIK, W. K., and BECKER, Y.: *J. Mol. Biol., 13:*511, 1965.
20. HENSHAW, E., REVEL, M., and HYATT, H.: *J. Mol. Biol., 14:*241, 1965.
21. WARNER, J. R.: *J. Mol. Biol., 19:*383, 1966.
22. BALTIMORE and GIRARD: Personal communication.
23. CARTER, W. A., and LEVY, H. B.: *Science,* In press.
24. CARTER, W. A., and LEVY, H. B.: *Arch. Biochem. Biophys.,* In press.
25. JOKLIK, W. F., and MERIGAN, T. C.: *Proc. Natl. Acad. Sci., 56:*558, 1966.
26. RAMPERSAL, O. R., and WOOL, I. G.: *Science, 149:*1102, 1962.

INTERFERON INDUCTION BY NONVIRAL AGENTS

L. BORECKÝ

T HE DISCOVERY OF YOUNGNER and Stinebring (1964) that certain bacteria possess an interferon inducing capacity similar to viruses, opened the door for exploration of an immense family of microorganisms with widely differing properties. The interest in non-viral interferon stimulation was soon justified when interferon inducers were found in microbial products such as lipopolysaccharides (Youngner and Stinebring, 1964; Ho, 1964) or, Statolon (Kleneschimidt *et al.*, 1964) etc.

As a result of intense activity in several laboratories, important informations is now available which contributes to a progressive understanding of the problem of non-viral interferon induction.

At present the following questions seem to be of obvious interest in this field:

1) The distribution of interferon-inducing ability among various groups of microorganisms.

2) The nature of microbial cell constituents possessing the interferon inducing capacity.

3) The character of cells which respond with interferon formation on non-viral stimulation.

4) The mechanism which distinguishes non-viral interferon formation from viral induction.

Our recent observations can be summarized as follows:

Comparison of interferon inducing ability of various bacterial species and serologically different strains in mice, as shown in Figure 1, seems to indicate that, based on interferon production dynamics, the microorganisms tested may be divided into following types:

I. Gram-negative bacteria such as *E. coli* of various serotypes (0128:B12:H2, 0111:B4, 086:B7, 055:B5), *Salmonellae typhi*

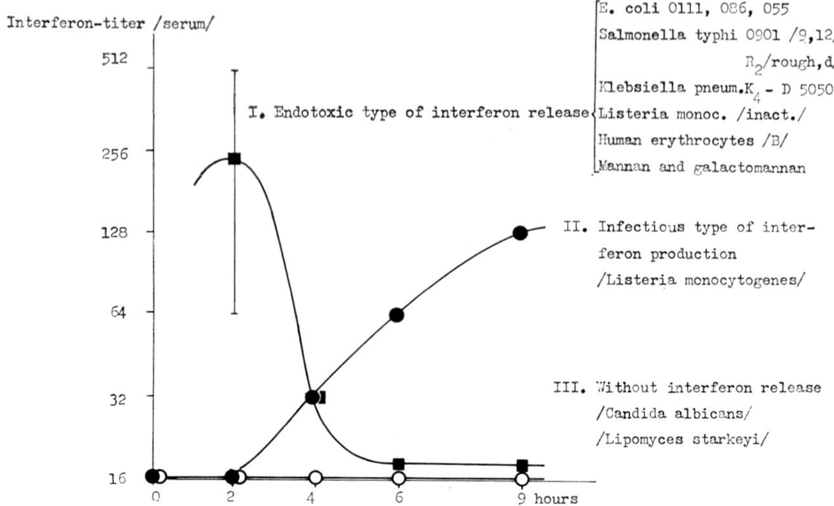

Figure 1. The dynamics of interferon release in mouse serum after intra-venous injection of various microorganisms, microbial products, and human erythrocytes.

(0901-9,12 and the R_2-rough, d strains), Klebsiellae pneumoniae (strains K_3-D 5050, K_3-C 5046) — whether live or (heat) killed, as well as endotoxic extracts from them — induce an "early" appearance of interferon in mice, which has been described by Youngner and Stinebring in 1964 and recently named "endotoxic" by the same authors (Youngner *et al.*, 1966 — personal communication).

A similar type of response has been observed when mice were injected with killed (but not with live) *Listeria monocytogenes* (strain 76.00), or, *Bordetella pertussis* (phase I, Evans-vaccine strain).

The yield of interferon produced in mice in response to injection of an O-antigen devoid *S. typhi* (strain R_2-rough, d) was comparable to that obtained after intravenous infection with a strain equipped with complete O-antigen (S. typhi 0901). This finding suggests that basal core lipopolysaccharides (Lüderitz *et al.*, 1966) provide a sufficient stimulus for interferon induction. However, as seen in Table I, treatment of bacterial extracts obtained from 2 S. *typhi*

TABLE I

SENSITIVITY OF INTERFERON INDUCING ABILITY OF EXTRACTS FROM 0901 AND R_2 (D) SALMONELLA TYPHI STRAINS TO TREATMENT WITH SODIUM-DEOXYCHOLATE

Bacterial Extract	Interferon Titer in Mouse Serum[†]					
(Corresponding to 2-3 x 10^9 Cells/ml)	Salmonella Typhi 0901 (9,12) Extract			Salmonella Typhi R_2 (Rough,d) Extract		
Control bacterial extract	512[†]	128[†]	128	128	256	128[†]
Sodium-deoxycholate treated bacterial extract[‡]	512	256	128[†]	16	32	32[†]

[†]Serum was obtained 2 hours after intravenous injection of mice.

[‡]Bacterial extracts were mixed with sodium-deoxycholate (final concentration 1%) incubated 2 hours at 37°C, dialyzed before injection.

strains with sodium-deoxycholate seems to impair the interferon inducing capacity of R_2-rough strain while leaving intact that of 0901 strain.

TABLE II
INTERFERON INDUCING ABILITY OF E. COLI 055, 086 AND 0111 STRAINS
IN IMMUNIZED MICE AND "NORMAL" HENS

E. coli		*Interferon-titer (Serum)*†		
Serotype	*Blood Group Activity*‡	*Control Mice*	*Immune Mice*§ (*E. coli 086*)	*"Normal" Hens*
055	O	512†	128	128
086	B	512	32	32
0111	AB	512	256	32

The bacterial suspensions contained 9 x 10^9 cells per mouse, and 7,5 x 10^{10} cells per hen, respectively.
†Serum samples for interferon titration were obtained 2 hours after intravenous infection of mice or hens.
‡Possible group activity according to Springer *et al.* (1961).
§Mice were immunized with 5 intraperitoneal doses of E. coli 086.

This finding indicates differences in vulnerability of interferonogenic capacity of R and/or O lipopolysaccharides, respectively. No substantial differences have been observed when 4 antigenically distinct *E. coli* strains were compared for interferon inducing capacity in mice. On the other hand, differences in interferon response were noticed repeatedly when immunized mice or "normal" hens were infected with various *E. coli* serotypes (Table II). This observation offers an immunological explanation for difficulties encountered in interferon induction in hens when certain *E. coli* strains or endotoxic extracts were used for injection (Stinebring and Youngner, 1964).

The possibility that blood group activity shared by various Gram-negative bacteria (Springer *et al.*, 1961) might be involved in process of interferon induction, led us to investigate the interferon inducing capacity of human erythrocytes (Table III). Surprisingly, it has been found that several samples of erythrocytes (especially those of B group) when injected intravenously into mice stimulate an "early" type of interferon appearance in serum. However, the interferon inducing capacity of positive erythrocyte sample after 48-72 hours of storage, usually, disappeared, and, moreover, interferon induction with erythrocytes from the same person was not a regular phenomenon. This observation suggests that adsorbed endotoxic material might be responsible for interferon induction. If true, interferon induction by human erythrocytes might signalize a transitory endotoxemia. However, experimental testing of this possibility did not exclude the other possibility that antigenic surface polysaccharides of erythrocytes may, in certain not clear circumstances, function as interferon inducers.

TABLE III
INTERFERON INDUCING ABILITY OF HUMAN ERYTHROCYTES IN MICE

Blood Group	No. of Samples Tested	No. of Samples Positive	Interferon Titre in Sera (Range)
O	5	1	16
A	7	3	16 – 64
B	9	6	16 – 512
AB	1	0	–

Mice were injected with a 20% suspension of erythrocytes. (0,3 ml per mouse i.v.). Blood samples from plexus orbitalis were obtained 2 hours later. The serum was tested against EMC virus in L-cells.
Characteristics of human erythrocytes as interferon inducers:
1. Loss of interferon inducing ability after storage at 4°C during 24-72 hours.
2. Irregular reproducibility by repeated testing of cells from the same person.

For a long time, we lacked a cell system which would permit us to study bacterial interferon induction in cells kept *in vitro*. Recently, we found that peritoneal cells of mice release interferon when treated with bacteria or their products up to 12 hours after explanation. Later, the interferon releasing ability disappeared completely, although such cells still responded with interferon formation of viral stimulation (Lackovič and Borecký, 1965). A similar observation has been made recently by Wheelock (1966 — personal communication) who studied interferon induction by NDV in human leukocytes. In mouse peritoneal cells, infected within 2 hours after explatation with various bacterial agents, we found that in the case of *E. coli* a similar amount of interferon was released when the input multiplicity of infection varied between 0.1-10. Higher multiplicity was needed in order to achieve a similar interferon response when *Salmonellae, Klebsiellae,* or, heat killed *Listeria monocytogenes* were tested (Table

IV) . This finding resembles the data of Youngner *et al.* (1966 — personal communication) who calculated the dose dependence of interferon production in response to NDV.

TABLE IV

INTERFERON PRODUCTION IN MOUSE PERITONEAL CELLS INFECTED
In Vitro WITH VARIOUS MULTIPLICITIES OF MICROBIAL AGENTS

Interferon Inducing Agent	Number of Bacteria Per 1 Peritoneal Cell						
	100	10	1	0,1	0,01	0,001	0,0001
	Interferon Titer						
E. coli 0111:B_4	4	32	64	64	32	16	8
—"— 086:B_7	16	32	32	32	16	16	—
—"— 055:B_5	8	32	64	64	16	8	—
Salmonella typhi R_2 (d)	128	32	16	4	<4	<4	—
—"— typhi 0901 (9, 12)	32	8	4	<4	<4	—	—
Klebsiella pneum. K_4-D 5050	4	<4	<4	—	—	—	—
Listeria monocytogenes	8	4	<4	—	—	—	—
Candida albicans (Berkhout)	<4	<4	—	—	—	—	—
Lipomyces starkeyi	<4	<4	—	—	—	—	—

The samples for interferon titration were taken 6 hours after infection of peritoneal cells with microorganisms.

II. A different picture of interferon production in serum results when mice or hens are infected with live Listeria monocytogenes. The curve resembles that obtained by Youngner and Stinebring (1964) when Brucella abortus was used as interferon inducer and which is by authors called "live virus type." (The designation "infectious type" is used in Figure 1.)

It is interesting to note that 48 hours after infection of hens with live Listeria, monocytogenes and "endotoxic type" of interferon response could be elicited in their sera when reinfected with Listeria monocytogenes.

III. A third group of microorganisms seems to be unable to elicit interferon formation in mice. (Candida albicans Berkout and Lipomyces starkeyi — from Czechoslovak type culture collection.) However, purified polysaccharides from these microorganisms such as a mannan of about 5500-8000 mol. weight isolated from Candida albicans (Šikl *et al.*, 1965), and, a galactomannan of about 20,000 m.w., isolated from Lipomyces starkeyi (Masler *et al.* — not yet published), proved to be capable of interferon induction both *in vivo* (in mice serum) and *in vitro* (in mouse peritoneal cells), although in different degree. The interesting feature of interferon induction by mannans *in vivo* is the presence of interferon in serum while absent in the spleen, as well as an "early-endotoxic" type of appearance. *In vitro* studies showed that so far only mouse peritoneal cells, infected soon after explantation, responded to mannan with interferon production. No interferon release has been observed in mannan treated continuous mouse "L"-cells, or, primary chick embryonic cells (Tables V and VI).

TABLE V
INTERFERON INDUCTION BY MANNAN FROM CANDIDA ALBICANS

In mice (100 ug/i.v.):

serum	+
spleen	—
Time of maximal interferon titers (serum)	between 2-6 hours
Time of complete disappearance of interferon (serum)	approx. 24 hours
In vitro:	
Mouse peritoneal cells	
(100 ug/2.5 x 10^6 cells)	
up to 12 hours after explanation	+
48 hours after explanation	—
Time of maximal interferon titer in growth medium	between 6-12 hours
Time of complete disappearance of interferon from growth medium	approx. 48 hours
Minimal interferon inducing amount of mannan	approx. 10 ug/2.5 x 10^6 cells
Mouse "L"-fibroblasts (48 hours old)	—
Chicken embryonic cells (24 hours old)	—

TABLE VI
DOSE AND CELL DEPENDENCE OF INTERFERON INDUCTION BY MANNAN
FROM CANDIDA ALBICANS IN MOUSE PERITONEAL CELLS *In Vitro*
DYNAMICS OF INTERFERON RELEASE

Amount of Mannan in ug	Number of Cells Per ml	Hours After Treatment	Interferon Titer
300	2.5 x 10^6	6	64
100	— " —	— " —	32
10	— " —	— " —	4±
1	— " —	— " —	<4
100	2.5 x 10^6	6	32
— " —	1.2 x 10^6	— " —	16
— " —	0.6 x 10^6	— " —	8
— " —	0.3 x 10^6	— " —	4
100	2.5 x 10^6	0	<4
— " —	— " —	3	16
— " —	— " —	6	32
— " —	— " —	12	32
— " —	— " —	24	8
— " —	— " —	48	<4

It has been calculated that about 25,000 peritoneal cells represent the minimal number of cells producing detectable amount of interferon on mannan stimulation.

Resuming, our experiments seem to indicate that:

1) Differentiation of interferons into "early-endotoxic" and "late-infectious, or, live virus" types (as introduced by Youngner and Stinebring, 1964) provides a basis for distinguishing different mechanisms of release, different cells involved in production, probably different qualities of the interferon molecules;

2) The finding that a homopolysaccharide (mannan) with relatively low molecular weight functions as an efficient interferon inducer is at variance with the hypothesis, that the polyanionic character of a molecule should be responsible for interferon induction (Kleinschmidt *et al.*, 1964).

Interferon release followed after infection of mice with a strain of *Salmonella typhi* devoid of O-antigen suggests that side chain sugars of lipopolysaccharides are not essential for interferon induction.

On the other hand, our experiments seem to indicate that the presence of specific antibodies directed against polysaccharide component of cell surface may hinder the expression of interferon inducing capacity of *E. coli* strains in an immune environment of mice or in "normal" hens.

Further investigations are needed to explain the irregular interferon inducing ability of human erythrocytes in mice since their functioning as temporary carriers of endotoxic substances resorbed in certain ill defined circumstances from intestines might help to explain the existence of "preformed" interferon in RE cells (Kono and Ho, 1964).

This work has been done in collaboration with Doctors V. Lackovič, D. Blaškovič, L. Masler and D. Šikl. The authors are indebted to Dr. Jola Borecká from the Institute of Epidemiology and Microbiology, Bratislava, for providing the bacterial strains.

REFERENCES

1. Ho, M.: Interferon-like viral inhibitor in rabbits after administration of intravenous endotoxin. *Science, 146:*1472, 1964.

2. KLEINSCHMIDT, W. J., CLINE, J. C., and MURPHY, E. B.: Interferon production induced by Statolon. *Proc. Natl. Acad. Sci. (U.S.), 52:*741, 1964.
3. KONO, Y., and HO, M.: The role of the reticuloendothelial system in interferon formation in the rabbit. *Virology, 25:*162, 1965.
4. LACKOVIČ, V., BORECKÝ, L.: The reticuloendothelial system and virus infection II. Production of interferon- and antibody-like substances in mouse peritoneal cells infected with myxoviruses *in vivo. Arch. Virusforsch., 17:*619, 1965.
5. LÜDERITZ, O., STAUB, A. M., and WESTPHAL, O.: Immunochemistry of O and R antigens of Salmonella and related Enterobacteriaceae. *Bacteriol. Rev., 30:*192, 1966.
6. SEDLAK, J., and RISCHE, H.: *Enterobacteriaceae — Infectionen.* Leipzig, G. Thieme Verlag, 1961.
7. ŠIKL, D., MASLER, L., and BAUER, Š.: The polysaccharides of yeasts and yeast-like microorganisms. I. The extracellular surface mannan of Candida albicans berkhout. *Chem. Evesti (Bratisý-ava) 19:*21, 1965.
8. SPRINGER, G. F., WILLIAMSON, BRANDES M.: Blood group activity of gram-negative bacteria. *J. Exper. Med., 113:*1077, 1961.
9. STINEBRING, W. R., and YOUNGNER, J. S.: Patterns of interferon appearance in mice injected with bacteria or bacterial endotoxin. *Nature, 204:*712, 1964.
10. WHEELOCK, E. F.: Virus replication and high titered interferon production in human leucocyte cultures inoculated with Newcastle disease virus. Personal communication via IEG / 6, 1966.
11. YOUNGNER, J., and STINEBRING, W.: Interferon induction in chickens injected with Brucella abortus. *Science, 144:*1022, 1964.
12. YOUNGNER, J., HALLUM, J., and STINEBRING, W.: Comparison of interferons appearing in the circulation of mice injected with different viral and nonviral stimuli. Personal communication via IEG / 6, 1966.

INTERFERON AS A POTENTIAL ANTIVIRAL DRUG

N. B. FINTER

A MOTIVE UNDERLYING MANY studies on viral interference has been the hope that they might lead to ways of controlling clinical virus infections. Thus, when Isaacs and Lindenmann (1957) discovered interferons, it was natural to consider whether these could be used as antiviral drugs. From this point of view, interferons have some particularly attractive properties (Table I). Firstly, they are active against a whole range of quite unrelated viruses, and in this respect they differ widely from those few synthetic antiviral drugs which are currently available. Secondly, they are remarkably non-toxic, both in tissue cultures and in the whole animal. A further point is that, although they are proteins, interferons are very poor antigens. Indeed, since, for example, human interferon must be formed many times in the body of an individual during natural virus infections, it seems hardly likely that it will be antigenic in man.

TABLE I

PROPERTIES OF INTERFERON

I Broad antiviral spectrum
   Non-toxic
   A poor antigen
II Maximum antiviral effects only after some hours
   Species-specific

Unfortunately, interferons have two less favourable properties. Firstly, cells must be treated with interferon for some hours before they become fully resistant to virus infection. This may seem to suggest that interferons could only be used for the prophylaxis of virus infections, but this is not necessarily so. In an animal, interferon could theoretically still have beneficial effects when administered some time after infection with a virus, perhaps even after the onset of symptoms, provided that, at that time, there are still

"

a significant number of virus-susceptible cells which have not yet been infected. A relevant point is that it is not necessary to prevent the growth of a virus completely in order to get clinically useful results. Secondly, interferons are markedly species-specific in their antiviral effects. For this and other reasons, interferon intended for use in man is much best obtained from human cells, and this makes it relatively difficult to prepare large amounts for clinical use.

In a recent review, data from many laboratories concerning effects of interferons *in vivo* have been considered (Finter, 1966b). Here, therefore, I shall present in particular data from my own laboratory, since this is an aspect of interferons in which I have been interested for some years.

In early experiments, Isaacs and his colleagues (Lindemann *et al.*, 1957; Isaacs and Westwood, 1959) showed that intradermal injection of an interferon preparation into the skin of a rabbit could prevent the growth of vaccinia virus injected into the same site 24 hours later. In my laboratory in 1962, we were interested to see for how long a single injection of interferon could thus protect the skin. In one such experiment, the skin of a rabbit's back was shaved and marked out in squares. On three different days, two such squares were inoculated intradermally with a preparation of rabbit interferon, and a further two with control material. Subsequently, all the sites were challenged with vaccinia virus on the same day, and at the different sites this was seven or four days or one day after the interferon injections. At the sites which were challenged with virus one day after injection of interferon, the development of a primary vaccinial response was completely suppressed. With an interval of four days between interferon and challenge, vaccinial lesions developed, but they were considerably smaller than the corresponding control lesions. With a seven day interval, there was no discernible effect from the interferon treatment.

Comparable experiments with interferon have been carried out in the skin of monkeys and in man. Protection of the skin of human volunteers by rhesus monkey kidney interferon was reported by the Scientific Committee on Interferon in 1962.

Such demonstrations of local antiviral effects confirm that

interferons are active in the animal and human body. Also, there are some special situations where such local applications of interferon could be of practical value. In particular, infections of the eye with herpes simplex virus, and much less frequently with vaccinia virus, are encountered in clinical practice, and these can cause serious damage to the cornea. Cantell and Tommila (1960) showed that the eyes of rabbits could be protected against experimental infections with vaccinia virus by repeated instillations of rabbit interferon. My former colleague, Dr. Weston Hurst, and I were similarly able to protect the eyes of rabbits against herpes simplex virus (Hurst and Finter, 1962, unpublished). The rabbits were treated with eye drops of interferon prepared from rabbit kidney cells, or comparable control material, given hourly during the day for six days. Twenty-four hours after the start of this treatment, the eyes were infected with herpes virus, and those treated with the control material developed a severe herpetic keratitis. In contrast, the eyes of the rabbits treated with interferon remained entirely free from clinical signs of infection. Jones *et al.* (1962) reported that interferon appeared to have beneficial effects in the treatment of vaccinial infections of the human eye.

At about this time, we made a number of attempts in my laboratory to protect rabbits against vaccinia and other viruses by injecting interferon at a different site in the body. These were unsuccessful, and it was obvious that this could have resulted merely from lack of potency in the interferon preparations which we had available. Therefore, as a preparatory step to studies in which the effects of interferon on mice with systemic virus infections were investigated, we looked for relatively potent sources of mouse interferon. Eight sources of interferon were studied (Finter, 1965), and of these, the brains of mice infected with the arbovirus, West Nile, contained the highest concentration, with a mean titre of approximately 10,000 units of interferon per ml. The next richest source was serum from mice inoculated intravenously with Newcastle disease virus (Baron and Buckler, 1963) which had a mean titre about four times lower. The best of the tissue culture sources studied was media collected from L cells infected with Newcastle disease virus; the mean titre of this material was about 1/40th of that in the brains of mice infected with West Nile virus. The other materials studied contained relatively small amounts of interferon. It should be pointed out that the absolute values obtained for the potency of these preparations have no particular significance in themselves, since interferon can at present only be measured in terms of arbitrary units. There are, nevertheless, some criteria by which to assess the size of a dose of mouse interferon, as will be considered later.

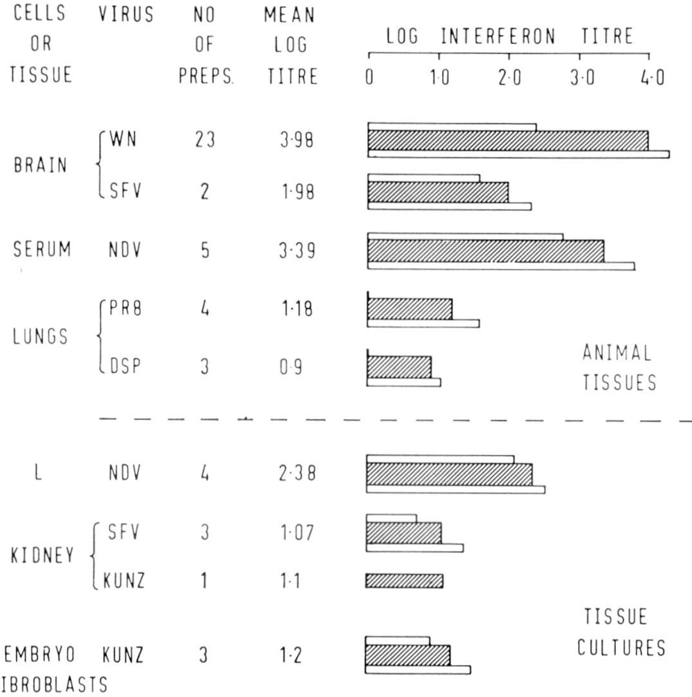

Figure 1. Yields of mouse interferon obtainable from different sources.

On the basis of these results, interferon obtained from the brains of mice infected with West Nile virus has been used in a number of mouse protection experiments (Finter, 1964a, 1966a), which have all been of the same general type. In these, groups of 10-20 mice were inoculated intramuscularly or intravenously with interferon, or with comparable control material made from the brains of uninfected mice. At the chosen times before or after these injections, the mice were infected with mouse encephalomyocarditis virus, or more often with Semliki Forest virus (SFV), usually injected intraperitoneally. Each morning and evening for 14 days after infection, the mice were inspected and the deaths recorded. The results of early experiments showed that adequate amounts of interferon could protect mice to a striking extent against these experimental infections. Later experiments have measured the amounts of interferon needed to protect under various circumstances, and have also studied the effects of injecting interferon at different times before or after virus infection.

In experiments in which mice were inoculated with different amounts of interferon, and were challenged after 24 hours with 110 LD_{50} of SFV, it was found that the numbers of mice ultimately surviving infection appeared to be directly proportional to the log dose of interferon injected, up to at least 30,000 units per mouse (Finter, 1966a). Under these conditions, there were no survivors among mice treated with less than about 1,000 units of interferon.

In other experiments, groups of mice were treated with 56,000 units of interferon, and were challenged with different amounts of SFV. All the mice survived challenge with 20 LD_{50}, but with larger doses of virus, there were fewer survivors, and none survived infection with 1280 LD_{50}. However, even with this largest dose of challenge virus, the interferon treatment had some effect as shown by a statistically significant prolongation of the mean survival time.

In most of these experiments, comparatively large doses of challenge virus were used, usually 100 LD_{50} or more, which killed nearly all the mice even in the groups treated with interferon. This allowed an efficient method of statistical analysis to be used, based on the survival times of the individual mice, so that the maximum amount of information concerning effects of interferon treatments could be derived. Such doses of virus are, however, much larger than those involved in natural virus infections of man or animals. We have, therefore, recently carried out some further protection experiments, using much smaller doses of challenge virus. The results obtained (Finter, 1967) have shown certain features of particular interest. Firstly, they have emphasized the remarkable potency of interferon as an antiviral agent

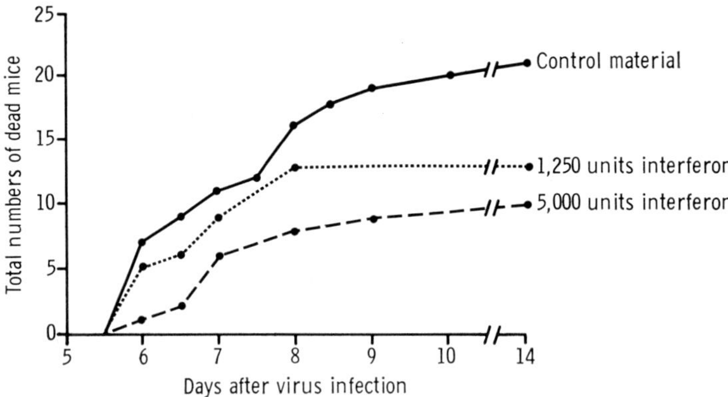

Figure 2. Effects of interferon in mice already infected with Semliki Forest virus. Groups of 30 mice were treated 18 hours after infection with interferon, or with comparable control material.

in vivo. Intramuscular or intravenous injection of even 125 units of interferon per mouse, 24 hours before infection with SFV, led to a highly significant increase in the number of surviving mice. Secondly, there was a significant increase in the number of mice surviving, when 5,000 units of interferon were injected intravenously even as late as 18 hours after infection (Fig. 2). Thirdly, it has been found that the prophylactic effect of a single injection of 5,000 units of interferon lasts for between 5-7 days. These amounts of interferon should be considered in relation to total amount of 17,000 units found in the brain of a mouse infected with West Nile virus (Finter, 1964b), and the estimated total amount of 100,000 units (or 750,000 units, in terms of the units used in our laboratory) which Baron *et al.* (1966b) have calculated to be formed in a mouse during the course of a viraemia (Fig. 3).

Workers in other laboratories have shown that rats, mice, chicks and rabbits can similarly be protected against a variety of virus infections by inoculation of adequate amounts of interferons (experiments reviewed by Finter, 1966b): an interesting recent observation is that concentrated preparations of mouse interferon inhibited the development of splenomegaly in mice infected with Friend leukaemia virus (Gresser *et al.*, 1966).

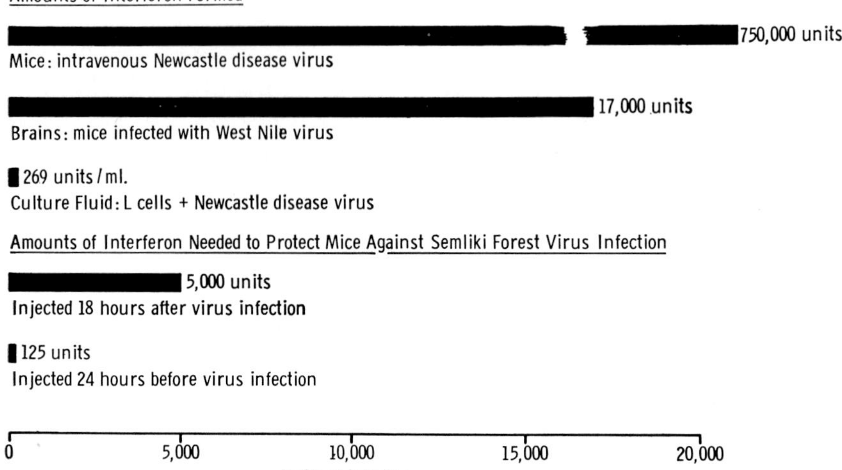

Figure 3. Amounts of interferon needed to protect mice against infection with Semliki Forest virus related to amounts of interferon obtainable from three sources.

Instead of injecting preformed interferons for the control of virus infections as in the experiments just described, an alternative is to administer a substance which stimulates the active production of interferon in the body. Baron and Buckler (1963) showed that large amounts of Newcastle disease virus injected intravenously into mice led to the appearance of high concentrations of interferon in the serum between about 2 and 9 hours later. Since interferon is rapidly cleared from the blood (Baron *et al.*, 1966b; Finter, 1966a; Subrahmanyan and Mimms, 1966), the total amounts of interferon formed after such an injection are very much higher than the serum concentrations themselves suggest. It has been shown by Baron and his colleagues (Baron *et al.*, 1966a) and in our own laboratory (Finter, 1966a), that mice thus injected with Newcastle disease virus are well protected against subsequent experimental virus infections. Wheelock and Dingle (1964) similarly demonstrated interferon in the serum of a leukaemic patient to whom they had administered large doses of viruses intravenously in an attempt to control his leukaemia. Wheelock and Sibley (1965) and Petralli *et al.* (1965) showed that interferon appeared in the sera of subjects injected subcutaneously with yellow fever virus and measles virus vaccines, respectively. The latter authors showed that the appearance of interferon in the blood coincided with a period during which the subjects were partially or totally resistant to cutaneous vaccination with vaccinia virus. Unfortunately, because of the formation of specific antibodies, a given virus can only be used once in each individual to provoke formation of interferon, as directly demonstrated by Wheelock and Dingle (1964).

A number of substances other than viruses have been shown to lead to the appearance of interferon in the body of animals. One of the most interesting of these is a complex polysaccharide derived from the mould *Penicillium stoloniferum*. A filtrate from a fermentation broth of this mould was shown by Powell *et al.* (1952) to protect mice against subsequent injections of otherwise lethal amounts of SFV or MM viruses. Later, Kleinschmidt and Probst (1962) isolated and partly purified an active principle, which they termed statolon, from another strain of the same mould. Statolon has been found to have a wide antiviral spectrum both in tissue cultures and in animals, and this appears to result from the ability of statolon to induce the formation of interferon, as shown by Kleinschmidt *et al.* (1964). When injected into mice, a single injection of 50 mg/Kg of statolon protected mice for up to 29 days against infection with a dose of MM virus which killed 63% of control mice (Kleinschmidt and Murphy, 1967).

Helenine (Shope, 1953), the product of another mould, *Penicillium funiculosum*, has also been shown to stimulate formation of interferon. It may be possible one day to utilize one of these two substances, when they have been sufficiently purified, or another such inducer of interferon, for the control of virus infections in man. The advantages of using a patient's own cells to produce an autologous interferon are obvious. However, such substances are not at present available, and in any event, injections of preformed interferon may be found to have some practical advantages. Preparations of human interferon have been obtained from tissue cultures of human amnion and foreskin cells, but only relatively small amounts can be obtained from such primary tissue cultures of human cells. However, workers in two laboratories

(Strander and Cantell, 1966; Falcoff *et al.,* 1966) have recently shown that very considerable amounts of interferon can be obtained from human leucocyte cultures, as first demonstrated by Gresser (1961). Alternative sources of human interferon are tissue cultures of diploid human cell strains, and of transformed human cell lines. Before interferon made from such sources, especially the latter, could be injected into man, a considerable degree of purification would be obligatory.

REFERENCES

BARON, S., and BUCKLER, C. E.: (1963) *Science, 141:*1061.

BARON, S., BUCKLER, C. E., FRIEDMAN, R. M., and McCLOSKEY, R. V.: (1966a) *J. Immunol., 96:*17.

BARON, S., BUCKLER, C. E., McCLOSKEY, R. V., and KIRSCHSTEIN, R. L.: (1966b) *J. Immunol., 96:*12.

CANTELL, K., and TOMMILA, V.: (1960) *Lancet, ii:*682.

FALCOFF, E., FALCOFF, R., FOURNIER, F., and CHANY, C.: (1966) *Ann. Inst. Parteur (Paris), 111* (5) *:*562.

FINTER, N. B.: (1964a) *Brit. Med. J., ii:*981.

FINTER, N. B.: (1964b) *Nature, 204:*1114.

FINTER, N. B.: (1965) *Nature, 206:*597.

FINTER, N. B.: (1966a) *Brit. J. Exper. Pathol., 47:*361.

FINTER, N. B.: (1966b) In: FINTER, N. B., Ed., *Interferons.* Amsterdam, North-Holland Pub. Co., p. 232.

FINTER, N. B.: *J. Gen. Virol.,* 1967, In press.

GRESSER, I.: (1961) *Proc. Soc. Exper. Biol. & Med., 108:*799.

GRESSER, I., COPPEY, J., FALCOFF, E., and FONTAINE, D.: (1966) *Proc. Soc. Exper. Biol. & Med., 124:*84.

ISAACS, A., and LINDENMANN, J.: (1957) *Proc. Roy. Soc. Ser. B, 147:*258.

ISAACS, A., and WESTWOOD, M.: (1959) *Lancet, ii:*324.

JONES, B. R., GALBRAITH, J. E. K., and AL-HUSSAINI, M. K.: (1962) *Lancet, ii:*875.

KLEINSCHMIDT, W. J., and MURPHY, E. B.: (1967) *Bacteriol. Rev.* In press.

KLEINSCHMIDT, W. J., and PROBST, G. W.: (1962) *Antibiot. Chemotherap., 12:*298.

KLEINSCHMIDT, W. J., CLINE, J. C., and MURPHY, E. B.: (1964) *Proc. Nat. Sci. U.S., 52:*741.

LINDENMANN, J., BURKE, D. C., and ISAACS, A.: (1957) *Brit. J. Exper. Pathol., 38:*551.

PETRALLI, J. K., MERIGAN, T. C., and WILBUR, J. R.: (1965) *Lancet, ii:*401.

POWELL, H. M., CULBERTSON, C. G., McGUIRE, J. M., HOEHN, M. M., and BAKER, L. A.: (1952) *Antibiot. & Chemotherap., 2:*432.

SCIENTIFIC COMMITTEE ON INTERFERON: (1962) *Lancet, i:*873.

SHOPE, R. E.: (1953) *J. Exper. Med., 97:*601.

STRANDER, H., and CANTELL, K.: (1966) *Ann. Med. Exper. Fenn., 44:*265.

SUBRAHMANYAN, T., and MIMS, C.: (1966) *Brit. J. Exper. Pathol., 47:*168.

WHEELOCK, E. F., and DINGLE, J. H.: (1964) *New England J. Med., 271:*645.

WHEELOCK, E. F., and SIBLEY, W. A.: (1965) *New England J. Med., 273:*194.

CHEMICAL AND BIOLOGIC CHARACTER-
ISTICS OF VERTEBRATE INTEFERONS

THOMAS C. MERIGAN, M.D.

INTRODUCTION

I N 1963, THE WORK OF Lampson *et al.* (1963) demonstrated that
chick interferon could be purified many thousandfold by the
usual techniques of protein chemistry including metal precipitation and ion exchange column chromatography. These studies
stimulated both ourselves and others such as Dr. Fantes (see
also Chap. 2) and Dr. Bodo (personal communication, 1966) to
purify this material further. Newer methods which have been
involved include gradient ion exchange column chromatography
and preparative polyacrylamide gel electrophoresis. Because of
their similar molecular weights and charge, we have found that
methods for purifiication of interferon from one species (the
chicken) could be applied to that obtained from other species,
including mouse and man. We have recently been primarily occupied with human interferon purification and characterization
because of the obvious dividend of potential clinical applicability.
However, it is clear that yields are best with the chick type as it
is more stable when purified. Later, I will deal with several of
the experiments that were possible because of the availability of
these purified interferons.

In the past two years, it has become clear that a variety of
nonviral agents can induce interferon production (Ho *et al.*, 1966),
including both extracts of living materials and other micro-organ-
isms. These include bacteria, rickettsia, pleuro-pneumonia-like organisms, trachoma inclusion conjunctivitis agents, endotoxin, phytohemagglutinin, fungal products (statolon and helinine), and
cyclohexamide. Recently, both Rytel and Smith and ourselves have
independently described the largest agent to induce interferon,
the intracellular protozoan, Toxoplasma. In conclusion, I will
mention some of the current work going on in our laboratory

with the newly discovered synthetic co-polymer interferon inducers. In general, all these nonviral inducers produce somewhat lesser levels of interferon compared to viruses. The materials produced by all these inducers have a similar antiviral action, but differ somewhat in terms of molecular characteristics.

PROPERTIES

It is clear that the major species of interferon induced by viruses in a variety of animal cell systems are of approximately neutral charge and 30,000 in molecular weight. They appear to have disulfide bridges within their structure. The purified material has an extremely high potency of antiviral action. In fact, Hilleman (1965) compared the activity of interferon against viruses on a weight basis to the most active antibiotics when they are assayed against sensitive bacteria. Both Dr. Fantes and we have purified chick interferon over 10,000-fold, and one can estimate a maximum of 4,500 molecules per cell are required to protect. Examination of the material at that stage in both our laboratories still reveals a significant degree of residual contamination by other proteins. Amino acid analysis and delineation of the number of polypeptide chains in the molecule has been difficult because of these contaminants and the small amount of interferon available. Studies with metabolic antagonists such as Actinomycin and studies of the structure of interferon induced by different agents indicate that the virus acts only as an inducer and that the structure of the interferon is directed by the genome of the cell in which it is produced. For example, the same virus can induce different interferons, depending on which type of animal cell it infects. Conversely, multiple viruses can induce the same type of interferon if they infect the same animal species. We recently observed that the interferon induced by Newcastle disease virus from human cells in tissue culture behaves chromatographically similarly to the serum interferon induced by measles virus in infants (Merigan *et al.*, 1966).

Recent studies, however, have demonstrated, especially with the newly described nonviral inducers, that interferons of higher molecular weight and even multiple species of interferons can be observed in the serum (see also Youngner, this Symposium). In fact, in our own laboratory, even rabbit serum interferon induced by virus has been found by Ke (1966) to be composed of two distinctly separable molecular species. It appears that there are at least three molecular weight classes of interferons, one of around 90,000 to 100,000, another of 40,000 to 60,000, and a third of approximately 20,000 to 30,000. The pattern of production of these molecular species depends on the nature of the inducer, the animal species studied, the site examined, and the time after induction.

TABLE I

MOLECULAR WEIGHT AND MOLECULAR CHARGE OF VARIOUS MOUSE INTERFERONS.
MOLECULAR WEIGHT WAS MEASURED BY G-100 SEPHADEX COLUMN CHROMATOGRAPHY,
AND MOLECULAR CHARGE MEASURED IN POLYACRYLAMIDE GELS AT pH 4.3 WITH
RESPECT TO A METHYL GREEN MARKER AS DESCRIBED ELSEWHERE

Type of Interferon	*Inducing Agent*	*Molecular Weight*	*Electrophoretic Rf*
Tissue culture	Chikungunya Virus	26,000	0.35
Serum	Newcastle Disease Virus	26,000*	0.35
Tissue culture	Statolon	34,000	0.35
Splenic extract	Statolon	30,000	0.36
Serum	Statolon	85,000	0.63

*Although this may be the predominant molecular species in serum obtained 6 hours after NDV injection, recent more detailed analysis of serum reveals two more species, one of 45,000 and another of 80-90,000 which may predominate in sera drawn under different conditions (Hallum, J. V., Merigian, T. C., and Youngner, J. S. In preparation, 1967).

For example, although we find that statolon, an anionic polysaccharide derived from a penicillium, induces a 90,000 m.w. species in the serum of the mouse, a 30,000 m.w. species is formed in the spleen of the statolon injected mouse which is indistinguishable in molecular weight or electrophoretic properties from viral induced mouse interferon seen *in vivo* or in tissue culture (Merigan and Kleinschmidt, 1965). It is clear with both viruses and nonviral agents such as statolon that the molecular nature of the interferon within the circulation varies depending on the time after injection. In general, the heavier molecular weight species appear more promptly after injection of the interferon inducer. We have observed a 110,000 m.w. species in the circulation of the chicken three hours after Statolon injection. On the other hand, we find a 30,000 species at 16 hours which may represent newly synthesized and released splenic interferon.

Prior studies of both Youngner and Ho and their collaborators indicate that certain of these heavier interferons are not blocked by agents known to arrest protein synthesis and appear to be preformed within the animal (see also Chapter 4). At present, the exact interrelationships between these multiple molecular species are not clear. We have recently carried out an experiment clearly demonstrating the separate mechanism for production of these species within the same animal. We find in cyclohexamide treated mice that the heavier serum interferon appears in normal yield confirming Youngner's prior observation, whereas we find the lighter spleen interferon is markedly diminished. It is intriguing that Youngner *et al.* (1965) have observed the 28,000 m.w. mouse serum interferon induced by virus can also be inhibited by cyclohexamide blockade of protein synthesis.

However, it appears likely to us that all these species have the same active center since they all appear to have a similar specificity toward cell and virus and a similar mode of action. The obvious possibility is either that they represent an active material on different carrier proteins or are related through being isozymic forms of one another.

Recent studies by Nagano (1966) suggest that a rabbit antiviral material which he feels to be interferon is composed of two parts. One of these, a protein moiety, is necessary for demonstration of its action in tissue culture and confers the cell species specific aspects of its action. However, the other portion of the molecule is a non-species specific antiviral carbohydrate of approximately 3,000 m.w. This carbohydrate can only be demonstrated to have an antiviral action *in vivo,* where perhaps it can combine again with the protein carried available in the organism. This schema is conceivable as no studies have yet been carried out which absolutely exclude a significant carbohydrate moiety within the interferon molecule. However, these studies sorely need confirmation by other workers, especially in animal species such as the mouse, chicken, or human, in which more information is available as to chemical and biological properties of the interferon.

Interferon has been found to be highly cell species specific in all situations in which purified interferons have been studied (for example, thousand-fold excesses are not active on heterologous cells). Previous studies utilizing crude interferon preparations indicated some lack of species specificity, and recent investigations have suggested some species overlap in interferon action if phylogenetically closely related species are studied.

Interferon is active against a wide range of viruses. In collaborative studies with us, Lavelle Hanna and Ernest Jawetz of the University of California find it even acts in tissue culture against a member of a group of intracellular parasites not usually classed with the viruses, the Psittacosis-Lymphogranuloma Venereum-Trachoma group (PLT agents) or chlamydiac group (Hanna *et al.*, 1966). These agents are intracellular parasites that appear to replicate, at least in part, by binary fission and contain both RNA and DNA in their structure. They are sensitive to the action of antibiotics and are currently thought to be more closely related to bacteria than viruses. These are the largest agents against which interferon appears to act. The LB1 inhibitor produced in L cells by Newcastle Disease Virus is pH stable, nonsedimentable by 100,000 G, and is inactivated by heat and trypsin in a way identical to mouse interferon. Column chromatographically purified interferon is also active against the LB1 agent, and we have recently observed this inhibitory action against two other TRIC agents, although the inhibition is cell species specific (Hanna, *et al.*, 1967). It is of interest that the organisms in this class have their own ribosomes (Tamura, 1965), yet it is not known whether all of the proteins required for their intracellular replication are synthesized on their own rather than the host's ribosomes. There have been several as yet unpublished attempts to demonstrate interferon against higher intracellular parasites, including rickettsia, bacteria, and protozoa which were unsuccessful. It is not surprising that the latter two groups of organisms seem to be insensitive to interferon action because both have been reported to support virus replication. Therefore, their biosynthetic pathways might be expected to resemble those of the host cell in their apparent lack of change in the presence of interferon.

The first illustration demonstrates the failure of interferon to effect certain host functions. In collaborative studies with Baron

---

1) Purified interferon does not influence *cell division rates* in rapidly growing cells (Baron *et al.*, 1966) .

2) Purified interferon does not influence the ability of the cell to synthesize *RNA, DNA* or *protein* (Levy and Merigan, 1966; Marcus and Saab, 1966; Cocito, Schonne and DeSomer, 1965) .

3) Crude interferon does not influence rate of *antibody* production *in vivo* (Anderson, 1965) and *in vitro* (Klesius and Lockhart, 1966) . Personal communication; Mazzur *et al.*, 1967.

4) Purified chick interferon does not block induction of *interferon* in tissue culture (Isaccs, Rotem and Fantes, 1966) .

Figure 1. Selective nature of interferon's action. (Failure to Influence Host's Own Biosynthesis.)

(1966) and Levy (1966) , we have noted that interferon is highly selective in its action. Our purified interferon, in contrast to crude interferon, has no measurable effect on the host cell in any of its biosynthetic functions including the synthesis of RNA, DNA, and protein. Cells have been demonstrated to divide at a normal rate following treatment with purified interferon, and after interferon treatment others have found antibody synthesis normal both when studied *in vivo* and *in vitro*. Isaacs, Rotem, and Fantes (1966) have recently shown that a material which they call "blocker" can be removed for crude interferon preparation during purification, and that it is responsible for an inhibiting effect of crude chick interferon preparations on interferon production (Fig. 2) .

In other studies with purified interferon, it has been shown that several initially proposed mechanisms of action are not tenable. Interferon does not appear to be capable of influencing oxidative phosphoralation in cells, nor is it a nuclease. It does not induce ribonuclease production in cells, nor is it an inhibitor of viral RNA polymerase.

SITE OF ACTION (FIG. 3)

Interferon does not interact extracellularly with either virus or its infectious nucleic acid. It has been demonstrated, in studies including those employing radioactive virus, that virus is absorbed to and penetrates the interferon protected cell in normal fashion. In fact, at least a few of the early host genome dependent events in viral infection appear to occur in interferon protected cells. It seems likely that the uncoating activity, or the activity responsible

are infected with a large DNA virus, vaccinia. It also appears that the messenger RNA of this DNA virus is synthesized in the interferon protected cells. However, the specific polyribosomal complexes between the messenger RNA and host ribosomes do not form due to changes in one of the reactants, that is, either the messenger or the ribosome. The lack of formation of the viral specific polysomes blocks synthesis of viral coated proteins at the translational level. Levy reports similar observations in lack of formation of viral specific polyribosomal complexes working with Mengo virus (see Levy, Chapter 5). All later events in viral biosynthesis then do not occur. For example, neither the viral DNA dependent, DNA polymerase noted with DNA viruses nor the viral RNA dependent, RNA polymerase observed with RNA viruses appear. The usual increase in enzymes such as thymidine kinase are not observed during vaccinia infections in interferon protected cells, and the viral coat proteins or the viral structural DNA or RNA do not appear to be synthesized in any quantity; hence, no complete virions are formed or released in the interferon protected cell. However, this cell, because of the destruction of its own polyribosomal complexes, goes on to die. It therefore is sacrificed and aborts the virus, preventing its spread to the uninfected, healthy neighboring cells.

Very recent *in vitro* studies by Marcus and Saab (1966) indicate a host ribosomal level of action for interferon. They found that ribosomes taken from interferon treated cells behave differently from controls when incubated with viral messenger RNA in that they do not bind or translate the viral messenger as well.

Recently, Regelson has described antiviral activity in the serum of mice injected with anionic co-polymer of defined composition (Regelson, personal communication, 1966). In conclusion, we wish to report confirmation of his finding, identification of the active material as interferon, and our results on the structural requirements for co-polymer to act as interferon inducers. The significance of these compounds is heightened by his report of this material as inhibition of Friend virus leukemia in mice and their present trials in the therapy of human neoplasms.

The characteristics of the maleic acid/vinyl ethyl ether co-polymer induced mouse serum interferon indicate it to be a protein of 70,000 m.w. Its antiviral protection is achieved indirectly through the cells rather than on direct incubation with the virus. Its heat inactivation characteristics are like mouse interferon and Actinomycin treatment blocks the antiviral activity of this substance in a fashion similar to virus induced interferon. It is not active on heterologous cells.

We have studied the structural variation in the maleic acid copolymer. As Regelson has noted, such antiviral activity can be induced by co-polymers from 17,500 thousand in m.w. Although we find co-polymers from 40,000 to $1\frac{1}{4}$ million m.w. are less active, and require high doses to active full response. Maximum interferon levels varied between 1/250 to 1/1,200 in different experiments. Methyl or ethyl analogues behaved similarly, but a styrene side chains significantly altered the activity. These styrene derivatives were somewhat toxic and even at doses near their LD_{50} dose no antiviral activity was noted. Methyl esterification of the carboxyls to extent of 50% did not significantly alter the interferon inducing capacity of the material. However, similar amidation decreased its activity somewhat.

In general, these materials resemble the anionic polysaccharide statolon in their ability to induce interferon and in the pattern of production with peak levels being reached between 18 hours after injection. However, the level of interferon induced is between three to eight-fold lower.

At present, we are engaged in collaboration with Dr. Regelson in an evaluation of one of these materials as an interferon inducer in man. A possible drawback of these materials is their failure to be metabolized in the animal and hence possible complications of their disposition in the reticulo-endothial system.

REFERENCES

1. ANDERSON, S. G.: *Austral. J. Exp. Med. Sc., 43:*345, 1965.
2. BARON, S., MERIGAN, T. C., and McKERLIE, M. L.: *Proc. Soc. Exper. Biol. & Med., 121:*50, 1966.
3. COCITO, C., SCHONNE, E., and DESOMER, P.: *Life Science, 4:*1253, 1965.
4. FANTES, K. H., and FURMINGER, I. G. S.: *Nature (London), 206:*928, 1965.
5. FANTES, K. H., O'NEILL, C. F., and MASON, S. J.: *Biochem. J., 91:*20P, 1964.
6. HANNA, L., MERIGAN, T. C., and JAWETZ, E.: *Amer. J. Ophthal., 63:*1115, 1967.
7. HANNA, L., MERIGAN, T. C., and JAWETZ, E.: *Proc. Soc. Exper. Biol & Med., 122:* 417, 1966.

8. HILLEMAN, M. R.: *J. Cell. and Comp. Physiol., 62*:337, 1963.
9. HO, M., FANTES, K., BURKE, D. C., and FINTER, N. B.: In *Interferons.* Ed., N. FINTER. Philadelphia, W. B. Saunders Company, 1966.
10. ISACCS, A., ROTEM, Z., and FANTES, K.: *Virology, 29*:248, 1966.
11. JOKLIK, W. K., and MERIGAN, T. C.: *Proc. Nat. Acad. Sc., 56*:558, 1966.
11a. KE, Y. H., and HO, M.: *Nature (London), 211*:541, 1966.
12. LAMPSON, G. P., TYTELL, A. A., NEMES, M. M., and HILLEMAN, M. R.: *Proc. Soc. Exper. Biol. & Med., 112*:468, 1963.
13. LEVY, H. B.: *Virology, 22*:575, 1964.
14. LEVY, H. B., and MERIGAN, T. C.: *Proc. Soc. Exper. Biol. & Med., 121*:53, 1966.
15. MARCUS, P. I., and SAAB, J. M.: *Virology, 30*:502, 1966.
16. MAZZUR, S. R., ELLSWORTH, B., and PAUKER, K.: *J. Immunol., 93*:683, 1967.
17. MERIGAN, T. C., GREGORY, D., and PETRALLI, J.: *Virology, 29*:515, 1966.
18. MERIGAN, T. C., and KLEINSCHMIDT, W.: *Nature (London), 212*:1383, 1966.
19. MERIGAN, T. C., WINGET, C. A., and DIXON, C. B.: *J. Molec. Biol., 13*:679, 1965.
20. MINER, N., RAY, W. J., JR., and SIMON, E. H.: *Biochem. and Biophys. Research Commun., 24*:268, 1966.
21. NAGANO, Y., KOJIMA, Y., HANEISHI, T., and SHIRASAKA, M.: *Jap. J. Exper. Med., 36*:535, 1966.
22. SONNABEND, J. A., MARTIN, E. M., MECS, E., and FANTES, K. H.: *J. Gen. Virol., 1*:41, 1967.
23. TAMURA, J.s *J. Molec. Biol., 11*:91, 1965.
24. YOUNGNER, J. S., STINEBRING, W. R., and TAUBE, S. E.: *Virology, 27*:541, 1965.

INTERFERON PRODUCTION IN MICE INJECTED WITH VIRAL AND NONVIRAL STIMULI

JULIUS S. YOUNGNER

DURING RECENT YEARS, an increasing number of non-viral stimuli which cause the appearance of interferon in a variety of cell systems have been reported. These stimuli include organisms such as bacteria (1), rickettsiae (2), mycoplasmas (3), and agents of the psittacosis-lymphogranuloma-trachoma group (4); and substances prepared from bacteria (5), molds (6, 7), and higher plants (7a). This report is intended to summarize our experiences with the production of interferon by viral and non-viral stimuli in a single system, the intact mouse, which has been intensively studied in our laboratory.

METHODS

The methods employed in these studies can be briefly summarized. The different interferons were prepared by the intravenous injection of 0.1 ml volumes of the different stimuli into mice, using at least 10 animals per sample. At appropriate times, blood was obtained by cardiac puncture using heparin-rinsed syringes and pooled plasmas were tested for interferon production by the plaque reduction method. Cultures of chick embryo or L-cells were exposed for 20 hours to dilutions of interferons and then challenged with 40 to 60 plaque-forming units (PFU) of vesicular stomatitis virus (VSV). The inhibitory titers of the plasmas were expressed as the reciprocal of the dilution which reduced the plaque count to 50% of that of control cultures. The details of these methods have been published elsewhere (1, 5, 8).

The molecular weights of the different interferons were determined by a modification of the method described by Merigan (9, 10). Since interferons produced in cells of a given species are effective only in cells of the same or closely related species (9, 11), assays of mixtures of mouse and chicken interferons are possible. Therefore, mouse inhibitors were mixed with a standard chicken interferon and co-chromatographed on columns of G-100 Sephadex so that the well-characterized chicken interferon could be used as an internal standard. The G-100 Sephadex columns were calibrated before and after each interferon run by chromatographing a mixture of blue dextran, phenol red, and solutions of the following three crystalline proteins: bovine serum albumin, ovalbumin, and chymotrypsinogen. The elution volume of each protein was taken as the difference between the volume of appearance of the blue dextran peak and the absorption peak at 280 mμ for the protein standards. A plot of $\log_{10}$ molecular weight of the standards against the elution volume gave a straight line. The molecular weight of the inhibitor was obtained by comparing the elution volume of peaks of interferon activity to the plot obtained with the standard proteins.

The properties which were used to identify an inhibitor as interferon were species specificity, trypsin sensitivity, acid stability, intracellular site of action, lack of direct effect on virus, and lack of virus specificity. These properties were identical for all the inhibitors produced in mice by the viral and non-viral stimuli reported in this paper.

STIMULI OF INTERFERON PRODUCTION IN MICE

Table I lists the various viral and non-viral stimuli which have been reported to produce circulating interferon in mice. In addition to many viruses, infectious agents such as TRIC agents, rickettsiae, mycoplasmas, and many gram-negative bacteria have been found to produce interferon in mice. It is interesting that we

TABLE I
STIMULI OF INTERFERON PRODUCTION IN MICE

Viruses (Many)
Trachoma-inclusion conjunctivitis (TRIC) agents
Mycoplasmas (M. pneumoniae)
Bacteria (*Brucella abortus, Salmonella typhimurium, H. pertussis, Escherichia coli, Serratia marcescens*)
Bacterial lipopolysaccharides (*E. coli, B. abortus*)
Mold products
    Statolon (*Penicillium stoloniferum*)
    Helenine (*Penicillium funiculosum*)
    Cycloheximide (*Streptomyces griseus*)
    Acetoxycycloheximide (*Streptomyces griseus*)

have failed to stimulate interferon with the gram-positive bacteria we have tested. The intravenous injection of large numbers of viable *Staphylococcus aureus* or *Bacillus subtilis* did not produce any circulating virus inhibitors (5). We have found that the production of interferon by living bacteria is not dependent upon replication of the organisms in the tissues or body fluids of the mouse (Stinebring and Youngner, unpublished data).

PROPERTIES OF INTERFERONS PRODUCED IN MICE INJECTED WITH VARIOUS VIRAL AND NON-VIRAL MATERIALS

We have reported previously that the early interferon which appears 2 hours after the injection of mice with *E. coli* endotoxin has a molecular weight of 89,000 and that the interferon produced 6 to 12 hours after the injection of living *Brucella abortus* shows two peaks of activity by Sephadex filtration which correspond to molecular weights of 77,000 and 54,000 (10). In contrast, the intravenous injection of NDV in mice results in the appearance of peak interferon titers 10 to 12 hours after injection of the virus and the molecular weight of this inhibitor is 28,000 (9, 10).

These experiments were extended using several additional bacterial species and an endotoxin prepared from *B. abortus*. Results were also obtained with several mold products (statolon, cycloheximide, and acetoxycycloheximide). In addition to studies of molecular weights, procedures known to alter the reactivity of animals to the lethal or pyrogenic effects ef endotoxin were investigated for their influence on the appearance of interferon in the circulation of mice injected with the different stimuli. The procedures employed were: a) pretreatment of mice with endotoxin to produce a state of reduced responsiveness (hyporeactivity) to a subsequent dose of endotoxin (12), and b) infection of mice with attenuated tubercle bacilli to produce an enhanced state of responsiveness to the lethal effects of endotoxin (13). Previous results from our laboratory have shown that the appearance of interferon in the circulation of mice can be affected by procedures which alter responsiveness to the lethal and pyrogenic effects of endotoxin (14).

Table II summarizes the results obtained in mice injected with the different stimuli. In addition to giving the molecular

TABLE II

PROPERTIES OF INTERFERONS PRODUCED IN MICE INJECTED WITH VARIOUS VIRAL
AND NON-VIRAL MATERIALS

Stimulus-producing Circulating Interferon after Intravenous Injection of Mice	Molecular Weight of Interferon (Sephadex G-100)	Time Maximum in Plasma (Hours)	Inhibited by Blockade of Protein Synthesis	Inhibited by Pretreatment with E. coli Endotoxin	Enhanced by BCG Infection
Viruses					
NDV (infective)	28,000	8–12	+	+ (95%)	0
Bacteria & Products					
Brucella abortus	77,000; 54,000	6–12	+	+ (100%)	0
Salm. typhimurium	85,000; 33,000	2	0	+ (100%)	+
Serr. marcescens	87,000; 33,000	2	0	+ (100%)	+
E. coli endotoxin	89,000	2	0	+ (100%)	+
B. Abortus endotoxin	n.d.	2	0	+ (100%)	+
Mold Products					
Statolon	90,000	8	0	0	0
Cycloheximide	33,000	12	0	+ (80%)	0
Acetoxycyloheximide	n.d.	12	0	n.d.	n.d.

+ = yes; 0 = no.
n.d. = not determined.

weights of the interferons present at the time of peak titer, the influences of prior treatment of mice with cycloheximide, endotoxin, or infection with *M. tuberculosis* (BCG) also are described. For purposes of comparison, data are included for the interferon produced by infective NDV.

Several interesting patterns emerge from the data presented in Table II. Firstly, although a wide variety of molecular weights are recorded for interferons produced in response to different stimuli, they can be separated into two broad categories: light (M.W. 28,000 to 33,000) and heavy (M.W. 54,000 to 90,000) interferons. Secondly, interferons which reach their peak titers early tend to have heavier molecular weights than those which peak later. This conclusion is only roughly applicable since the heavy interferon (M.W. 90,000) which appears after inoculation of mice with statolon reaches peak titers between 8 and 12 hours after inoculation (15, 8). Thirdly, blockade of protein synthesis inhibits only the interferons produced in response to NDV and *Brucella abortus,* indicating that these stimuli induce the synthesis of interferon, whereas the other stimuli cause the release of preformed virus inhibitor.

As far as the hyporeactive state produced by prior inoculation of endotoxin is concerned, it can be seen that all the stimuli tested were affected by this procedure, except statolon. These results are interpreted to mean that, although the interferons produced in mice by statolon and endotoxin are both present in a preformed state in the animal and both have similar molecular weights of 90,000, they are released from different cell populations and/or by different mechanisms. The finding that the other stimuli employed are all affected by pretreatment of mice with endotoxin indicates the likelihood that the same cells are involved in the release of interferon, despite the different time patterns of appearance of interferon in the circulation. The cellular, rather than humoral, nature of this hyporeactivity has been clearly indicated (14, and unpublished data).

In the case of hyperreactivity elicited by infection of mice with tubercle bacilli, it is clear that an increased capacity to produce interferon is seen only with stimuli which exert their effects by virtue of their endotoxin content (*S. typhimurium, S. marcescens,* and the endotoxins of *E. coli* and *B. abortus*). It is likely that infection with BCG results in the appearance of a cell type or types which can release heavy molecular weight interferon when stimulated by endotoxin; these cells fail to react in this manner with other stimuli. It is interesting to note that interferons with identical molecular weights appear in BCG-infected and in uninfected mice injected with *E. coli* endotoxin (10).

TABLE III
CHARACTERISTICS OF TWO TYPES OF INTERFERON PRODUCTION IN MICE

	Response	
	Endotoxin-type	*Live Virus-type*
Stimulus	Bacterial lipopolysaccharide *Serratia marcescens* *Salmonella typhimurium*	Newcastle Disease Virus *Brucella abortus*
Time interferon maximum in plasma	2 hours	8–12 hours
Molecular weight of major interferon component	80,000–90,000	28,000-77,000
Requirement for protein synthesis	no (Interferon is preformed in host)	yes (Interferon is newly synthesized)
Heterologous hyporeactivity	yes	yes
Enhancement by infection with tubercle bacilli	yes	no

From the results which have been described, it is possible to identify at least two separate and distinct patterns of interferon production in mice injected intravenously with different stimuli. These patterns are described in Table III. One of these can be termed the "endotoxin-type" response and is characterized by the appearance in the blood of a heavy molecular weight (85,000 to 90,000) interferon which is released from a preformed state and which reaches maximum levels about 2 hours after the intravenous injection of the stimulus. In addition, in the "endotoxin-type" response, there is a marked enhancement of the amount of interferon released when animals infected with attenuated tubercle bacilli are utilized.

The other clearly definable interferon response can be termed the "live virus" response. Intravenous injection of infective NDV or live *Brucella abortus* results in the appearance of interferon which is newly synthesized and which reaches peak titers 6 to 12 hours after injection. The molecular weight of the interferon produced by this response is variable and depends to a great extent on the stimulus employed and the time of bleeding. This type of response is not enhanced in mice infected with tubercle bacilli but hyporeactivity is observed when mice pretreated with endotoxin are employed.

It is of interest to note that in the case of *B. abortus,* either of the two types of response described above can be elicited depending on whether one uses the live, intact organism or the separated endotoxin fraction of the cell wall. The intact bacterium produces the "live virus" response while the fractionated lipopolysaccharide gives the "endotoxin-type" of response.

The pattern obtained with statolon is somewhat different than the two response patterns described above. Statolon releases into the circulation a preformed, heavy molecular weight interferon which peaks later than the usual "endotoxin-type" response, and is not enhanced by using mice infected with tubercle bacilli. No hyporeactivity can be demonstrated with statolon in mice previously treated with endotoxin. These differences have led us to conclude that the interferons produced in response to endotoxin and statolon are released from different cell populations and/or by different mechanisms (8).

MODIFICATION OF INTERFERON RESPONSES IN ANIMALS

Attempts were made to compare the ability of different procedures to modify the two patterns of interferon response which have been described. Differentiation of the "endotoxin" and "live virus" responses was carried out using heterologous hyporeactivity, splenectomy, and cortisone pretreatment as test procedures.

Heterologous Hyporeactivity. Studies have been reported (14) which showed that prior treatment of mice with endotoxin completely eliminated the characteristic appearance of interferon which followed endotoxin inoculation of untreated mice. With NDV challenge, mice treated 48 hours previously with endotoxin showed a markedly depressed interferon titer compared to the untreated controls. Data were also obtained which showed that pretreatment of mice with NDV decreased interferon response to injection of endotoxin and that normal responsiveness to endotoxin did not reappear until about 6 days after the original injection of NDV. This experiment demonstrated that decreased appearance of interferon in hyporeactive animals is not limited to the materials used to produce this state.

Table IV summarizes the results of another experiment in which mice were challenged with viral and non-viral stimuli 48 hours after pretreatment with these materials. The patterns of homologous and heterologous hyporeactivity showed that in mice pretreated with NDV, reduced interferon titers followed second injection of NDV, endotoxin, or statolon. Pretreatment with endotoxin or statolon produced homologous hyporeactivity and a markedly depressed interferon response to NDV. In contrast, endotoxin and statolon did not produce reciprocal hyporeactivity (8). The lack of reciprocal hyporeactivity between endotoxin and statolon points to the probable difference in the cell populations affected by these substances and again emphasizes the cellular (rather than humoral) basis of this reduced responsiveness.

TABLE IV

PATTERNS OF HOMOLOGOUS AND HETEROLOGOUS HYPOREACTIVITY IN MICE PRETREATED AND CHALLENGED WITH DIFFERENT STIMULI OF INTERFERON

Mice Pretreated with	*Per Cent Inhibition of Interferon Titer in Plasma after Challenge 48 Hours Later with:*		
	Endotoxin	*NDV*	*Statolon*
Endotoxin (250 µg)	100	95	0
NDV (2 x 10⁸ PFU)	100	98	94
Statolon (1,000 µg)	0	75	84

Influence of Splenectomy on Interferon Formation. Fruitstone *et al.* have recently shown that splenectomy markedly depresses the interferon titers produced by NDV in mice (16). De Somer and Billiau, working with rats, found the spleen to be involved in the production of interferon in response to *E. coli* endotoxin, but not necessary for interferon production induced by Sindbis virus (17).

Experiments were carried out to determine the influence of splenectomy on the "endotoxin" and "live virus" interferon response patterns in mice. Mice were splenectomized or sham-operated 2-3 days prior to stimulation with either NDV, *E. coli* endotoxin, or statolon. There was no difference in results when mice were used for interferon production 2 days or 2 weeks after splenectomy. After injection of the different stimuli, groups of 10 mice were bled at the appropriate time intervals. The results (Table V) showed that splenectomy drastically reduced interferon production by NDV and statolon, while having little or no effect on endotoxin-stimulated interferon. Further differentiation of the two different types of interferon response in the mouse is provided by these findings. The "live virus" type of response is highly dependent upon the presence of the spleen, whereas this organ seems not to be involved in the "endotoxin" type of response.

In the case of statolon, which occupies an anomolous position in regard to the two types of interferon response, it is interesting to note that Merigan and Kleinschmidt have reported that two interferons with different molecular weights are produced in mice by this material (15). An interferon with a M.W. of 90,000 appears in the circulation, while an inhibitor with a M.W. of 30,000 can be found in the spleen. Further, in contrast to the preformed nature of the circulating, heavy interferon (8), the lighter spleen interferon is inhibited by pretreatment of mice with doses of cycloheximide which block protein synthesis in the intact animal (Merigan, personal communication). This information may provide a basis for understanding the inhibiting effect of splenectomy on interferon production in mice by statolon.

TABLE V

INFLUENCE OF SPLENECTOMY ON INTERFERON PRODUCTION IN
MICE BY NDV, E. *coli* ENDOTOXIN, AND STATOLON

Stimulus (i.v.)	Treatment*	Interferon Titer of Plasma at:					
		2 hr	4 hr	5 hr	6 hr	8 hr	10 hr
NDV	Sham-operated	880†	2,700		3,800		21,000
$(1.5 \times 10^8$ PFU)	Splenectomized	720	880		940		130
Endotoxin	Sham-operated	500		270		20	
(100 μg)	Splenectomized	400		98		80	
Statolon	Sham-operated	1,500		6,400		10,000	
(1,000 μg)	Splenectomized	800		1,100		1,150	

*Splenectomized or sham operated 2-3 days prior to stimulation.
†Pool of plasma from 10 mice.

Influence of Pretreatment with Cortisone on Interferon Production. The influence of steroid hormones on interferon production has been studied in different systems by several groups of workers (18, 19; Postic and Ho, personal communication). The experiment summarized in Table VI was done to test the influence of pretreatment with cortisone on the different types of interferon response. Mice were inoculated subcutaneously with either saline or 7.5 mg of cortisone. Twenty-four hours later, viral and non-viral stimuli were injected intravenously into cortisone-treated and control mice. Bleedings were made at the appropriate times for the different stimuli and the interferon titers of the plasma pools are recorded in Table VI. It can be seen that pretreatment with cortisone inhibited in varying degrees the production of interferon by all of the viral and non-viral stimuli which were employed. No clear-cut differentiation of the two types of interferon response in mice was obtained by treatment with this steroid hormone. These results again point out the complex and diverse effects cortisone has in the intact animal.

TABLE VI
INFLUENCE OF PRETREATMENT WITH CORTISONE ON INTERFERON
PRODUCTION IN MICE BY VIRAL AND NON-VIRAL STIMULI

Stimulus *(i.v.)*	*Time of* *Bleeding*	*Interferon Titer of Plasma* *of Mice Pretreated with:*		*Inhibition*
		Saline[a]	*Cortisone*[a]	
NDV $(1 \times 10^7$ PFU)	10 hr	14,000	2,800	80%
B. abortus $(3.2 \times 10^8$ viable organisms)	8 hr	1,600	32	98%
S. marcescens $(4.6 \times 10^7$ viable organisms)	2 hr	1,000	70	93%
E. coli endotoxin $(100 \ \mu g)$	2 hr	800	250	69%
Statolon $(1,000 \ \mu g)$	8 hr	9,000	1,800	80%

[a]Mice injected subcutaneously with either saline (0.15 ml) or cortisone (7.5 mg) 24 hours prior to interferon production.

DISCUSSION

The presence in the mouse of interferons with different molecular weights raises the possibility that these materials may represent discrete molecules, polymers of smaller interferon subunits, or complexes of interferon with other materials. Attempts have been made to distinguish the different interferons by a variety of chemical and physical treatments but no significant differences in behavior have been found (Hallum and Youngner, unpublished data). At this time, we do not have enough information to choose between the various alternatives. The variations of inhibitors which appear in mice inoculated with different stimuli — in molecular weight, in time of appearance, and in dependence on protein synthesis — must be due to the wide range of cell types which are available in the intact animal and to the activation of different mechanisms of synthesis and release.

A possibility exists that all interferons produced in the intact mouse are preformed and that their release is effected by the interaction of given stimuli with certain cells in which the precursor interferon is present. The seeming requirement for new protein synthesis in the case of the "live virus" type of response may be for an enzyme or enzymes which are capable of reducing the heavy molecular weight precursor to smaller subunits. The state of the inducer or the type of damage to cells which it produces may have a marked influence on the form in which the interferon is released. This hypothesis is compatible with the known facts about interferon production in the mouse and cannot be ruled out without additional information which will bear on this question.

SUMMARY AND CONCLUSIONS

The characteristics of the interferon responses of mice injected with viral and non-viral stimuli have been described. Two main patterns of response, designated the "endotoxin" and "live virus" types of response, were distinguishable and have been described in detail. The variations in the molecular weight, time of appearance, dependence on protein synthesis, and the role of the spleen in the production of the inhibitors which appear in mice inoculated with different stimuli must be due to the wide range of cell types which are available in the intact animal and to the activation of different mechanisms of synthesis and release.

ACKNOWLEDGMENTS

The active participation of the following associates in various aspects of the work reported is gratefully acknowledged: Dr.

Warren R. Stinebring, Dr. Jules V. Hallum, Mrs. Sheila E. Taube and Miss Chuen-Mei Liu.

This study was supported by Public Health Service research grant AI-06264 from the National Institute of Allergy and Infectious Diseases.

REFERENCES

1. YOUNGNER, J. S., and STINEBRING, W. R.: Interferon production in chickens injected with *Brucella abortus. Science, 144*:1022-1023, 1964.

2. HOPPS, H. E., KOHNO, S., KOHNO, M., and SMADEL, J. E.: Production of interferon in tissue cultures infected with *Rickettsia tsutsugamushi. Bact. Proc.,* p. 115, 1964.

3. STINEBRING, W. R., and YOUNGNER, J. S.: Unpublished data.

4. MERIGAN, T. C., and HANNA, L.: Characteristics of interferon induced *in vitro* and *in vivo* by a TRIC agent. *Proc. Soc. Exper. Biol. & Med., 122*:421-424, 1966.

5. STINEBRING, W. R., and YOUNGNER, J. S.: Patterns of interferon appearance in mice injected with bacteria or bacterial endotoxin. *Nature, 204*:712, 1964.

6. KLEINSCHMIDT, W. J., CLINE, J. C., and MURPHY, E. B.: Interferon production by statolon. *Proc. Natl. Acad. Sc. (U.S.), 52*:741-744, 1964.

7. RYTEL, M. W., SHOPE, R. E., and KILBOURNE, E. D.: An antiviral substance from *Penicillium funiculosum.* V. Induction of interferon by helenine. *J. Exper. Med., 123*:577-584, 1966.

7a. WHEELOCK, E. F.: Interferon-like virus inhibitor induced in human leukocytes by phytohemagglutinin. *Science, 149*:310-311, 1965.

8. YOUNGNER, J. S., and STINEBRING, W. R.: Comparison of interferon production in mice by bacterial endotoxin and statolon. *Virology, 29*:310-316, 1966.

9. MERIGAN, T. C.: Purified interferons. Physical properties and species specificity. *Science, 145*:811-813, 1964.

10. HALLUM, J. V., YOUNGNER, J. S., and STINEBRING, W. R.: Interferon activity associated with high molecular weight proteins in the circulation of mice injected with endotoxin or bacteria. *Virology, 27*:429-431, 1965.

11. BUCKLER, C. E., and BARON, S.: Antiviral action of mouse interferon in heterologous cells. *J. Bacteriol., 91*:231-235, 1966.

12. LANDY, M., and BRAUN, W., (Ed.) : *Bacterial Endotoxins.* New Brunswick, N. J., Rutgers University Press, 1964.

13. SUTER, E.: Hyperactivity to endotoxin in infection. *Trans. New York Acad. Sci., Ser. II, 24*:281-290, 1962.

14. YOUNGNER, J. S., and STINEBRING, W. R.: Interferon appearance stimulated by endotoxin, bacteria, or viruses in mice pretreated with *Escherichia coli* endotoxin or infected with *Mycobacterium tuberculosis. Nature, 208*:456-458, 1965.

15. MERIGAN, T. C., and KLEINSCHMIDT, W. J.: Different molecular species of mouse interferon induced by statolon. *Nature, 208*:667-669, 1965.

16. FRUITSTONE, M. J., MICHAELS, B. S., RUDLOFF, D. A. C., and SIGEL, M. M.: Role of the spleen in interferon production in mice. *Proc. Soc. Exper. Biol. & Med., 122*:1008-1011, 1966.

17. DE SOMER, P., and BILLIAU, A.: Interferon production by the spleen of rats

after intravenous injection of Sindbis virus or heat-killed *Escherichia coli.*
*Arch. f. d. ges. virusfors., 19:*143-154, 1966.

18. RYTEL, M. W., and KILBOURNE, E. D.: The influence of cortisone on experimental virus infection. VIII. Suppression by cortisone of interferon formation in mice injected with Newcastle disease virus. *J. Exper. Med., 123:*767-776, 1966.

19. DE MAEYER, E., and DE MAEYER, J.: Two-sided effect of steroids on interferon in tissue culture. *Nature, 197:*724, 1963.

PREPARATION AND ELECTROPHORESIS OF HIGHLY PURIFIED CHICK INTERFERON

K. H. FANTES

Soon after the discovery of chick interferon by Isaacs and Lindenmann (1957), it became apparent that interferons induced in other animals were, like chick interferon itself, essentially species specific. It was thought for some time that cells from each species produced their own unique and uniform interferon. Since then, however, it has been shown that even cells from the same species can produce different interferons, depending on the nature of the stimulus, and that even the same stimulus can induce in the same cells, though after different time intervals, interferons with somewhat differing properties. Moreover, the same stimulus can also induce, even at the same time and in the same animal, different interferons in the serum and in the spleen.

All these diverse antiviral substances have similar biological properties, but they show great variations in their molecular weights, isoelectric points and stabilities. Whether some of them should be called "interferon-like" substances instead of "interferons," depends at present on the view of the individual worker, since a rigid, generally accepted definition of "interferon" does not exist.

Work in our laboratories on these antiviral substances has been concerned mainly with only one of them, namely, virus-induced chick interferon. A short account of its purification is given below, followed by a discussion of recent results obtained when the most highly purified interferon so far prepared by us was subjected to electrophoresis in acrylamide gel.

The first attempt to purify chick interferon was made by Burke (1960, 1961). He purified chorioallantoic membrane interferon about twenty-fold, mainly by chromatography on DEAE- and SM-celluloses.

A great step forward was taken when Lampson *et al.* (1963) were able to purify allantoic fluid interferon 1830 times — or 4,500 times after electrophoresis — by precipitation of inert protein with perchloric acid, adsorption of the active material by zinc hydroxide and chromatography of the liberated interferon on carboxymethylcellulose.

Merigan and his colleagues (1964a, 1964b, 1965) took the purification of allantoic fluid interferon even further. Using CM-sephadex C-25 instead of CM-cellulose, and pH-gradient instead of stepwise elution, they achieved 6,500-fold purification without the help of electrophoresis.

My colleagues and I (1964, 1965) first worked with allantoic fluid interferon, but more recently we have also purified chick embryo tissue culture interferon by the same methods. The crude tissue culture interferon had a higher titre and a higher specific activity, and was thus a better starting material. Briefly, this series of steps has been found to be the most useful:

1) Adsorption of active material from the starting fluid by "Doucil" (a sodium aluminum silicate).

2) Elution of activity with a smaller volume of 0.5 M K-thiocyanate at pH 7.5.

3) Consecutive rejection of precipitated inert protein after acidification of the eluate to pH 3.5 and pH 2.0.

4) Addition of methanol (5 volumes) to the supernatant fluid at pH 2 thereby precipitating further inactive protein.

5) Precipitation of interferon by neutralising the acidic aqueous methanol solution.

6) Extraction of interferon from the precipitate with 0.01 M phosphate buffer at pH 7.5.

7) Passage of the extracts in this buffer through DEAE-cellulose.

8) pH gradient chromatography in 0.1 M phosphate buffer on CM-sephadex-C50 (sometimes twice).

An example of the method as applied to the purification of tissue culture interferon is presented in Table I.

The most highly purified interferon obtained, from both tissue culture and allantoic fluids, had a specific activity of 1.6 x 10^6 units per mg protein, which, for allantoic material, represented nearly 20,000-fold purification. Since such a specific activity was considerably higher than any reported elsewhere, we wished to verify that the difference was not due to different sensitivities of the assay methods. Assays of freeze-dried samples (kindly supplied by Dr. T. C. Merigan), carried out concurrently in Dr. Merigan's and in our laboratories, showed that both methods gave very similar results.

TABLE I

PURIFICATION OF TISSUE CULTURE INTERFERON

Sample	Volume (ml)	Total Units (x 10⁶)	Specific Activity*	Purifi- cation Factor**	Overall % Recovery of Activity
Crude culture fluid	22,500	5.40	470		100
pH 7.5, 0.5M KSCN Doucil eluate	3,000	4.80	1,980	4.2	89
pH 3.5 supernatant fluid	3,000	3.60	5,180	11	67
pH 2.0 supernatant fluid	3,000	2.37	15,500	33	44
Redissolved methanol (pH 7.5) precipitate	225	1.80	19,300	41	33
Combined DEAE cellulose eluate fractions	960	1.14	132,000	281	21
CM-sephadex gradient eluate (combined peak fractions)	25	0.40	1,600,000	3,400	7

*Specific activity = interferon units per mg protein.

**Purification factor = $\dfrac{\text{specific activity of sample}}{\text{specific activity of starting fluid}}$

The highly purified interferon has recently been subjected by us to acrylamide gel electrophoresis, work done in collaboration with my colleague Dr. I. G. S. Furminger. Earlier results obtained by Williams and by us (Fantes, 1965; Fantes 1966), using acrylamide gel electrophoresis at an alkaline pH (Davies, 1964) and by Polson (1966) who used sucrose gradient electrophoresis, with interferon now known to be somewhat less pure than the present samples, suggested that it was substantially homogeneous.

Our technique (Fantes, 1966) consists of slicing the gel lengthwise, after electrophoresis, staining one half with amido black and cutting the other half into 20 equal (about 2 mm) portions; these are individually eluted and assayed for interferon content. Early experiments using Davis's method produced a rather diffuse protein zone, but it seemed to coincide with antiviral activity; the recovery of interferon was, however, always low. We subsequently found that a fairly high proportion of the total activity was retained in the sample and spacer gels. These were, therefore, omitted and the samples, in sucrose, were applied directly to the small-pore gel.

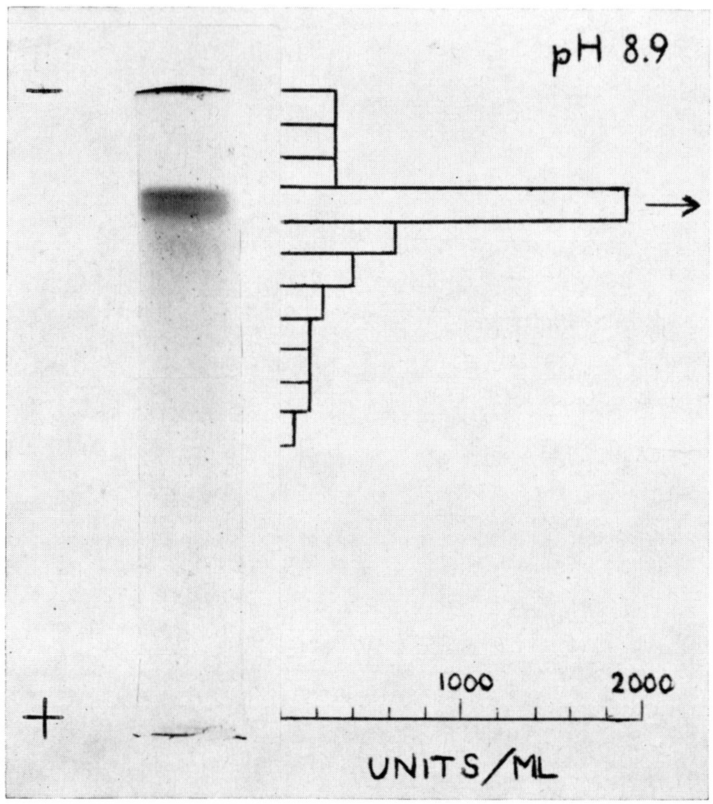

Figure 1. Polyacrylamide gel electrophoresis at pH 8.9 of highly purified chick interferon. 38,400 units (specific activity 1.6 x 10^6 units/mg protein), in 1 ml 0.01 M phosphate buffer pH 7.5 were mixed with 1 ml 40% sucrose and applied to a gel column (40 x 6.5 mm). After electrophoresis (6 mA/column), the gel was cut lengthwise; one-half was stained with amido black, the other was cut into twenty 2 mm portions, each of which was suspended in 1 ml 0.5 M phosphate buffer pH 7.5 containing bovine plasma albumin (500 μg/ml) and Tween 80 (20 μg/ml). After 24 hours at 4°, the supernatant fluids were assayed for interferon content. The arrow indicates that the interferon content of the fraction was greater than that shown.

Recovery of interferon was now much better, and a very sharp double-band of protein was superimposed exactly over the activity peak (Fig. 1). The sharpness of the protein zone and the much wider spread of activity suggested that the protein bands might not be interferon.

In order to get more evidence, electrophoresis was carried out at an acid pH (Reisfeld *et al.*, 1962); but under these conditions, all or nearly all activity was invariably lost. One of the materials used by both Davis and Reisfeld to catalyse the polymerisation of the small-pore gels was persulphate, and it was suspected that this material, at least at an acid pH, might destroy interferon. Crude interferon was therefore incubated (18 hours: 4°C) at pH 4.3 and at pH 8.9, with and without persulphate (0.006M, the concentration used by Reisfeld for the polymerisation of the gels).

Complete loss of activity occurred in the acidified sample containing persulphate, but none in the other three preparations. Highly purified interferon was similarly stable on incubation at pH 4.3, but only in the absence of persulphate. It seems possible that acid persuphate may, like performic acid, disrupt disulphide bonds, known to be essential for interferon activity (Fantes and O'Neill, 1964; Merigan, 1964b; Merigan *et al.*, 1965). If this is a general effect, electrophoresis of other proteins in persulphate-catalysed gels could lead to the formation of artifacts.

When the small-pore gels were polymerised with riboflavin instead of persulphate, the recovery of interferon after electrophoresis at pH 4.3 was good, but the activity peak now appeared to lie between two protein bands and was probably not associated with either. The distribution of stained protein and activity after electrophoresis at pH 4.3 is shown in Figure 2.

If the interpretation of these findings is correct, the specific activity of pure interferon will be found to be much greater than 1.6 x 10^6 units per mg protein. Although the activity does not coincide with any obvious protein zone, there are not sufficient reasons for supposing that it is associated with a non-protein antiviral factor, similar to the one postulated by Nagano *et al.*, (1966), since that material is active only *in vivo*.

It seems more likely that the generally accepted view that interferon is a protein is correct. Because of its high specific activity, however, it is probable that much larger concentrations will have to be applied to the gels before interferon-protein bands become visible. The very low level of protein associated with appreciable activity even in our present material could perhaps explain why interferon was considered a poor antigen (Paucker

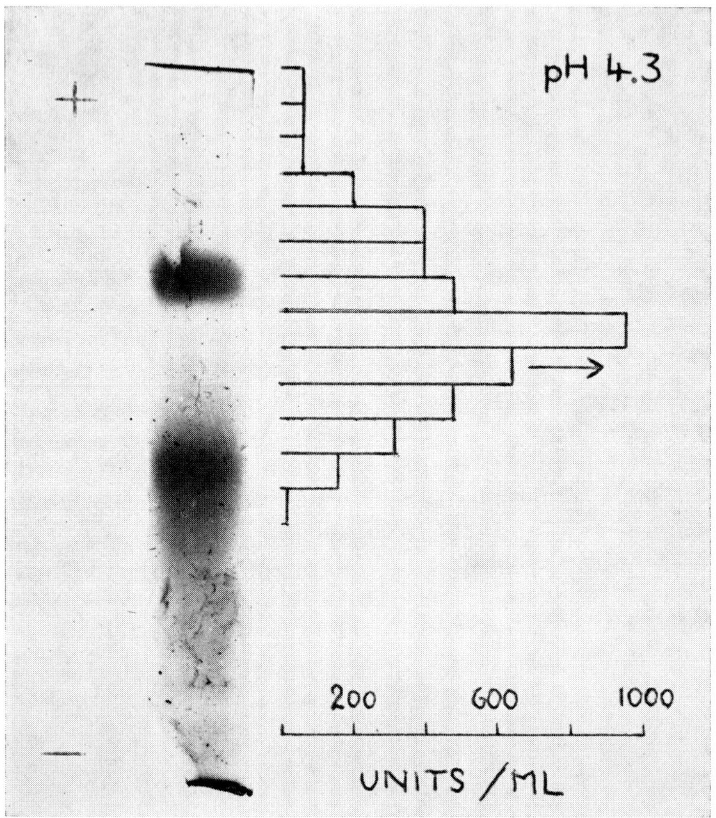

Figure 2. Polyacrylamide gel electrophoresis at pH 4.3 of highly purified chick interferon. 38,400 units (specific activity 1.6 x 10⁶ units/mg protein), in 1 ml 0.01 M phosphate buffer pH 7.5 were mixed with 1 ml 40% sucrose and applied to a gel column (40 x 6.5 mm). After electrophoresis (6 mA/ column), the gel was cut lengthwise; one-half was stained with amido black, the other was cut into twenty 2 mm portions, each of which was suspended in 1 ml 0.5 M phosphate buffer pH 7.5 containing bovine plasma albumin (500 μg/ml) and Tween 80 (20μg/ml). After 24 hours at 4°, the supernatant fluids were assayed for interferon content. The arrow indicates that the interferon content of the fraction was greater than that shown in the figure.

and Cantell, 1962) , and it might also have been the reason why Burke and Walters (1966) were unable to find a messenger RNA that codes for interferon.

REFERENCES

Burke, D. C.: (1960) *Biochem. J., 76*:50P.

Burke, D. C.: (1961) *Biochem. J., 78*:556.

Burke, D. C., and Walters, S.: (1966) *Biochem. J., 101*:25.

Davis, B. J.: (1964) *Ann. New York Acad. Sci., 121*:404.

Fantes, K. H., O'Neill, C. F., and Mason, P. J.: (1964) *Biochem. J., 91*:20P.

Fantes, K. H., and O'Neill, C. F.: (1964) *Nature, 203*:1048.

Fantes, K. H.: (1965) *Nature, 207*:1298.

Fantes, K. H.: (1966) *Interferons.* Ed. by N. B. Finter. Amsterdam, North Holland Publishing Co., p. 142.

Isaacs, A., and Lindenmann, J.: (1957) *Proc. Roy. Soc., B, 147*:258.

Lampson, G. P., Tytell, A. A., Nemes, M. M., and Hilleman, M. R.: (1963) *Proc. Soc. Exper. Biol. & Med., 112*:468.

Merigan, T. C.: (1964a) *Science, 145*:811.

Merigan, T. C.: (1964b) *International Symposium on Nonspecific Resistance to Virus Infection, Interferon and Viral Chemotherapy.* Bratislava, Czechoslovakia.

Merigan, T. C., Winget, C. A., and Dixon, C. B.: (1965) *J. Mol. Biol., 13*:679.

Nagano, Y., Kojima, Y., and Haneishi, T.: (1966) Personal communication; I.E.G., #6.

Paucker, K., and Cantell, K.: (1962) *Virology, 18*:145.

Polson, A.: (1966) *Interferons.* Ed. by N. B. Finter. Amsterdam, North Holland Publishing Co., p. 135.

Reisfeld, R. A., Lewis, U. J., and Williams, D. E. (1962) *Nature, 195*:281.

Williams, J.: See Fantes, K. H., 1965.

INFLUENCE OF WHOLE-BODY IRRADIATION, CORTISOL TREATMENT AND ADRENALECTOMY ON INTERFERON INDUCTION IN VIVO IN RATS

P. De SOMER, E. De CLERCQ and A. BILLIAU

INTRODUCTION

Interferon-like substances can be induced *in vivo* in different animal species by injection of various viral and non-viral stimuli. Literature on this subject has recently been reviewed by Finter (1).

In a previous day, we determined the kinetics of interferon-appearance in rats after administration of Sindbis virus and endotoxin (2). It was found that the injection of Sindbis virus or endotoxin is followed by a period during which the interferon response to a second challenge with either virus or endotoxin is reduced. This finding was reported several times by other authors and the term "hyporeactivity" has since currently been used in this respect (3, 4).

In another study, we found that the injection of Sindbis virus or endotoxin 24 to 48 hours prior to a viral or non-viral antigen markedly reduced the formation of antibodies against the latter (5). This inhibitory effect was absent when the Sindbis virus or endotoxin were given a certain time after the antigen, and thus resembled the inhibition of antibody formation by irradiation or cortisol treatment. Hypothesizing that the observed inhibition of antibody production in animals treated with interferon inducers would be related to the hyporeactive state, we thought it worthwhile to investigate whether x-irradiated or cortisol-treated animals would also show reduced interferon responses.

To our knowledge, no studies have as yet been published dealing with the influence of irradiation on in vivo interferon induction. Corticoids have been reported to inhibit the interferon responses to both viral and endotoxin challenges (6, 7, 8). However, some workers found the virus induced-interferon in rabbits to be difficultly inhibited by cortisol pretreatment (8). The same authors also reported that adrenalectomy potentiated the effect of endotoxin while it left the virus-induced interferon yields unchanged.

MATERIALS AND METHODS

Male and female Wistar R-rats weighing about 150 g (range from 120 to 180 g) were used throughout our experiments. Endotoxin was prepared from *Proteus* rettgeri according to the method of Roberts (9). A standard dose of 0.3 mg per rat was used throughout this study. Sindbis virus was propagated and titrated as described previously (2).

The first part of our irradiations was performed with a Maxitron 250 x-ray therapy unit. The rats were placed in lucite restraining cages which were continuously rotated during exposure. The radiation factors were as follows: operating voltage 250 KV; tube current 25 mA; filtration through 0.25 mm copper and 1 mm aluminum; the distance from the target of the x-ray tube to the lead plate on which the cages were placed was 80 cm. At this distance, a sublethal dose of 400 r, measured in air, was delivered in 9 min 20 sec. When allowance is made for the retrodiffusion on the lead plate, the rats received strictly a total body dose of 488 rad. For practical reasons, a 2000 curies cesium-137 unit was used in the second part of our experiments. The rats were placed in stationary lucite cages, put on a lucite support, at a distance of 50 cm to the target of the bomb. We obtained a rate of 39 r/min, measured in air with a Siemens ionisation Chamber. A dose of 550 r delivered by the telecesium-137 unit responded to 400 r of the 250 KV-unit (relative biological efficiency).

For lethal irradiation, 1000 r were administered. This treatment killed 50% of the animals after 7 days.

The preparation of spleen homogenate, the dialysis procedures and the plaque-inhibition technique for interferon determinations have been described in a previous report (2). Dialysis was omitted for the samples derived from endotoxin-treated animals.

Most of the interferon determinations were carried out with vaccinia as a challenge virus, but occasionally VSV was used.

RESULTS

1) Influence of Sublethal X-irradiation on the Interferon Yield after Injection of Endotoxin in Rats

From a series of preliminary studies not reported herein, it appeared that irradiation would enhance the interferon production after endotoxin administration in rats. In order to assess more precisely this effect, the following experiments were carried out. Five pairs of rats were given a sublethal x-ray dose, while five others remained untreated. Twenty-four hours later the animals received an intravenous dose of endotoxin; serum and spleen samples were taken 2 hours after this injection and were pooled per pair of rats. Interferon yields obtained in this experiment are shown in Table I. A 4- to 5-fold higher interferon concentration was noted in irradiated than in control animals.

TABLE I
INFLUENCE OF SUBLETHAL WHOLE-BODY IRRADIATION
ON ENDOTOXIN-INDUCED INTERFERON YIELDS IN RATS

	Interferon Titers	
	Serum U/ml	*Spleen U/gm*
X-ray +	2115	40.000
endotoxin	1360	25.000
	2040	32.000
	1480	35.000
	2080	45.000
	m = 1815	m = 35.400
Endotoxin	288	8900
	500	8000
	300	7000
	410	6500
	540	4000
	m = 414	m = 6880

1. X-ray administration 24 hours before endotoxin.
2. Figures represent values obtained on paired pools of sera or spleen homogenates.

The next experiments were designed to investigate the time relationship of the enhancing effect of x-ray administration on interferon production by endotoxin. Groups of three rats were used. While control groups received only endotoxin, all the others were also sublethally irradiated at times indicated in Table II.

Serum samples and spleens were taken two hours after endotoxin administration. The interferon determinations were carried out on pools of sera and spleen homogenates of each group. It appears from the data in Table II that, when the animals were given the x-rays simultaneously with, or 1 hour after endotoxin, the interferon yields were fairly comparable to those obtained in the control animals. On the contrary, when the irradiation was applied earlier, i.e., 4, 14, 24, or 96 hours before endotoxin injection, the interferon yield was increased. X-ray administration 7 days before endotoxin had no effect any more. Conclusively, the enhancing effect can be located in a period ranging approximately from 4 hours to 4 days after the administration of x-rays.

TABLE II
INFLUENCE OF SUBLETHAL WHOLE-BODY IRRADIATION ON
ENDOTOXIN-INDUCED INTERFERON YIELDS IN RATS.
TIME RELATIONSHIPS.

Exp. nr	Pretreatment	Time Lapse Between X-ray and Endotoxin	Interferon Titers	
			Serum U/ml	Spleen U/ml
1	Control	—	230	7.500
	X-ray after endotoxin	1 hour	185	5.400
	X-ray and endotoxin simultaneously	0	125	5.650
	X-ray before endotoxin	4 hours	340	17.800
		14 hours	430	19.900
		24 hours	655	28.600
2	Control	—	140	12.000
	X-ray before endotoxin	24 hours	430	12.000
		4 days	425	48.000
		7 days	170	14.000

Each figure represents 1 interferon determination carried out on a pool of 3 sera, respectively spleen homogenates.

The interferon response to endotoxin is characterized by its short duration; circulating interferon reaches its peak level 2 hours after endotoxin injection and subsequently decreases to zero in 2 to 3 more hours. In view of the potentiating effect of irradiation on the interferon production, it remained to be determined whether the interferon response would be of longer duration in irradiated animals than in controls. Five groups of 3 animals were sublethally irradiated, while 5 others were left untreated. Endotoxin was administered 40 hours later. Serum samples and spleens were taken respectively $\frac{1}{2}$, 1, 2, 4, and 8 hours after the endotoxin injections. The results of interferon determinations are represented in Figure 1. It appears that the shape of the time-response curve after endotoxin administration was not changed by irradiation; only the height of the peak value was significantly increased.

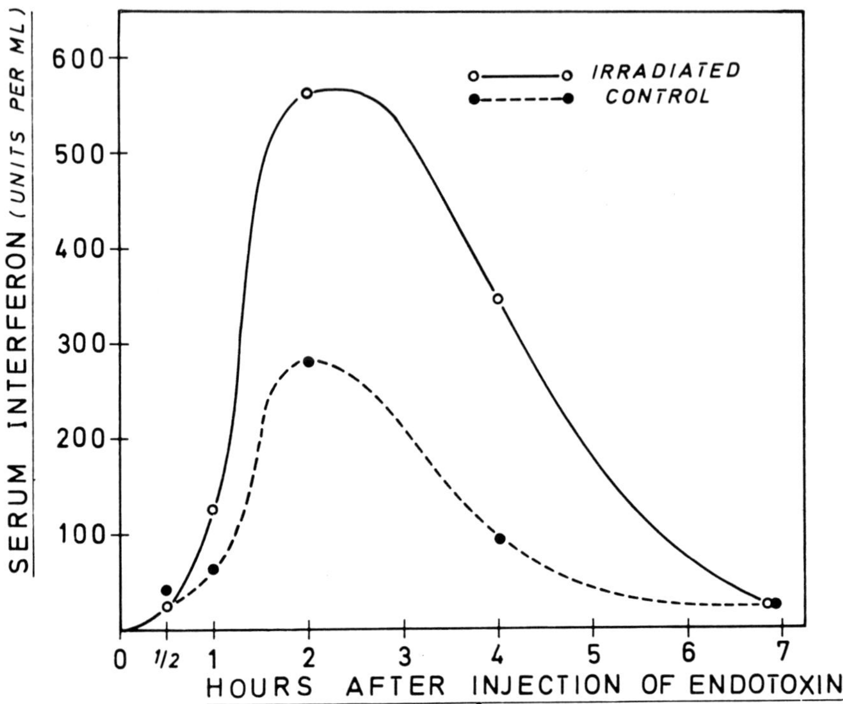

Figure 1. Time response curve of endotoxin induced interferon in control rats and rats sublethally irradiated 40 hours before endotoxin injection. Each determination was carried out on a serum pool from 3 animals.

2) Influence of Lethal Irradiation on Endotoxin-induced Interferon

In a next experiment, we determined the amounts of interferon induced by endotoxin in rats which had been lethally irradiated at different times before endotoxin administration. Five groups of rats were given a lethal dose of x-rays while the control group remained untreated. Endotoxin was injected after 24, 48, 72, and 96 hours respectively. In order to avoid daily variations in interferon yields, the time schedule of irradiation was so arranged that the endotoxin administrations could be given to all animals on the same day. The results are represented in Figure 2. It appears that a lethal dose also enhanced interferon production when given 24 or 48 hours before endotoxin. However, 3 to 4 days after lethal irradiation, a certain mortality occurred and the interferon response of the remaining animals gradually decreased. In the group receiving endotoxin 4 days after irradiation, the serum

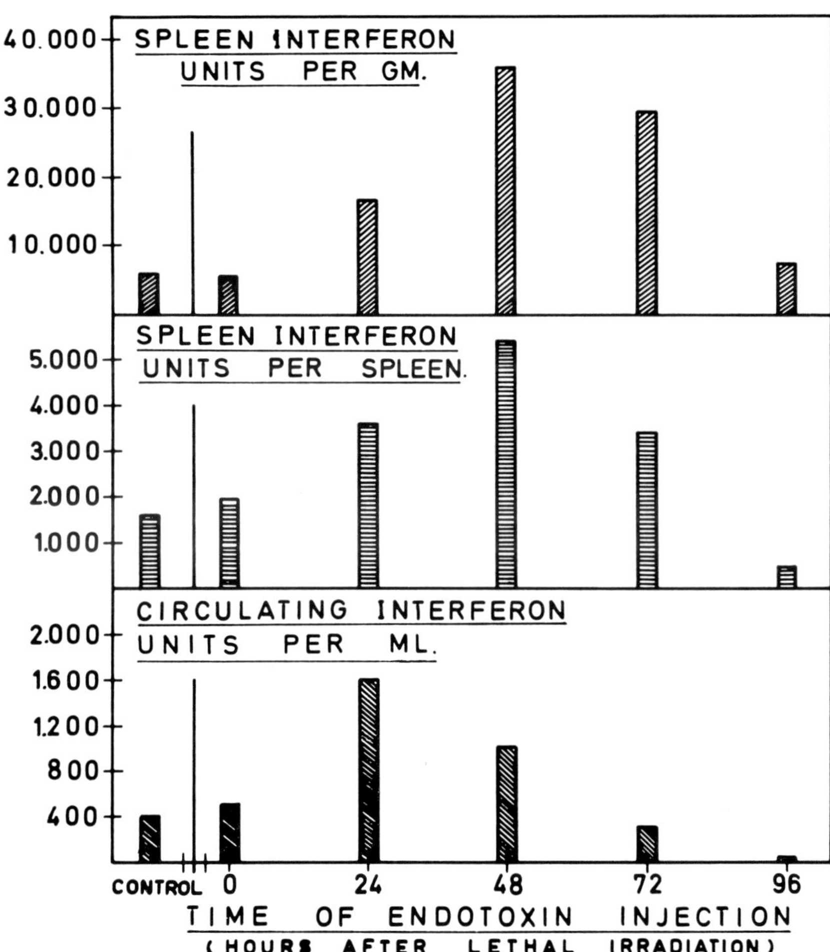

Figure 2. Influence of lethal whole-body irradiation on endotoxin-induced interferon yields. Each group consisted of 3 rats. Determinations were carried out on pooled sera, respectively spleen homogenates. Blood samples and spleens were taken 2 hours after endotoxin injection.

interferon concentration was almost zero. In the spleens of these rats, on the contrary, the interferon concentration still equalled that of the control group, but in view of the reduced size of the spleen, it only represented 1/3 of the total amount of interferon present in the spleens of control animals.

TABLE III
INFLUENCE OF SUBLETHAL WHOLE-BODY IRRADIATION ON
SINDBIS-VIRUS-INDUCED INTERFERON YIELDS IN RATS

Exp. nr.	*Pretreatment*	*Time Lapse Between X-ray and sindbis Virus*	*Interferon Titers*	
			Serum (U/ml)	*Spleen (U/gm)*
1	Control	—	3.410	54.000
	X-ray after Sindbis virus	2	675	40.000
	X-ray and Sindbis virus simultaneously	0	930	23.500
	X-ray before Sindbis virus	4	905	14.700
		14	2.930	45.000
		24	1.600	36.000
		48	1.205	18.000
2	Control	—	5.875	>80.000
	X-ray before Sindbis virus	14	8.000	64.500
		24	2.240	44.000
		48	5.600	21.400

Each figure represents 1 interferon determination carried out on a pool of 3 sera, respectively spleen homogenates.

3) Influence of Sublethal Irradiation on Virus Induced Interferon

Groups of 3 rats were sublethally irradiated; non-irradiated controls were included. At times indicated in Table III, the rats received 10^9 pfu of Sindbis virus intravenously. Spleens and blood samples were taken 8 hours after virus injections. Sera and spleen homogenates were pooled per group. The results of interferon determinations are summarized in Table III. Both in the serum and the spleen the interferon yields of irradiated animals were lower than those of control animals. As to the time distribution of this effect, the number of observations is still too small to permit definite conclusions.

4) Influence of Cortisol Pretreatment on the Interferon Response Elicited by Endotoxin Injection

Groups of three rats were used in three experiments in which we tested the effect of different doses of a cortisol-hemisuccinate solution (Roussel, France) and of cortisol-acetate suspension (Hydroricortex, R.I.T., Belgium), administered either at 1 or at 24 hours before endotoxin. It appears from the data of Table IV that intramuscular administration of a dose as high as 10 mg (i.e., about 1/3 of the acute lethal dose), had no effect when administered 24 hours before endotoxin, but reduced the interferon yields in both the serum and spleen when it was given 1 hour before the endotoxin challenge. The minimum dose given 1 hour before endotoxin, which exerted still an inhibitory effect was 5 mg per rat i.m. Doses of 10 and 20 mg of the cortisol-acetate suspension given 24 hours before endotoxin had no inhibitory effect.

TABLE IV
INFLUENCE OF CORTISOL PRETREATMENT ON THE INTERFERON RESPONSE
INDUCED BY ENDOTOXIN INJECTION

Exp. nr.	*Pretreatment*	*Time of Endo- toxin Injection (Hours After Cortisol)*	*Interferon Titer†*	
			Serum (U/ml)	*Spleen (U/gm)*
1	Cortisol-HS (10 mg i.m.)	24	130	4.000
	Cortisol-HS (10 mg i.m.)	1	32	2.250
	———	—	200	5.600
2	Cortisol-HS (10 mg i.m.)	1	12	2.500
	Cortisol-HS (5 mg i.m.)	1	89	3.200
	Cortisol-HS (1 mg i.m.)	1	900	14.000
	———	—	540	20.000
3	Cortisol-Ac (10 mg i.m.)	24	200	1.800
	Cortisol-Ac (20 mg i.m.)	24	200	2.000
	———	—	320	2.700

Cortisol-HS = cortisol-hemisuccinate solution.
Cortisol-Ac = cortisol-acetate suspension.
†Exp. nrs. 1 & 2 = interferon determinations carried out with VSV as challenge virus.
Exp. nr. 3 = interferon determinations carried out with vaccinia as challenge virus.
Blood and spleens taken 2 hours after endotoxin injection.
Each figure represents 1 interferon determination carried out on a pool of 3 sera, respectively spleen homogenates.

5) Influence of Cortisol Pretreatment on Virus Induced Interferon Production

Similarly, we also tested the effect of cortisol on the production of interferon after Sindbis virus injection. Again groups of three rats were used. It appears from the data in Table V that in our hands cortisol did not inhibit interferon production. This failure contrasts with the data of Mendelson *et al.* (6) and of

Rytel *et al.* (7). Therefore, we tried to reproduce the results of these authors. Groups of 5 mice were subcutaneously injected with cortisol-hemisuccinate or cortisol-acetate. A control group received no cortisol. At times indicated in Table V, 10^9 pfu of Sindbis virus were injected intravenously. Sera were collected 8 hours later. A dose of 5 mg of cortisol-hemisuccinate killed 2 of the 5 mice within 24 hours. The remaining 3 animals in this group died after having received the Sindbis virus. From the data obtained in the other groups, it appeared that neither a dose of 5 mg of cortisol-acetate administered 24 hours before or a dose of 1 mg of cortisol-hemisuccinate injected 1 hour before Sindbis virus were able to inhibit the production of circulating interferon.

TABLE V

INFLUENCE OF CORTISOL PRETREATMENT ON THE INTERFERON RESPONSE
INDUCED BY SINDBIS VIRUS INJECTION

Exp. nr. Test Animal	Pretreatment	Time of Sindbis Virus Injection (Hours After Cortisol)	Interferon Titer	
			Serum (U/ml)	Spleen (U/gm)
1	Cortisol-HS (10 mg i.m.)	24	10.000	64.000
Rat	Cortisol-HS (10 mg i.m.)	1	2.500	71.000
	————	—	2.500	35.000
2	Cortisol-Ac (5 mg s.c.)	24	10.000	n.d.
Mouse	Cortisol-HS (5 mg s.c.)	24	lethal	lethal
	Cortisol-HS (1 mg s.c.)	1	10.000	n.d.
	————	—	10.000	n.d.

Cortisol-HS = cortisol-hemisuccinate solution.
Cortisol-Ac = cortisol-acetate suspension.
Blood and spleens taken 8 hours after Sindbis virus injection.
Each figure represents 1 interferon determination carried out on a pool of 3 sera, respectively spleen homogenates.

6) Influence of Adrenalectomy on Circulating Interferon Induced by Endotoxin

The inhibitory effect of cortisol on interferon induction by endotoxin should be paralleled by a potentiating effect of adrenalectomy. Postic *et al.* reported the existence of such an effect in rabbits (8). We were able to obtain similar results in rats, as is shown in Table VI. Six rats were adrenalectomized through a median abdominal incision; 3 control rats were left intact. Seven days after surgery all animals were injected with endotoxin. Three adrenalectomized rats received 0.2 mg, while the remaining 3 were given 0.05 mg. Sera were taken 2 hours after endotoxin injection and were pooled per group. Interferon administrations revealed that the adrenalectomized animals produced about 4 times more circulating interferon than did the controls, in spite of the fact that they had received smaller doses of endotoxin.

TABLE VI
INFLUENCE OF ADRENALECTOMY ON CIRCULATING INTERFERON LEVELS
INDUCED BY ENDOTOXIN INJECTION IN RATS

Pretreatment	*Dose of Endotoxin*	*Circulating Interferon (U/ml of Serum)*
Adrenalectomy	0.2 mg	1460
Adrenalectomy	0.05 mg	1740
Control	0.3 mg	400

Time lapse between adrenalectomy and endotoxin injection: 7 days.
Figures represent values obtained on pools of three serum samples.
Blood samples taken 2 hours after endotoxin injection.

DISCUSSION

1) Endotoxin-induced Interferon

Sublethal x-irradiation, 14 hours to 4 days before endotoxin injection increased the interferon-response to the latter. Cortisol-treatment, on the other hand, inhibited the interferon production, while adrenalectomy had a potentiating effect.

A sufficient explanation for these results is difficult to provide, since the mechanism by which endotoxin acts as an interferon inducer is poorly understood. Although indirect evidence has lead to the spleen as a major interferon producer in endotoxin-treated animals, no morphologically well-determined cell type could as yet be designated as the principal source of interferon (2, 10, 11). Neither is it clear whether endotoxin acts directly on the cell or through a systemic reaction in which intermediary humoral factors would play a role. The recent observation of Wagner that peritoneal macrophages are able to produce interferon when stimulated *in vitro* by endotoxin seem to be conclusive in this respect (12). These observations also suggest that the macrophages in the spleen and in other tissues should be held responsible for the *in vivo* interferon production. Wagner (12) also found that the *in vitro* production of interferon by endotoxin-stimulated macrophages is blocked by actinomycin and consequently is to be considered as synthesized *de novo*. This is in strong contrast with the *in vivo* observations of Ho *et al.* (13), and of Youngner *et al.* (14), and either tends to invalidate conclusions drawn from *in vivo* experiments with inhibitors of metabolic pathways or indicates that the *in vitro* production of interferon after incubation with endotoxin does not reflect the situation *in vivo*. It remains still to be questioned whether circulating interferon induced by endotoxin is synthetized *de novo* or results from release of preformed interferon.

In view of these many imponderabilia in the mechanism of interferon induction by endotoxin, an explanation of the noted effects of irradiation, cortisol-treatment and adrenalectomy must necessarily be tentative. It seems to be well established that irradiation, and adrenalectomy increase the lethal effect of endotoxins, while corticoids protect against it (15, 16). Apparently, there seems to exist a certain parallelism between lethal effect and interferon production. Although the mechanism of endotoxin lethality is rather obscure, this seems to indicate that *in vivo* induction of interferon by endotoxins must be explained in terms of cell damage.

However, if the same cell damage is to be held responsible for both interferon release and subsequent hyporeactivity to a second endotoxin challenge, then the inconsistency has to be taken into account that cortisol is unable to prevent hyporeactivity (8).

Irradiation could act either by sensitizing the interferon producing cells, by increasing the absolute number of the cell-type which is responsible for interferon-release, or by increasing the amount of accumulated preformed interferon in the cells. To our knowledge, no evidence supporting either of these hypotheses is present in literature. An accurate study of morphology, interferon production, and distribution of different cell types in hematopoietic organs and in the circulation, under conditions that irradiation stimulates endotoxin-induced interferon seems to be indicated.

After lethal irradiation, the yields of endotoxin-induced interferon initially paralleled those in sublethally irradiated animals, in that they were supranormal. From the third to the fourth day after irradiation, the response returned to normal and then further decreased in the lethally irradiated animals. This terminal decrease probably corresponded to an over-all depletion of reticuloendothelial and blood cells, due to the inhibition of mitosis.

2) Sindbis-induced Interferon

At first sight, the induction of interferon *in vivo* by injection of viruses, may seem less puzzling, because an *in vitro* model of this phenomenon was described long before the *in vivo* counterpart was observed. However, the same questions which still remain unanswered concerning the mechanism of interferon induction by endotoxin are also unsolved in the case of virus-induced interferon. Neither the intricate mechanism, nor the cell-species involved have as yet been determined. Furthermore, whereas evidence exists that virus-induced interferon is synthetized *de novo,* the possibility still remains that all interferons are preformed and that the apparent requirement of protein synthesis and transcriptive functions would be for an enzyme or enzymes involved in the processing of precursor interferon into detectable inhibitor. Owing to this lack of basic knowledge on the mechanism of interferon production, the failure of x-irradiation and cortisol to influence the virus-induced interferon yields with respect to their divergent effect on endotoxin-induced interferon, can hardly be explained in terms of the cell-population or metabolic pathway invloved.

Yet, our observations support the assumption that the virus- and endotoxin-induced interferons are released from a different cell-population and/or by different mechanisms.

Our data on the influence of cortisol are in good agreement with those of Ho *et al.* (9) obtained in rabbits, but contrast with the results of other authors who found that cortisol does inhibit the yield of virus-induced interferon in mice (6, 7). This discrepancy, presumably resulting from the use of another test animal and/or viral challenge should warn against premature generalizations in this field. The functional lesions and alterations caused by corticoids and radiation are indeed complex and often unpredictable. Therefore, further elaboration of the observations seems worthwhile; conceivably the study of interferon production by different cell types *in vivo* would permit a new approach to the intricate disturbances caused by irradiation.

SUMMARY

1) Sublethal whole-body irradiation potentiated endotoxin in inducing circulating and spleen interferon in the rat. The effect lasted from 4 hours to 4 days after irradiation.

2) Lethal irradiation initially had the same potentiating effect as sublethal treatment, but 4 days after x-ray administration, the animals began to die and in the remaining rats the serum interferon response declined to almost zero. However, the concentration of interferon in the spleen of these animals still equalled the control values, but in view of the reduced size of the spleen, this represented only 1/3 of the total amount of interferon present in control spleens.

3) Preliminary data concerning the influence of sublethal irradiation on the induction of interferon by Sindbis virus seemed to indicate a slight inhibitory effect.

4) Cortisol-pretreatment reduced the interferon response to endotoxin in rats. In contrast, the interferon yields in Sindbis virus-injected rats and mice seemed not to be influenced by cortisol pretreatment.

5) Adrenalectomy potentiated the interferon inducing capacity of endotoxin in rats.

The cortisol-hemisuccinate preparation used in this study contained N-methylacetamide as a solvent. Experiments with this substance have indicated that it is not without influence on the production of interferon in vivo. Nevertheless it was confirmed that corticoids inhibited endotoxin-induced interferon formation in the rat while they fail to do so with Sindbis virus-induced interferon. A report on these data is being prepared.

ACKNOWLEDGMENTS

The authors are indebted to Dr. A. Dunjic and Dr. D. Jovanovic from the Institute for Radiotherapy of the St. Rafaës-kliniek, Director: Prof. Dr. H. Maisin, University of Louvain, Belgium, for their collaboration in irradiating the test animals.

The skillful technical assistance of Mrs. G. Reghers-Aelvoet, Miss J. Costermans, A. Focan, F. Cornette, and Mr. R. Conings is acknowledged with gratitude.

REFERENCES

1. FINTER, N. B. (Ed.): *Interferons.* Amsterdam, North Holland Publishing Co., 1966.
2. DE SOMER, P., and BILLIAU, A.: Interferon production by the spleen of rats after intravenous injection of Sindbis virus or heat-killed E. coli. *Arch. Ges. Virusforsch., 19:*143, 1966.
3. YOUNGNER, J. S., and STINEBRING, W. R.: Interferon appearance stimulated by endotoxin, bacteria, or viruses in mice pretreated with Escherichia coli endotoxin or injected with Mycobacterium tuberculosis. *Nature (Lond.), 208:*456, 1965.
4. HO, M., and KONO, Y.: Tolerance to the induction of interferons by endotoxins and virus. *J. Clin. Invest., 44:*1059, 1965.
5. DE SOMER, P., BILLIAU, A., and DE CLERCQ, E.: Inhibition of antibody production in rats and mice by intravenous injection of interferon-inducing amounts of Sindbis virus or E. coli. *Arch. Ges. Virusforsch.,* In press.
6. MENDELSON, J., and GLASGOW, L.: The in vitro and in vivo effects of cortisol on interferon production and action. *J. Immunol., 96:*345, 1966.
7. RYTEL, M. W., and KILBOURNE, E. D.: The influence of cortisone on experimental viral infection. VIII. Suppression by cortisone of interferon formation in mice injected with Newcastle disease virus. *J. Exper. Med., 123:* 767, 1966.
8. POSTIC, B., DE ANGELIS, C., BREINIG, M., and HO, M.: The effect of cortisol and adrenalectomy on the induction of interferon by endotoxin and viruses in the rabbit. *Proc. Ann. Meet. Am. Assoc. Immunologists,* 1966.
9. ROBERTS, R. S.: Preparation of endotoxin. *Nature (Lond.), 209:*80, 1966.
10. KONO, Y., and HO, M.: The role of the reticuloendothelial system in interferon formation in the rabbit. *Virology, 25:*162, 1965.
11. VAN ROSSUM, W., and DE SOMER, P.: Some aspects of the interferon production in vivo. *Life Sci., 5:*105, 1966.
12. WAGNER, R. R.: Communication on the PAHO/WHO Conference on Vaccines against viral and rickettsial diseases of man, Washington, November 1966.
13. HO, M., and KONO, Y.: Effect of actinomycin D on virus and endotoxin induced interferon-like inhibitors in rabbits. *Proc. Natl. Acad. Sci. (U.S.A.), 53:*220, 1965.
14. YOUNGNER, J. S., STINEBRING, W. R., and TAUBE, S. E.: Influence of inhibitors of protein synthesis on interferon formation in mice. *Virology, 27:*541, 1965.
15. SMITH, W. W., ALDERMAN, I. M., SCHNEIDER, C., and CORNFIELD, J.: Sensitivity of irradiated mice to bacterial endotoxin. *Proc. Soc. Exper. Biol. & Med., 113:* 778, 1963.
16. SCOTT, W. J. M.: The influence of the adrenal glands on resistance. II. The toxic effect of killed bacteria in adrenalectomized rats. *J. Exper. Med., 39:* 457, 1924.

MECHANISM OF ACTION OF INTERFERON: A DISCUSSION

IRVING GORDON

T HE DATA PRESENTED BY Levy and Carter (1) add further strong support to the hypothesis that interferon suppresses the multiplication of sensitive viruses by interrupting translation of virus mRNA or input virus RNA. As Levy and Carter point out, Joklik and Merigan (2, 3) have shown that in interferon-treated cells subsequently infected with vaccinia virus, formation of virus-specific polyribosomes is greatly reduced. They also showed that the disaggregation of host cell polyribosomes effected by Mengovirus infection is followed by withdrawal of ribosomes from the 74S pool in normally infected cells but not in cells previously treated with interferon. The data from both laboratories, which were obtained with mouse cells treated with mouse-derived interferon, indicate that the binding of ribosomes to mRNA is defective. In contrast, Marcus and Salb (4), examining the interaction of Sindbis virus RNA with chick ribosomes in a cell-free system, found that polyribosome formation did occur at 2/3 the normal efficiency when the ribosomes came from cells treated with high concentrations of chick-derived interferon. The fact that the polyribosomes were incapable of translation, however, was demonstrated by several functional criteria. The discrepancy between these findings and the others might be attributed to quantitative or qualitative differences in the action of interferon, or might be the result of dissimilarities in experimental conditions, including the interferons employed and their concentration.

Normal chick (4) or mouse (1) rapidly-labeled cell RNA binds equally well to ribosomes from normal or interferon-treated cells. It is reasonable to postulate, therefore, as has been done by these authors, (1-4) that there is a point or region of the virus mRNA that does not exist on cell mRNA or vice versa, and that ribosomes from interferon-treated cells can recognize the difference.

There are great discrepancies in the sensitivity of different viruses to interferon. The multiplication of some is markedly inhibited by doses of interferon that do not inhibit the replication of others, and certain viruses, the adenoviruses being noteworthy examples, are highly insensitive to interferon. Furthermore, the interferons themselves are not completely categorized. The translational-inhibitory protein made by a cell after exposure to virus-induced interferon conceivably could interrupt translation at a different efficiency than the translational-inhibitory protein induced by statolon- or endotoxin-induced interferon, and result in partial impairment of virus replication that might extend over several steps.

My colleagues, J. E. Officer, Douglas Stevenson, and R. R. Wycoff, and I have satisfied ourselves that the virus-inhibitory effect of the latter two types of non-virus induced interferon mentioned is indeed mediated by a translational-inhibitory protein. When statolon- and endotoxin-induced interferons are employed against both a moderately-sensitive virus (Mengovirus) and a highly sensitive virus (vesicular stomatitis virus) their suppressive effect is reversed by prior treatment of the cells with actinomycin D (Fig. 1; the statolon was a gift of Dr. Walter Kleinschmidt, and the endotoxin-induced interferon was a gift of Dr. Warren Stinebring). Note evidence of blocking with the low dilutions of endotoxin-induced interferon. The difference in the pattern of reversal of inhibition by actinomycin in Mengovirus as compared with VSV infection needs verification.

Mengovirus-specific RNA is not made in L cells treated with statolon-induced interferon, nor is an RNAse-resistant replicative form demonstrable (Fig. 2). It appears probable that polyribosomes are not made in cells infected with these viruses and treated with these interferons, but more data need to be collected. We can tentatively predicate that the differences observed in the degree of protection conferred by different interferons on cells infected with different viruses are likely to involve the efficiency with which interferon can induce defective translation of virus mRNA. Whether there might be differences in the efficiency with which interferon(s) inhibit the translation of different parts of the virus messenger remains to be determined.

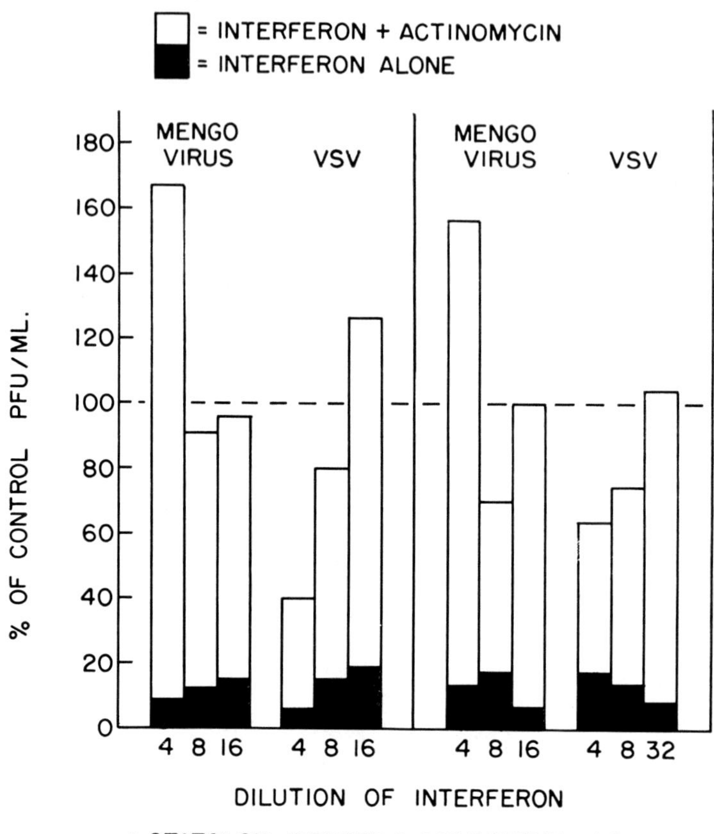

Figure 1. The effect of actinomycin D on inhibition of virus multiplication by statolon-induced and endotoxin-induced interferon. L cells were respectively treated with control fluid, interferon, and interferon plus 3 μg/ml of actinomycin D 18 hours before infection with either Mengovirus or vesicular stomatitis virus (VSV). One hour after infection, all cultures were treated with 3 μg/ml of actinomycin D. They were harvested at 7 hours and virus output determined.

Although in neither of the *in vitro* experiments cited above (1, 4) did normal cell messenger show any diminution in its ability to form polyribosomes after interferon treatment, nor have prior examinations by other approaches revealed any effect of interferon on cell messenger, it is conceivable that there are exceptions which would not be recognized when total macromole-

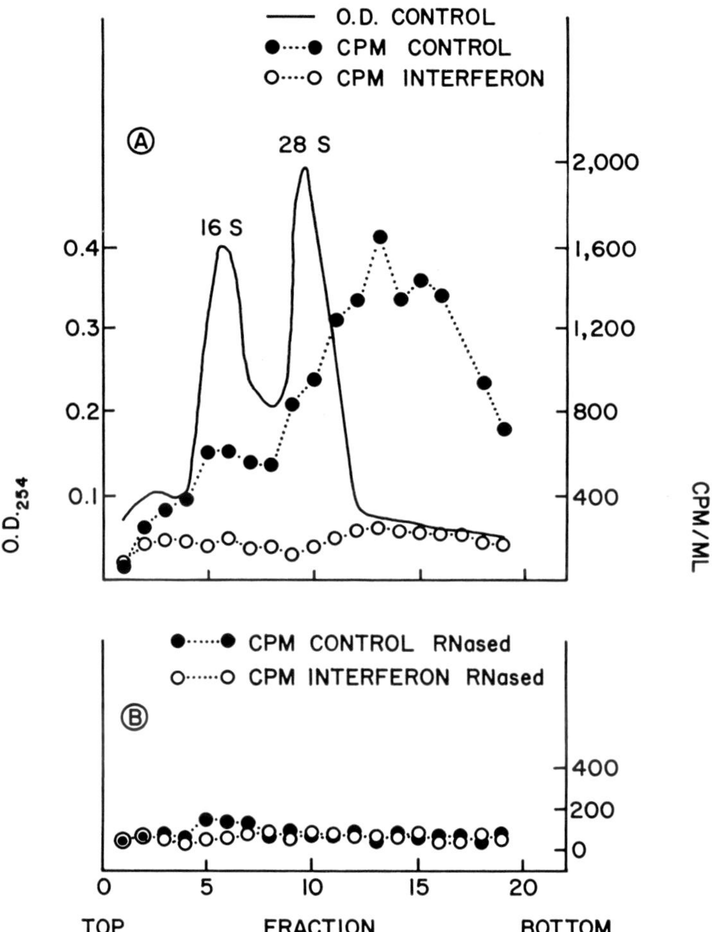

Figure 2. Zonal sucrose density centrifugation analysis of Mengovirus-specific RNA extracted from L cells previously treated with statolon-induced interferon. Interferon or control fluid was added 18 hours before infection. Actinomycin D (3μg/ml) was added one hour after infection, the cultures were pulsed with uridine-H[3] (50 μc/culture) between the 5th and 6th hours after infection, and the RNA was extracted by the SDS-phenol technique as previously described (5) and analyzed in a 5 to 20% sucrose linear gradient. A portion of each fraction was treated with 10 μg/ml of ribonuclease at 0°C for 10 minutes.

cular synthesis is employed as the criterion, for example, incorporation of amino acids into peptides. We have performed some preliminary experiments on the effect of interferon on differentiation, since it can be used as an endpoint for interruption of translation under conditions of highly programmed regulation. With the aid of our colleague, G. Matioli, whose techniques we employed, we have prepared rabbit bone marrow preparations consisting of 90% erythroblasts and 10% myeloblasts. In conventional suspension culture these cells differentiate in 24 hours into reticulocytes or erythrocytes on the one hand, and granulocytes (principally eosinophils) on the other hand. We have demonstrated that one or both types of cell can effectively be derepressed to make translational-inhibitory protein by showing that VSV multiplication was inhibited by statolon-induced interferon. Fluorescent antibody studies indicated that the erythroblasts were infected. Both cell types differentiated as quickly and completely after interferon treatment as in the absence of it.

Additional experiments of this type seem worthwhile in order to complement data of the type already obtained. More complete examination of the effects of different interferons on translation by viruses differing in susceptibility to interferon also seems warranted.

REFERENCES

1. LEVY, H. B., and CARTER, W. A.: These Proceedings.
2. JOKLIK, W. K., and MERIGAN, T. C.: *Proc. Nat. Acad. Sci., 56:*558-565, 1966.
3. JOKLIK, W. K.: These Proceedings.
4. MARCUS, P. I., and SALB, J. M.: *Virology, 30:*502-516, 1966.
5. GORDON, I., CHENAULT, S. S., STEVENSON, D., and ACTON, J. D.: *J. Bacteriol., 91:* 1230-1238, 1966.

MASS PRODUCTION OF
HUMAN INTERFERON FOR
THERAPEUTIC TRIALS

E. FALCOFF, R. FALCOFF, AND C. CHANY

THE POTENTIAL THERAPEUTIC use of interferon or interferon-like substances in viral diseases is based on several observations. Interferon can be detected in the blood and tissues of patients and animals early in the course of viral disease (either natural or experimentally induced) and before the appearance of antibody. Interferon inhibits the multiplication of antigenically unrelated viruses both *in vitro* and *in vivo*. Since at the onset of viral infection only a fraction of the total number of susceptible cells are infected, administration of interferon at an opportune moment may delay or inhibit the evolution of the disease, thus permitting full mobilization of the host's defenses.

In order to evaluate the therapeutic usefulness of interferon in viral diseases of man, it is necessary to produce an interferon of high biologic activity and in large quantity. Since interferon is specific for the species, cells derived from human tissues must be utilized. Of several different possible cell types, we have investigated leukocytes and placental tissues. In this chapter, we will confine our remarks to the results obtained utilizing human leukocytes.

Proteins derived from blood have been utilized in the therapy of various diseases and blood transfusion itself is a well accepted procedure. Leukocytes may be easily separated from whole blood. Gresser has previously shown, and has been confirmed recently by others, these cells under appropriate conditions produce large amounts of interferon. We have attempted, therefore, to define a number of the optimal experimental conditions for the "en masse" production of interferon *in vitro*. Our results suggest the feasibility of producing interferon on a large scale for use in clinical trials. We will also present the results of experiments

pertaining to the purification of interferon and the physico-
chemical characterization of the active moiety.

1	CITRATED BLOOD (DIFFERENT DONORS) + DEXTRAN SEDIMENTS 1 HOUR - 25°C
2	UPPER 2/3 HARVESTED and CENTRIFUGED
3	LEUCOCYTE SEDIMENT INOCULATED with PARAINFLUENZA VIRUS I (M 100:1) —37°C - 3 HOURS
4	LYSIS of ERYTHROCYTES ENSUES, RECENTRIFUGATION and RESUSPENSION of LEUCOCYTE SEDIMENT in NUTRITIVE MEDIUM
5	INCUBATION 22 HOURS - 37°C
6	CENTRIFUGATION. SUPERNATANT TREATED at PH 2 24 HOURS and ADJUSTED PH 7
7	ULTRACENTRIFUGATION. FILTRATION
8	VARIOUS BIOLOGIC TESTS, i.e., ACTIVITY, STERILITY, etc.

Figure 1. Preparation of interferon (human leucocytes).

The first illustration summarizes our basic technique for the
preparation of interferon from leukocytes.

Citrated blood from several donors is mixed (without taking
into consideration the blood type). After sedimentation of most
of the erythrocytes by means of dextran, the leucocyte rich plasma
is centrifuged and the cellular sediment inoculated with parainfluenza virus I (Sendaï) at a multiplicity of 100. During infection,
the contaminant erythrocytes are agglutinated by the virus and
then completely lysed. After incubation for 3 hours, the leukocytes
are washed, resuspended in a medium containing human serum,
and incubated at 37°C for 22 hours. The medium is then centrifuged and the pH of the supernatant adjusted to 2 and then to
7 prior to ultracentrifugation and filtration. The filtrate consti-
tutes our interferon preparation.

Figure 2 illustrates the optimal conditions for the production
of interferon from leucocytes in this system.

In view of the possibility of utilizing interferon in clinical
trials, we have also developed a method of semi-purification of

interferon. This technique can be easily adapted to different volumes of interferon and provides high yields of biologically active material. Techniques of purification previously described utilizing CM cellulose or CM-Sephadex in chromatographic column are not suited for large scale production of interferon for several reasons:

1) The narrow relationship between the volume of the sample and the volume of the resin.
2) The necessity of slow elution.
3) The dilution of biologic activity of the initial preparation.

NUMBER OF LEUCOCYTES/ml of SUSPENSION	5×10^6
MULTIPLICITY of INFECTION	100
TIME of INCUBATION	22 HOURS
CONCENTRATION of HUMAN SERA	
in the NUTRIENT MEDIUM	5%
pH	7,4 – 7,5
TEMPERATURE of INCUBATION	37°C

Figure 2. Optimal conditions of interferon production.

Batch chromatography, the technique which we will describe, provides a partial purification (30-40 fold) and avoids these three limitations of column chromatography.

Crude interferon is adsorbed to carboxymethyl CM-Sephadex in a beaker containing a magnetic agitator. The resin is separated by centrifugation and washed in various phosphate buffers with increments of pH and molarity. By this technique, most of the extraneous protein is eliminated.

In evaluating the results of batch chromatography, it was noted that the biological activity of interferon was increased 40 fold with the loss of only 20 per cent of the initial total of interferon.

Figure 3 illustrates some of the physico-chemical properties of this semi-purified interferon.

We have administered the unpurified interferon to several patients with known viral and suspected viral diseases, considered to be incurable. In all, 28 patients have received interferon for prolonged periods of time:

> 11 acute myeloblastic leukemia
> 7 acute lymphocytic leukemia
> 2 cases of herpes zoster in leukemic patients
> 7 newborn with cytomegalic disease
> 1 newborn with generalized herpetic infection

MOLECULAR WEIGHT	25,000
ISO-ELECTRIC POINT	6,75
DENSITY in CESIUM CLORIDE GRADIENT	1,17
ACTINOMYCINE (1γ/ml)	ACTION INHIBITED
TRYPSIN (200γ/ml)	DESTROYED
RNA· ase (100γ/ml)	NOT DESTROYED
DNA· ase (100γ/ml)	NOT DESTROYED

Figure 3. Physico-chemical properties of interferon (human leucocytes) .

Figure 4, you can see the list of 11 patients with acute myeloblastic leukemia. Our preparation of crude interferon was essentially well tolerated by these patients, even when a total of more than 3 liters were administrated in the course of one year (patient 11) .

As to the survival of patients with leukemia treated with interferon, we cannot at this time draw any conclusions.

In 7 newborns, the intravenous injection of 40 ml per day was well tolerated for 3 weeks. In some of the other 20 patients, chilly sensations were experienced 30 minutes after injection of interferon. This constituted the only untoward reaction. In only one instance was administration of interferon for a prolonged period of time accompanied by any evidence of a significant allergic reaction. It is possible that partial purification of interferon will eliminate the pyrogens produced by viral infected leucocytes.

Although we have observed no case of serum hepatitis in our study, this danger remains. It is possible however that the techniques employed to separate the inducing virus from the interferon preparation have also eliminated potential contaminating viruses.

Patient	Age (Years)	Duration of Treatment (days)	Total Amount of Interferon (ml)	Survival (months)
1	55	80	270	>11
2	59	15	180	9
3	67	3	90	> 8
4	59	30	620	> 8
5	19	130	1280	> 7
6	14	50	660	> 9
7	11	16	330	7
8	15	310	2280	16
9	12	360	2200	>12
10	18	400	2000	16
11	6	400	3620	14

Figure 4. Administration of interferon to 11 patients with acute myeloblastic leukemia. Interferon inoculated intravenously (10-40 ml), daily or twice weekly.

As to the therapeutic value of interferon, it is too early to form any opinion. The preliminary results seem encouraging since the 7 newborns with cytomegalic disease have survived and this disease is usually fatal. Another newborn, with generalized infection due to herpes simplex virus, seemed to improve transiently after receiving interferon.

We must, however, also refrain at this time from commenting on the effect of interferon on patients with leukemia; although the survival of some of these patients has been prolonged.

The preliminary clinical trials, however, do show that intravenous administration of interferon derived from human leucocytes is feasible and without danger. It is hoped that these studies will form the basis for further trials with preparations of interferon of greater potency and greater purity.

DISCUSSION — INTERFERON

Dr. Baron: Dr. Isaacs has written that interferon was discovered in 1957, thus making it a 9-year-old child born of parents with a family name of interference. It is fitting that the parents are approximately 30 years old. As we know, the child was born into the skeptical scientific world and its legitimacy was questioned. However, relatively quickly, additional facts established its merit which led to acceptance. Since the interferon system apparently involved the virus and the cell, it promised better understanding of several mechanisms — viral reproduction, cell synthetic and regulatory mechanisms, pathogenesis of viral infection, diagnosis and, hopefully, prophylaxis and treatment of viral diseases. The papers which follow will report recent progress towards fulfilling these promises. I think it is fair to say that from the observations of this child to date, we may anticipate a remarkable young man but only if we continue to raise him with a balanced mixture of indulgence and skepticism.

Miss Truden: At the Variety Children's Research Foundation here in Miami, we have been studying the effect of crude and purified L cell interferon on the replication of Mengo virus in Ehrlich ascites tumor cells. Our method of analysis consists of MAK chromatography of P^{32}-labeled nucleic acids obtained from cells that have been infected in the presence of Actinomycin D after treatment with interferon. In this manner, we have determined that whereas crude interferon has no inhibitory effect on the synthesis of Mengo viral RNA, and in fact caused a 3-4 fold stimulation of its synthesis, interferon purified by CM-Sephadex column chromatography completely suppressed viral RNA replication.

We concluded that our crude preparations of interferon contained an antagonist capable of nullifying the antiviral action of interferon. This was substantiated by locating a substance, eluting before the interferon on CM-Sephadex, that was capable of reversing the viral inhibitory capacity of the purified interferon.

By enzyme inactivation studies, we have ascertained that the antagonist is proteinaceous in nature since it is inactivated by trypsin, but is unaffected by RNAase, DNAase, or lipase.

Furthermore, our antagonist exhibits a certain selectivity in that it is completely ineffective in the L cell, but is fully expressed in the ascites cell.

DR. REGELSON: Ascites sarcoma 180 tumor cells, with increasing concentrations of polysulfinate, have antitumor and antiviral activity. There is formation of a hydrophilic gel which is prevented by the oxyribonuclease. Budding of the cytoplasm and the swelling of the cell occur. The chemistry demonstration of these polyantimes is most evident for the polysulfinates and there is also polycarboxylates. They, of course, have antiviral activity forming electrostatic bonding between viruses, but, in addition, require the presence of living cells to demonstrate their antiviral effect and thus may act either by inducing the cellular interferon or through some other mechanism. Pyrancopolymers are a development of the Hercules people and are antiviral giving a polycarboxylates compound. They have a wide range of antitumor activity against the number of transplants that are going into man. There is significant inhibition of splenomegally which correlates with inhibition of virus titers within the spleen of the animals. Not all polyantimes inhibit. Heparin actually decreases host resistance. This demonstration by Kleinschmidt brought us together with Merigan demonstrating the presence of interferon in the serum of mice treated with a whole range of pyrancopolymer compounds in these mice and the difference in molecular weight. Induction of interferon is thus noted. The Pertussis Vaccine completely reverses the antiviral activity and Friend leukemia activity for pyrancopolymer.

DR. YOUNGNER: I would like to return to the properties of interferon which were discussed by Dr. Fantes. You have a characteristic listed that interferon is incapable of blocking the formation of interferon, and I wonder on what basis you make this classification and if you really meant that?

DR. FANTES: I would not say that I did that work. I provided the various interferon fractions and the main object of this work was to show that its action would be of further importance to interferon action.

DR. BARON: Then there is agreement that interferon itself, at least in high dosages, may exert a separate blocking activity on subsequent interferon production.

Can Dr. Merigan give us a time-course of development of interferon and resistance in the mouse treated with pyrancopolymer and what are the effects in tissue culture?

DR. MERIGAN: Very quickly, with a peak at 18 hours, we've done this with the pyrancopolymer and as far as *in vitro* we have not been able to demonstrate interferon induction with these materials working with L cells and tissue culture.

DR. SIGEL: I would like to amplify one statement of my associate, Miss Truden. In discussing blockers or antagonists, it is essential to go one step beyond the species specificity. In this instance, we are dealing with two cells derived from the same host, the mouse. Both cells are responsive to mouse interferon but the Ehrlich ascites cell recognizes not only interferon but also its antagonist, whereas the L cell ignores the antagonist. Thus, whatever the recognition mechanism of interferon by a cell, it must function as a determinant of specificity. In addition, the response of a particular cell is probably under the control of a second mechanism, namely, the one which determines whether a cell sees or ignores the interferon antagonist.

DR. HAYES: About two or three months ago, a paper was presented in the IEG #7 Group on the use of purified polychromide gels for the fractionation or separation of high moleculoid nucleic acids. This work was done by Dr. Looney from the University of Glasgow. I was wondering if Dr. Fantes has done anything with this type of polychromide system? Dr. Looney has shown that he can get the same type of fractionation pattern in isochromides that you can get in an electrophoretic pattern, and, needless to say, you can do it in a much shorter time, and also, these purified gels are now as useful in direct absorption readings at 265 millimicrons so that you must line them up and scan them or at 280 millimicrons to find out where the proteins are.

DR. DE SOMER: I would like to ask if no interferon is present in the copolymer. We examined some polymers that were undesirable but did not produce interferon and at first we thought this was due to interferon production. So separate your interferon from copolymer.

DR. MERIGAN: I think that one of the clear ways that you can separate the interferon from the copolymer is that it is in the characterization of the antiviral activity as it appears in the serum of the mouse. I stated previously that it was Trypsin-susceptible and it had a molecular weight of 70,000 as it appears in the serum. The inducer is actually a 17,000 molecular weight polyantime species.

Dr. Oxman: Dr. Gordon has reminded us that the remarkable property of the interferon system is apparent selective inhibitional viral function without detectable effect on host cellular functions. Work has been done by Drs. Baron, Rowe, Black, Tocomoto, and Habel at National Institutes of Health, and may be of some interest in focusing some attention on this question of recognition. Two sets of findings were obtained. The first is an example of the effect of interferon on the early function of the SV 40 virus. The function chosen was the induction of SV 40 T antigen which appears to be rather closely related or identical sensitive protein. It is not inhibited by FEDR and occurs in a system of abortive infection of mouse cells in which the transformation occurs but no significant synthesis of viral DNA, and no synthesis of infectious virus or structural viral proteins. We have found that SV 40 T antigen was only inhibited by interferon when cells were pre-treated with interferon and acutely infected with SV 40 virus. However, in the same cell virus system, transformed cells are produced which divide for an infinite number of generations thereafter always full of SV 40 T antigen, and the obvious question was whether these cells which appear to contain integrated SV 40 genome but no infectious virus and produce SV 40 T antigen all the time would have the same antigen, similarly sensitive interferon. To examine this mouse, interferon was used and these transformed mouse cells in propagative fashion in a continuous approximate log phase for more than 100 generations over four months in the continuous presence of high titer interferon. On monitoring, it was found at various times that the actual effect of interferon by infecting these cells with vesicular stomatitis virus was to produce an antiviral effect in these cells which was measured by the SV. Throughout, and at the end of this experiment, it was shown that there was no detectable depression of SV 40 T antigen by fluorescent antibody of CF in these cells, so that here is a situation where the same product presumably the product of at least a similar region of messenger RNA from the DNA viral geneome is strongly inhibited by interferon in acute infection. However, when the same DNA is integrated in a cell and produces a messenger RNA, this messenger RNA is apparently not inhibited in its function by interferon. The other somewhat related system chosen to look at was the utilization of the adeno-SV 40 hybrid virus. It is a useful point to know that adenovirus is approximately 50 times less sensitive to interferon than SV 40 either measured as CPE infectious virus or, in our case, SV 40 or adeno-T antigen production. The question then occurring to us was in these hybrid viruses which appear to consist of a linked single molecule of DNA-containing both SV 40 on each strand or double strand of DNA containing SV40 and adeno only bonded, whether the T antigen synthetic function of these two would be changed by the co-presence on the same molecule, so that we infected cells with SV 40 virus and noted the sensitivity of the T antigen formation interferon. Then, infected cells double with both SV 40 and adenovirus independently, bringing the genomes into the same cells. It was noted that in the same cells SV 40 T antigen production continued to approximately 50 times as sensitive interferon as was the co-present adeno-T antigen production. Then, using the hybrid virus to produce T antigens of both adeno and SV 40 in the same cells, we again examined the interferon sensitivity of the two and found, surprisingly, that the SV 40 T antigen brought in by the hybrid virus was no longer as sensitive to interferon as when brought in by SV 40 alone. In fact, it reverted to the relative insensitivity of adeno so that here in two situations, the integrated SV 40 genome and transformed cells, and the acute infection by combined adeno SV 40 and DNA molecule, we have an alteration in the function. The T antigen synthetic function of the DNA presumably mediated by messenger RNA and we would surmise that in both cases the messenger RNA is somewhat altered by its DNA by the association of the SV 40 DNA with host or with adeno DNA. Altered perhaps by the loss of a recognition region characteristic of the SV 40, or, we think more likely, the acquisition of in one case a host segment of messenger RNA and in the other adeno-segment of messenger RNA.

DR. MARCUS: I believe there is a firm basis for evidence that the antiviral protein that interferon induces is in fact a translation inhibitory protein that binds to ribosomes. Interferon interacts with the cell. We are now integrating data from the literature and our own data not always consistent with the literature. Interferon in some manner deeply processes the cell genome to make a new message. This message coding for the protein is antiviral protein. We presented evidence that this protein, antiviral protein, is, in fact, a translation inhibitory protein that we nickname "tip" and that this protein is synthesized in the cytoplasm, binds to ribosomes and the ribosome pool. If these ribosomes are sequestered into polysomes using host cell message there is normal binding and 100% efficiency of translation. However, if viral message binds to these ribosomes, the binding is somewhat reduced and what we think is more important even if binding does occur, there is no translation of the viral message. Evidence for the nature of the translation inhibitory protein has come primarily from the inference in the literature, and I would like to present our data that this protein is similar in kind to other translation inhibitors that we have described, one on the ribosomes from cells in metaphase and another similar to, probably, the same inhibitor that Hogland has described on the microsomes on normal liver cells. Both of these translation inhibitory proteins are sensitive to Trypsin, and I contrast their sensitivity with their insensitivity to Chymotrypsin, so that if we take ribosomes from metaphase cells or microsomes from normal liver cells and treat them with appropriate concentrations of Trypsin, one can restore the normal amino acid incorporating activity of these ribosomes with appropriate message. Then, we tried to demonstrate similar Trypsin sensitivity of the translation inhibitory protein on interferon ribosomes but much to our amazement the ribosomes from chick cells literally dissolved in the concentrations of Trypsin which were ample to show this effect with ribosomes and rat liver microsomes. Finally, by backing down on the concentration of Trypsin to very minute amounts, we were able to show differential effect. The normal ribosomes translate Sindbis viral message with 100% efficiency so that the polysomes formed are broken down at 37 degrees. However, at 0 concentrations of Trypsin, there is complete inhibition of translation of this Sindbis message and as we increase the Trypsin concentration of the ribosomes we finally reach a concentration a little beyond 0.3 and 0.4 gamma per ml which apparently destroys or removes this translation inhibitory protein and the very ribosomes that were completely refractory to translation of the message now function as do normal ribosomes. Polysomes bound at 0 degrees and 448 counts represents labeled Sindbis RNA bound in the polysome region 250 S region. If we raise the temperature of these complexes in the *in vitro* system to 37 degrees, 96% of the label is lost from that region and we consider this 96% translation efficiency. If we do the same experiment with interferon ribosomes from interferon treated cells, the number of counts bound is significantly less at 0 degrees. This is one of our lower values at 37 degrees; there is slight translation here. It is usually more than we get. This calculates out to be 15% translation. Now, if we mix equal numbers of normal and interferon ribosomes we find that the number of counts that are Sindbis-RNA bound, are relatively close to the normal, that is, 395 compared to the 448. However, if we incubate at 37 degrees and calculate what we would expect to be the translation efficiency, when translated independently, we find that we have significantly less. Although this is very preliminary evidence, it suggests that the presence of a ribosome containing the translation inhibitory protein in a run of normal ribosomes will block translation of the message or at least cut down its efficiency of translation. In view of the fact that binding was relatively close to normal, we would also suggest that in the presence of a mixture of two kinds of ribosomes the viral message can in fact recognize the difference and there is a greater efficiency in binding the normal ribosome.

Dr. Levy: Our results in a cell-free system are not totally in accord nor are they identical with what we see in the whole cell system. In our cell-free system, we see virtually no binding of messenger RNA to ribosomes or to polysomes. In the whole system, there is an inhibition of the order of magnitude that Dr. Marcus sees in association with his cell-free system. The important thing to recognize here at the moment is that there is an alteration in ribosomes induced by interferon and that the apparent differences are probably going to be worked out and when more experiments are done I think that an explanation for these differences will develop. I think the reason that we see a cut-off of RNA synthesis probably relates to the fact that there was cut-off of protein synthesis induced in these cells by these cells and polio-infected cells, which may not be related to any synthetic process initiated by the virus. There is something possibly growing with a number of these viruses that leads to a protein cut-off. As with Puramycin and with some other systems, I believe if you cut off protein cell synthesis there follows not too long after a cut-off of RNA synthesis.

Dr. Lockart: I would like to say something about the cell killing with respect to the L cells and Mengovirus. We have worked with L cells quite a bit and have actually incubated them with rather large concentrations of Actinomycin D and Puramycin at the same time so that all RNA and Protein synthesis was cut off and these cells actually remained quite normal looking as long as 48 hours under these conditions, certainly well by 24 hours. When these cells are infected with Mengovirus, they are destroyed in about 10 hours, and I am not sure that the other kinds of events that occur in these cells are a result of the shunted RNA or protein synthesis. I wonder if we are cutting off translation and if it may not be a little early to say that it is very complete. Perhaps one should think that it's a partial translation cut-off and that some kind of things can continue and others are cut off. I certainly think in terms that if what you do is to put a protein, or you offer the ribosomes, in such a way that they don't translate a given viral RNA such as one chooses to measure. How does one explain the kind of observation, for example, where the sensitivity of Herpes virus was compared with WEE in chick cells? If we put in sufficient interferon into a population of cells so that one sees absolutely no viral RNA synthesis and no destruction of any sort so that nothing occurs in this case, one then infects the same population of cells with Herpes virus and obtains essentially normal yields of Herpes virus so that we have altered everything with WEE but we haven't altered it at all with respect to Herpes. I think that both Widener and I have shown that you can add interferon to cells that are actively synthesizing virus at the time that you add it and still reduce viral synthesis by something like 90%. This means that everything that had to be synthesized for a new viral RNA synthesis to occur has happened. It is actually under way. Actually, a new virus was made and released and you can still shut it off.

Dr. Marcus: I think Dr. Lockart has raised a very important point, that there is an impressive bit of information about the spectrum of sensitivity to interferon which must be accounted for by any mechanism of action of interferon, and that is the striking fact that he mentioned, namely, that you can have cells treated with a given concentration of interferon. They are exquisitely sensitive to the arboviruses and very relatively insensitive to something low on the scale like Newcastle Disease or the Herpes and there is a whole spectrum in between. To account for these results, one visualizes that Sindbis virus may contain in its polycystchronic message more points that are sensitive to the action of this translation inhibitory protein.

Dr. Oxman: We have discussed most of the components of translation here, that is, the ribosome and the messenger RNA or the viral genome. However, I think that there is another component that we have not considered that might bring some of this information together and that is the nascent protein that is being translated from the viral genome. This is the component of the translational system too. If you could possibly take this nascent protein and perhaps react it with the regulatory molecule, twist this protein around to change its tertiary structure or later on perhaps a quatrinary structure so that it blocks translation, this would also keep the ribosome from the sub-ribosome unit from developing into a polyribosome. It would also prevent translation or read out, that is, the ribosome with the genome on it could no longer function and these findings might be brought together. Now, this hypothesis has been advanced by Klein and Bach at the University of Wisconsin and it has been published in the recent Coldspring Harbor Symposium and it is called *Control of Gene Expression at the Level of Translation.* I believe we might explain all of these findings by suggesting in well-known terms that this hypothetical antiviral protein might be a regulatory molecule that would bind in some way with, lets suggest the preliminaries that are being made by the virus, and in this way inhibit polysome formation and virus replication as well.

DR. REGELSON: In a discussion that I had with Bubel from Cincinnati, we talked about the non-specific action of various drugs and inducing hydrolysating enzymes which result in a tremendous increase in the microsomal content, for example, liver cells. There is a whole area of non-specific drug hydrolysis and detoxification that is very important to the clinical activity in a number of drugs, in the resistance it develops to them. Now, these enzyme inducing effects are readily related to a given drug that is administered to the animal with an associated increase in microsomal activity and blocks afforded by Actinomycin D. I ask the following question: "Has anyone studied interferon production with relation to the amount of microsomal activity in the cell, using this for a good model system for a study of interferon-like activity and blocking such?"

DR. MAHDY: We have been working with Rubella and it appears to us to present a very good system because in the actively working monkey kidney others as well as ourselves have failed to demonstrate detectable interferon. We were interested in determining the metabolic conditions for the production of the infection interferons. With the rubella systems, and using 0.1 micrograms Actinomycin D, we failed to show any significant effects of this treatment on the production of interferons by reacting monkey kidney cells to charge with ECHO 11 virus. Similarly, some of the times we have used vesicular stomatitis virus and the resistance in part of the cell that was treated with Actinomycin D prior and during infection did not get impaired by the Actinomycin D treatment. The phenomena of antiviral protein can be

continuous yield of interferon by the infected culture and the rapid passage of this low molecular weight product into the supernate fluid disturbed its accumulation in the cells. Flasks of the monolayer cultures of chick and mouse fibroblasts or Japanese B viruses were incubated as monolayers in the reverse of the usual position so that the cells are above and tissue fluid below. Every four hours the cells are covered for some seconds by tissue fluid and brought into the new position. Within 8 to 12 hours, the infected cells produced interferon in entirely distinctive concentrations, never observed by the usual techniques. The titer of the interferon in the cell extracts suspended in one tenth of fluid a matter of every two hours rose after six to twelve hours to the dilution of 1:8, 1:32. After 12 hours, the same monolayers were covered again by the fluid and interferon quickly disappeared from the cells during two to four hours, but was found again in increasing amounts in bottles that were turned and the fluid was brought into contact with the cells. This simple assay was especially effective in the bottles that were vertically saturated with the other cells of tissue fluid. This not only maintains the high moisture of the air but apparently effects the infected cells with minimum of nutrient compounds that supported the virus multiplication. This method of cellular interferon analysis may be useful for understanding the complicated correlation existing between interference and interferon production, especially in cells where there is a low level of interferon formation. The absence of interferon in the fluid does not mean its lack in the cells but often depends upon its critical quantitative level still providing an excellent protection of cells against super infection by various viruses such as equine stomatitis or Sindbis virus.

DR. BARON: Some of the interferon may be tightly bound to cells, or, a reaction leading to derepression of the cistron for Taylor's antiviral protein may have occurred at $4°C$. If a small amount of interferon is bound to cells, then I believe that it is less than 1% because 1% of many proteins will now specifically bind to cells. In our mouse cell system, we find that removal of interferon and replacement with medium or inhibitors of RNA and protein synthesis is not followed by loss of antiviral activity until more than 4 hours have elapsed. This recent finding suggests stability of Taylor's antiviral protein over this time.

MR. BARRY S. MICHAELS: We have been working with polyvinylpyrrolidone (PVP) and have found it to contain viral inhibitory activity against herpes simplex and vaccinia viruses *in vitro* without significant toxic effects to cells. No direct inactivating or neutralizing effects on virus infectivity were observed. The mode of action has not been determined except for the tentative evidence that this substance does not act by way of stimulation of the formation of an inhibitor such as interferon. This evidence was obtained by treating chick embryo fibroblasts with PVP and after a 24-hour incubation period rinsing cells of residual PVP. These cells were then fed with fresh medium, and after a further 24-hour incubation, the fluids were harvested and assayed for their inhibitory activity. No subsequent inhibitory activity was detected, and at this time it was concluded that PVP was the inhibitory agent demonstrated.

Presently, we have done some work on the tumor inhibitory activity of PVP of various average molecular weights on Rous sarcoma virus (RSV), in chickens. Studies have been carried out on the monomer, N-vinyl-2-pyrrolidone, four polymers of average molecular weights of 10,000, 40,000, 160,000, and 360,000, and N-methyl-2-pyrrolidone. All compounds were screened by simultaneous inoculation of both RSV and compound subcutaneously into the wing web of 3-day-old chickens. Activity was shown in all materials with the exception of PVP average molecular weight 160,000, as measured by reduction of tumor frequency, size, and by a delay in the latent period. Three of these compounds were virucidal and their action was exerted, at least in part, upon the virus. The PVP average molecular weight 40,000 was found to be the most effective inhibitor tested, being nonvirucidal and exerting its action on the host. This is the same material that was used in the early work with herpes simplex and vaccinia viruses. Little activity was demonstrated upon pretreatment of the chickens. Maximal activity was obtained by simultaneous inoculation of RSV and PVP, but, PVP also was effective when administered 6 hours after exposure to RSV.

Although we have not done any *in vivo* studies on the possible production of interferon in response to PVP, I can't help wondering whether some of the activity demonstrated may be due to an interferon phenomenon. I would like to question Dr. Regelson who mentioned briefly that he demonstrated activity of his pyran copolymers on a sarcoma system. I would like him to discuss more completely under what conditions this activity was demonstrated, and whether the activity of PVP could be attributed to interferon activity.

Dr. YOUNGNER: The results that were described today by several different people illustrate that a great deal of caution must be exerted in transposing conclusions and generalization in one species to another. This has been true in many other areas and is being emphasized more and more in the interferon area. I can point out to you several things said today which were diametrically opposite results on the effect of cortisone or splenectomy whether one uses rats or mice. Also, I would like to say that the fact that interferon is found in rabbit macrophages, whether or not they are stimulated with endotoxin, is related only to the rabbit, but in mice we have been unable to find the release of interferon spontaneously from mouse macrophages, nor, indeed, have we been able to get an interferon response when endotoxin is added. This further emphasizes the fact that each species has to be considered separately and broad generalization cannot be made.

Dr. MAHDY: I wonder if Dr. Falcoff has tested the leukocytes before infection with parainfluenza virus, for the presence of interferon or interferon-like substances? I ask this because we are conducting some work along this line, and in preparation of leukocytes either by phytohemagglutin or by dextran having molecular weight of 186,000, we have found that leukocytes from normal pools of blood do contain an interferon-like substance which is active against most vesicular stomatitis virus and polio.

Dr. FALCOFF: In these experiments, we have used controls. Especially in some experiments, we found inhibitory activity sometimes going up to fairly high titers. However, we are not sure that this has anything to do with interferon because in one test we have done we have no absolute definite data. In one test, Actinomycin did not inhibit action of the activity. Under this condition, if this is controlled in other experiments, we do not consider this spontaneously-appearing inhibitory type interferon.

Dr. DONOVICK: I am wondering how seriously we should take these molecular weights especially when we begin to classify different interferons. We all know that macromolecules tend to aggregate. They will shift with degree of hydration and with various strengths. In these reports on molecular weights, I wonder how many of them have been studied in more than one system?

Dr. Youngner: Dr. Merigan did the experiments first using density gradients to separate these interferons, and it is possible to do this so that the heavy endotoxin interferon and 90,000 and NDV interferon 28,000 can be distinguished by means other than sucodex filtration.

RATIONALE FOR THE DEVELOPMENT OF INHIBITORS OF VIRAL SYNTHESIS

RICHARD M. FRANKLIN

Depending on the amount of genetic information contained in the nucleic acid of a virus, several or many virus-specific macromolecules will be synthesized in virus-infected cells. For example, the poxviruses, containing DNA of total molecular weight 160×10^6 daltons in 1 or at the most 2 molecules (Joklik, 1962), can code for a great many proteins whereas poliovirus, with a single RNA molecule of molecular weight 2×10^6 (Anderer and Restle, 1964), can code for a maximum of ten proteins. Among these viral-specific macromolecules one may find nucleic acids, structural proteins, and enzymes. In some cases, there may also be viral-induced synthesis of cellular macromolecules.

One of the major approaches to inhibition of viral growth is through the use of selective agents which may inhibit the production or the activity of these viral-specific or viral-induced macromolecules without acting on cellular biosynthesis (Tamm and Eggers, 1965). A second avenue of approach, through the inhibition of virus attachment and penetration, although of great practical importance as the basis for the action of antibiotics, does not seem as promising a direction for developing antibiotic or synthetic chemotherapeutic agents. Therefore our present discussions will emphasize the viral-specific macromolecules and their inhibition.

In these introductory remarks, some general principles of viral replication will be reviewed and some of the features of viral biosynthesis which may be especially vulnerable to attack by selective inhibitory agents will be considered.

The first step in the synthesis of any virus is the adsorption to the cell surface and the penetration of the whole virion or some part thereof. Although the initial contact between virion and cell surface may be purely physical due to random motion of the virus

(Allison and Valentine, 1960); the next step, attachment, can be more specific, at least in certain cases (Dales, 1965). For example, the enteroviruses adsorb to specific receptors identified as protein components of the cellular membrane (Holland and McLaren, 1961). The RNA bacteriophage adsorb only to F$^+$ bacteria which have special pili found only in male bacteria (Brinton *et al.*, 1964) and these special pili are the site of adsorption (Crawford and Gesteland, 1964). Many examples are known of mutant bacteria resistant to bacteriophage because of changes in the bacterial cell surface resulting in lack of attachment of the bacteriophage. Poxviruses, however, are an example of a class of viruses which do not adsorb to specific attachment sites (Allison and Valentine, 1960).

Transfer of virus into mammalian cells is believed to be due to an active phagocytosis of the virion. Much evidence to support this concept arises from the extensive ultrastructural studies of Dales and coworkers (Dales, 1965) on the fate of many viruses such as poxviruses, adenoviruses, and reovirus. This phagocytic process must, however, be distinguished from the ultimate release of the viral nucleic acid into the cell, accompanied by the disintegration of the virion. This latter series of events should be considered to be the true penetration process (Dales, 1965). Penetration leading to activation of the viral genome may be quite different for the viruses infecting different organisms. In the case of some animal viruses, phagocytized virus may be partially degraded in lysosomes, thereby freeing the genetic material. This has now been observed in the case of poxviruses, adenoviruses, and reovirus. On the other hand the fate of the smaller animal viruses such as the Picornaviruses cannot be followed by ultrastructural techniques. Tailed bacteriophage possess a highly developed specialized mechanism for ejection of their genetic material into bacteria, but the spherical RNA bacteriophage and the rod-shaped and spherical bacteriophage having single-stranded DNA do not have a specialized ejection mechanism. Plant viruses are transmitted in a mechanical fashion by insect vectors. In some cases, the viruses are injected into phloem cells through the insect's stylet (Bradley, 1966).

Once the viral genome is released from the protective capsid, the actual replicative process commences. It is necessary to consider RNA and DNA viruses separately.

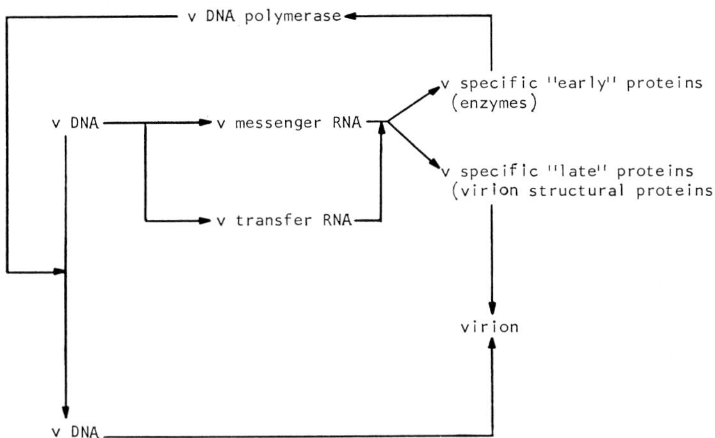

Figure 1. Generalized scheme for the molecular events occurring during the replication of a DNA virus. In the synthesis of viral DNA, there may be some complicated intermediate structures (replicative forms). This generalized scheme is valid for both double-stranded and single-stranded DNA viruses.

The DNA virus genome must replicate and the genetic information must also be translated via viral-specific messenger RNA in order to control the synthesis of viral-specific proteins (Fig. 1). Several types of viral-specific proteins can be postulated:

a) DNA-polymerase. It may be advantageous for a virus to have a specific DNA polymerase since viral DNA may multiply in a part of the cell which usually does not have DNA polymerase or the viral DNA may be quite different from cellular DNA and therefore may not serve as a primer for cellular DNA polymerase. One example of a viral-specific DNA polymerase is that induced in *E. coli* after infection with T-even bacteriophage (Kornberg, 1960).

b) Enzymes Involved in the Synthesis of Deoxyribonucleotides. In some cases, viral specific enzymes are produced and these may give some advantage to viral DNA replication during the period of rapid viral DNA synthesis. In the classic case of T-even bacteriophage, there are many viral specific enzymes concerned with the synthesis of 5-hydroxymethyl cytosine, a base found only in some viral DNA.

c) Enzymes Involved in Protein Synthesis. Particularly activating enzymes should be considered in cases where there may be new species of transfer RNA (Subak-Sharpe, Shepherd, and Hay, 1966).

d) Enzymes Which May Lyse the Cell. Lysozyme-like enzymes are often found in bacteria infected with DNA bacteriophage.

e) The Various Structural Proteins Which Comprise the Viral Coat. There may be one or several proteins comprising the capsid and other viral structures such as the tail of some bacteriophage.

Besides this class of macromolecules, we must consider viral-specific messenger RNA, transfer RNA's, and also the viral DNA which may have a unique structure.

The synthesis of viral-specific proteins implies the synthesis of viral-specific messenger RNA. This has been demonstrated for some animal viruses, such as poxvirus (cf. Joklik, 1966). In the case of T-even bacteriophage, different messenger RNA molecules are synthesized in the cell at different times after infection. This is at least one of the mechanisms whereby viral-specific protein synthesis is regulated.

The DNA of many viruses differs in nearest neighbour base sequence from that of the host cell (Subak-Sharpe *et al.*, 1966; Morrison *et al.*, 1967). Those which have now been studied include adenovirus 2, Pseudorabies, Equine rhinopneumonitis, Herpes simplex, and vaccinia. These are all relatively large DNA viruses with DNA molecules ranging in size from 23×10^6 to 160×10^6 daltons. In order to stimulate rapid rates of protein synthesis, it has been postulated that such viruses may be required to synthesize one or several new transfer RNA species which would correspond to the code of the particular DNA. The presence of different code words in viral DNA has been suggested by the very different nearest neighbour frequencies which have been found. At least one new transfer RNA has been demonstrated, that for arginine in cells infected with Herpes simplex virus (Subak-Sharpe, Shepherd, and Hay, 1966).

Some viruses containing DNA with nearest neighbour frequencies like those of host cell DNA have nucleic acid which differs structurally from that of the host cell. This viral DNA is circular and is present in the form of a supercoil. Examples are the virions of the papova group of animal viruses and replicative forms of bacteriophage Φ X 174 (Vinograd and Lebowitz, 1966).

Thus, there are many points of difference between viral and cellular DNA and between other cellular macromolecules and those synthesized in the DNA virus-infected cell.

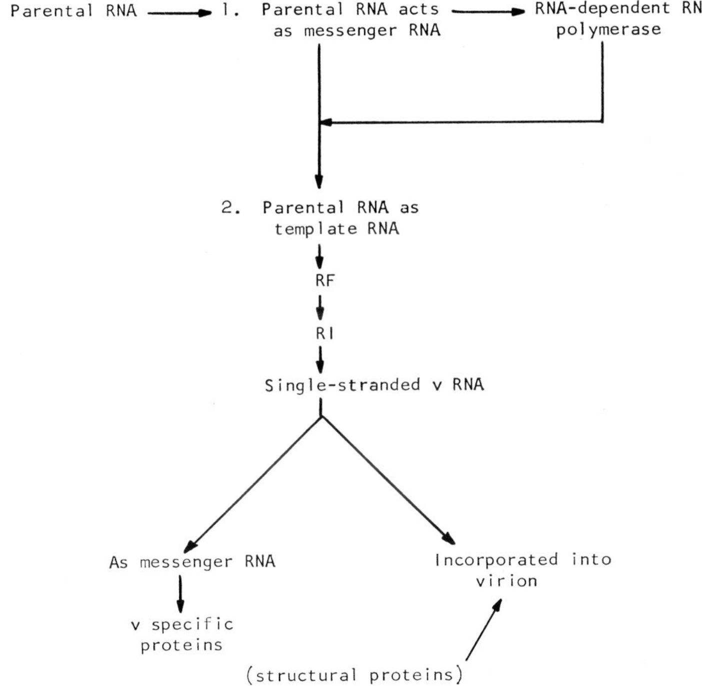

Figure 2. Generalized scheme for the replication of a RNA virus. Whereas most aspects of the scheme may be valid for all RNA viruses, the molecular details of viral RNA replication may differ for different groups of RNA viruses. In the case of reovirus, where the parental RNA is double-stranded, there could be an initial synthesis of single-stranded messenger RNA.

RNA viruses also control the synthesis of new proteins (Fig. 2). The viral coat protein may consist of one or several polypeptides. Besides this, there is at least one, perhaps two, enzymes involved in the synthesis of viral RNA. These form a class of enzymes known as RNA-dependent RNA polymerases. Most RNA viruses are small and contain relatively small amounts of genetic material $(1 - 3 \times 10^6$ daltons of RNA). In such cases, no more than 10 different polypeptides could be synthesized from the genetic information. A few groups of viruses such as those of the para-influenza group, and also reovirus, may have from 10 - 40×10^6 daltons of RNA and in these cases a larger number of viral-specific proteins may possibly be found in infected cells. Much work is still to be done in identifying these viral specific proteins.

The replicative mechanism of most RNA viruses involves a double-stranded RNA template (Erikson and Franklin, 1966). Parental RNA must first serve as messenger RNA to code for the synthesis of at least one type of RNA-dependent RNA polymerase. Once such an enzyme is made, it in turn catalyzes the polymerization of RNA using the parental or "+" RNA as template. This results in the synthesis of a RNA molecule complementary to the parental strand and therefore designated as "−" strand. The resulting double-stranded RNA is known as replicative form (RF). This structure may be used as a template for further synthesis of "+" strands, now utilizing the "−" strand as template. It is possible to isolate RF in the process of synthesizing new "+" strands. This complicated structure is partially single-stranded and partially double-stranded and consists of the double-stranded template with bound nascent "+" strands. In the infected cell, the structure is far more complicated, since polymerase must be associated with the growing point of the polynucleotide chain and the entire complex could be associated with polysomes or with other cellular structures. Thus, in cells infected with RNA viruses, one may find viral RNA of different size and physical properties from cellular RNA species, double-stranded RNA, and a complex containing replicative intermediate.

Although the mechanism of replication of RNA viruses outlined above may be valid for most RNA viruses, other mechanisms of replication must be considered for certain groups of viruses such as the para-influenza group, the avian tumor viruses, and the reoviruses. But even in these cases, the viral RNA may be of different size or structure from cellular RNA species. In the case of reovirus, both the double-stranded RNA of the virion and also single-stranded viral-specific RNA has been extracted from infected cells (Kudo and Graham, 1965).

As already mentioned, inhibitors of viral RNA synthesis can be considered potential chemotherapeutic agents only if they exhibit some selectivity. For example, actinomycin D is a very effective inhibitor of poxvirus synthesis (Reich *et al.*, 1962) but is a non-selective inhibitor of DNA-dependent RNA synthesis and is therefore highly toxic for cells and organisms. On the other hand, isatin-β-thiosemicarbazone is a highly specific inhibitor of the synthesis of certain DNA viruses such as poxviruses. This inhibitor only affects the synthesis of the proteins made late in infection in the case of poxvirus (Woodson and Joklik, 1965). It is already a useful chemotherapeutic agent for the treatment of smallpox whereas actinomycin D is of no value in treating the disease.

In considering the general schemes for synthesis of RNA and DNA viruses, several possibilities are evident for specific inhibition of the viral products (Figs. 3 and 4). But although the viral macromolecules are often different from normal cellular macromolecules, it must be remembered that, in general, the metabolic pathways leading to synthesis of monomer units (i.e., amino acids, nucleoside triphosphates) are no different from those in uninfected cells. In many cases, they may be exactly the same pathways, utilizing cellular enzymes. In other cases, similar pathways may be used but with viral-specific enzymes. An example of this latter case is the poxvirus-specific thymidine kinase (Kit *et al.*, 1962; McAuslan and Joklik, 1962).

Inhibitor specificity can occur even in normal or near-normal pathways. Several possible mechanisms can be suggested for such specificity:

1) In cases where there are viral-induced enzymes similar to pre-existing cellular enzymes, the physical and chemical properties of the viral-induced enzyme may be different from those of the normal one. The poxvirus thymidine kinase may be an example of such a case. There is a higher affinity of thymidine for poxvirus thymidine kinase than for cellular thymidine kinase (McAuslan, 1963; Salzman *et al.*, 1965) and this may be a factor in

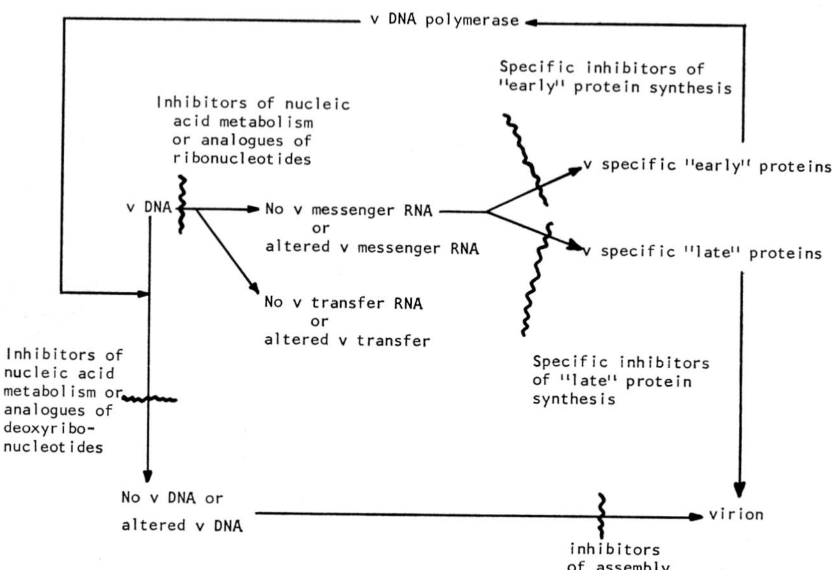

Figure 3. Some possible points in the synthesis of a DNA virus which may be susceptible to specific inhibitors.

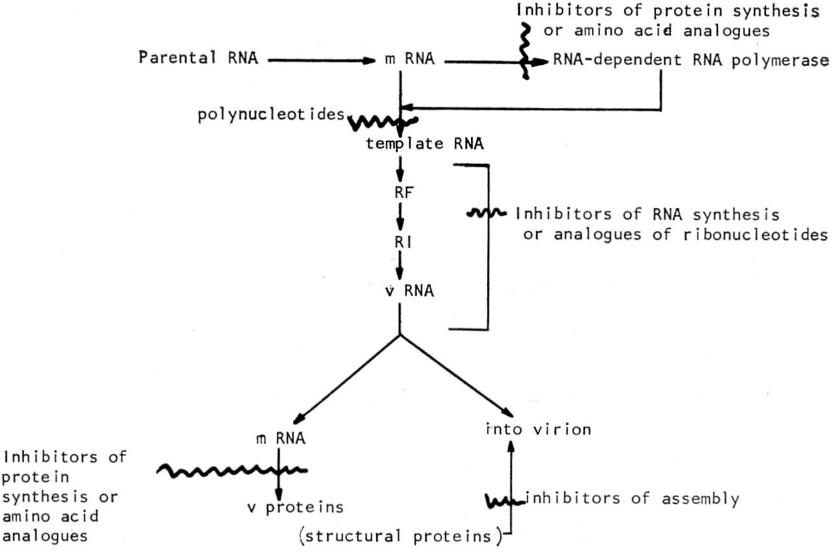

Figure 4. Some possible points in the synthesis of a RNA virus which may be susceptible to specific inhibitors.

enhancing the incorporation of 5-bromodeoxyuridine and 5-iodo-deoxyuridine into viral DNA, since these thymidine analogues must be phosphorylated prior to incorporation.

2) Synthesis may occur in an unusual part of the cell and a particular inhibitor may be confined to that part.

3) On the level of the organism, synthesis may be confined to a particular organ or region which may be particularly easy to treat without obtaining high levels of inhibitor in the body. Such a case is the treatment of Herpes keratitis with 5-iododeoxyuridine (Kaufman *et al.*, 1962).

One example of specific inhibition due to unique macromolecules involved in viral synthesis may be the inhibition of *in vitro* viral RNA synthesis by polynucleotides (Haruna and Spiegelman, 1966). The *in vitro* synthesis of bacteriophage Qβ RNA is inhibited by polynucleotides of either A or U (poly A or poly U) but not by polymers containing predominantly C or G. This may be due to a specific requirement for viral RNA synthesis in which a RNA-dependent RNA polymerase interacts with both ends of the viral RNA as an initial step. The effectiveness of poly A or poly U may be explained by the presence of A, U sequences near the ends of the viral RNA molecule.

Thus our present understanding of the molecular details of viral replication should open up vast new possibilities for chemotherapeutic attack on virus disease.

LITERATURE CITED

1. ALLISON, A. C., and VALENTINE, R. C.: Virus particle adsorption. II. Adsorption of vaccinia and fowl-plague viruses to cells in suspension. *Biochim. Biophys. Acta, 40:*393-399, 1960.

2. ANDERER, F. A., and RESTLE, H.: Untersuchungen über ein attenuiertes Poliomyelitis-Virus Type III, Reindarstellung und physikalische-chemische Eigenschaften des Virus. *Z. Naturforschg., 19b:*1026-1031, 1964.

3. BRADLEY, R. H. E.: Which of an aphid's stylets carry transmissible virus? *Virology, 29:*396-401, 1966.

4. BRINTON, C. C., JR., GEMSKI, P., JR., and CARNAHAN, J.: A new type of bacterial pilus genetically controlled by the fertility factor of E. coli K 12 and its role in chromosome transfer. *Proc. Natl. Acad. Sci., (U.S.), 52:*776-783, 1964.

5. CRAWFORD, E. M., and GESTELAND, R. F.: The adsorption of bacteriophage R-17. *Virology, 22:*165-167, 1964.

6. DALES, S.: Penetration of animal viruses into cells. *Prog. Med. Virol., 7:*1-43, 1965.

7. ERIKSON, R. L., and FRANKLIN, R. M.: Symposium on replication of viral nucleic

acids. I. Formation and properties of a replicative intermediate in the biosynthesis of viral ribonucleic acid. *Bact. Rev., 30:*267-278, 1966.

8. FRANKLIN, R. M.: Replication of bacteriophage RNA: Some physical properties of single-stranded, double-stranded, and branched viral RNA. *J. Virology, 1:*64-75, 1967.

9. HARUNA, I., and SPIEGELMAN, S.: Selective interference with viral RNA formation in vitro by specific inhibition with synthetic polyribonucleotides. *Proc. Natl. Acad. Sci., (U.S.), 56:*1333-1338, 1966.

10. HOLLAND, J. J., and McLAREN, L. C.: The location and nature of enterovirus receptors in susceptible cells. *J. Exper. Med., 114:*161-171, 1961.

11. JOKLIK, W. K.: The purification of four strains of poxvirus. *Virology, 18:*9-18, 1962.

12. JOKLIK, W. K.: The poxviruses. *Bact. Rev., 30:*33-66, 1966.

13. KAUFMAN, H. E., NESBURN, A. B., and MALONEY, E. D.: IDU therapy of herpes simplex. *Arch. Ophth., 67:*583-591, 1962.

14. KIT, S., DUBBS, D. R., and PIEKARSKI, L. J.: Enhanced phosphorylating activity of mouse fibroblasts (strain L-M) following vaccinia infection. *Biochem. Biophys. Res. Commun., 8:*72-75, 1962.

15. KORNBERG, A.: Biological synthesis of the deoxyribose nucleic acid of T2 bacteriophage. *Science, 131:*1503-1508, 1960.

16. KUDO, H., and GRAHAM, A. F.: Synthesis of reovirus ribonucleic acid in L cells. *J. Bact., 90:*936-945, 1965.

17. McAUSLAN, B. R.: The induction and repression of thymidine kinase in the poxvirus-infected HeLa cell. *Virology, 21:*383-389, 1963.

18. McAUSLAN, B. R., and JOKLIK, W. K.: Stimulation of the thymidine phosphorylating system in HeLa cells on infection with poxvirus. *Biochem. Biophys. Res. Commun., 8:*486-491, 1962.

19. MORRISON, J., KEIR, H., SUBAK-SHARPE, H., and CRAWFORD, L. V.: Nearest neighbor base sequence analysis of the deoxyribonucleic acids of a further three mammalian viruses: simian virus 40, human papilloma virus and adenovirus type 2. *J. Gen. Virology, 1:*101-108, 1967.

20. REICH, E., FRANKLIN, R. M., SHATKIN, A. J., and TATUM, E. L.: Action of Actinomycin on animal cells and viruses. *Proc. Natl. Acad. Sci., (U.S.), 18:*1238-1245, 1962.

21. SALZMAN, N. P., SHATKIN, A. J., and SEBRING, E. D.: The use of metabolic inhibitors in studies on the mode of replication of vaccinia virus. *Annals New York Acad. Sci., 130:*240-248, 1965.

22. SUBAK-SHARPE, H., BÜRK, R., CRAWFORD, L., MORRISON, J., HAY, J., and KEIR, H.: An approach to evolutionary relationships of mammalian DNA viruses through analysis of the pattern of nearest neighbor base sequences. *Cold Spring Harbor Symp. Quant. Biol., 31:*737-748, 1966.

23. SUBAK-SHARPE, H., SHEPHERD, W. H., and HAY, J.: Studies on sRNA coded by herpes virus. *Cold Spring Harbor Symp. Quant. Biol., 31:*583-594, 1966.

24. TAMM, I., and EGGERS, H.: Selective inhibition of viral reproduction. In *Viral and Rickettsial Infections of Man, 4th Ed.* Edited by F. L. HORSFALL, JR., and I. TAMM. Philadelphia, Lippincott, pp. 305-338, 1965.

25. VINOGRAD, J., and LEBOWITZ, J.: Physical and topological properties of circular DNA. *J. Gen. Physiol.,* In press.

26. WOODSON, B., and JOKLIK, W. K.: The inhibition of vaccinia virus multiplication by isatin-β-thiosemicarbazone. *Proc. Natl. Acad. Sci., (U.S.), 54:*946-953, 1965.

ALTERATIONS IN CELLULAR BEHAVIOR DURING ESTABLISHMENT OF ANTIVIRAL RESISTANCE

KURT PAUCKER AND MARTHA BOXACA

INTRODUCTION

A NUMBER OF RECENT OBSERVATIONS have suggested that exposure of tissue cultures to interfering agents may alter cellular activities in various ways. Thus, fragments of chorioallantoic membrane which had sustained prior contact with heated influenza virus were capable of producing interferon in response to live influenza virus at a time when no interferon would have been detectable in the absence of pretreatment (2, 8). On the other hand, cessation of interferon production induced by irradiated viruses caused cultures of L-cells (3) and of chick embryo fibroblasts (1) to resist restimulation of interferon production. The refractory condition thus observed was only transitory, and it was lost rapidly in dividing cultures of cells, well in advance of protection (3). However, in the absence of division, non-responsiveness to induction of interferon (Paucker and Boxaca, personal communication) as well as resistance to viral infection (9) appear to endure throughout the life of the cultures. Furthermore, the growth rate of resistant cultures of mouse fibroblasts, whether persistently infected with passenger viruses (6) or exposed to inactivated interfering virus (20) can be markedly reduced, although a state of antiviral resistance is not necessarily associated with a slow rate of division as compared with that of susceptible cells (15).

Only scant information is available on the question whether non-viral interfering agents, capable of functioning in tissue culture, bring forth similar changes in cellular behaviour. Statolon, an anionic polysaccharide which produces interferon in cultured cells (12) has been shown to depress the cellular response to subsequent stimulation of interferon formation by either the same

(11) or another inducer (17). Additional findings regarding the effects of statolon on cell cultures have recently accumulated in this laboratory, and they are included in this chapter.

PROTECTION AND PRODUCTION OF INTERFERON IN L-CELLS TREATED WITH VIRAL AND NONVIRAL INTERFERING AGENTS

A series of earlier publications may be consulted for all pertinent technical details regarding experimentation with suspended cultures of L-cells in studies on viral interference, interferon synthesis, and interferon action (3, 4, 18-20). As interfering agents, ultraviolet-irradiated Newcastle disease virus (NDVuv), at an input multiplicity of 100 EID_{50} before inactivation, and statolon, at a concentration of 20 microgram of active component per milliliter, were employed. These levels were found optimal under the prevailing experimental conditions. Vesicular stomatitis virus (VSV) was used throughout as challenge virus, either in the plaque-inhibition assay or by means of an immunofluorescent method to be described elsewhere (Boxaca and Paucker, in preparation).

In order to compare the rates at which protection against viral infection is established, cultures of L-cells were exposed to interfering preparations of NDVuv and of statolon. At periodic intervals, samples of cell suspension were withdrawn, inoculated with VSV, and infectible cells were enumerated 6 hours later in smears which had been stained with antiviral fluorescent gamma globulin. The results, presented in Figure 1, show that NDVuv was considerably more efficient than statolon in establishing protection in these cultures. A significant reduction in the number of fluorescent cells was noticeable already after 3 hours of incubation, and it increased linearly until at least the 9th hour when more than 99 per cent of the initial susceptible cell population was protected against VSV. By 24 hours, protection was reinforced further although the exact course of events was not followed in any detail. In cultures subjected to statolon, antiviral resistance did not become apparent until 6 hours after initial contact with the compound, increased progressively until about the 16th hour, and then slowed down until it was maximal after 22 hours had elapsed.

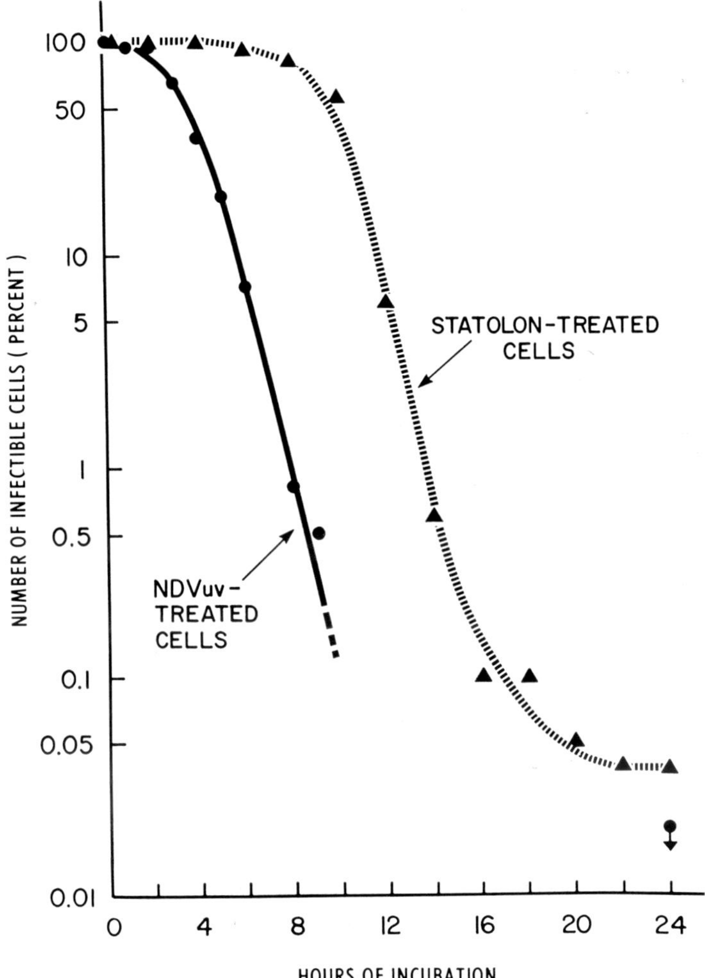

Figure 1. Development of protection against vesicular stomatitis virus in L-cells treated with irradiated Newcastle disease virus (NDVuv) or with statolon.

At some time during the establishment of protection, L-cells begin to release interferon into the medium. As seen in Figure 2, this event occurs also faster in NDVuv- than in statolon-treated cultures. In the first instance, interferon becomes detectable after 3 hours of incubation, and it increases rapidly until its final concentration is attained by the 18th hour. Statolon-induced interferon, on the other hand, becomes first measurable only after 14 hours in culture, and it reaches its ultimate level, which is generally low, some 10 hours later.

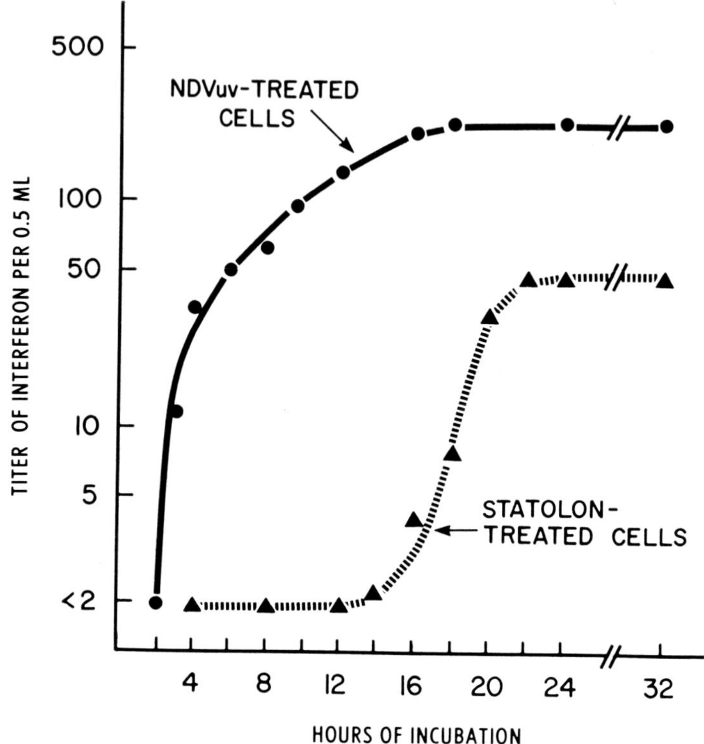

Figure 2. Production of interferon in L-cells treated with irradiated Newcastle disease virus (NDVuv) and with statolon.

It was interesting to note that in cell cultures exposed to NDVuv the advents of protection and interferon release were not separated in time. However, in statolon-treated cells, as shown in Figure 3, interferon did not become detectable until the time when the cultures had become nearly totally resistant to superinfection by challenge virus. In fact, it was noted on a number of occasions, especially with statolon preparations which had been stored for some time, that production of interferon could be irregular or be absent altogether. However, this erratic effect did not influence the regularity with which antiviral resistance developed in such cultures. The question would, therefore, seem warranted whether in this system at least interferon formation is prerequisite for the establishment of protection and whether these cellular events engendered by the same agent are more than coincidentally related.

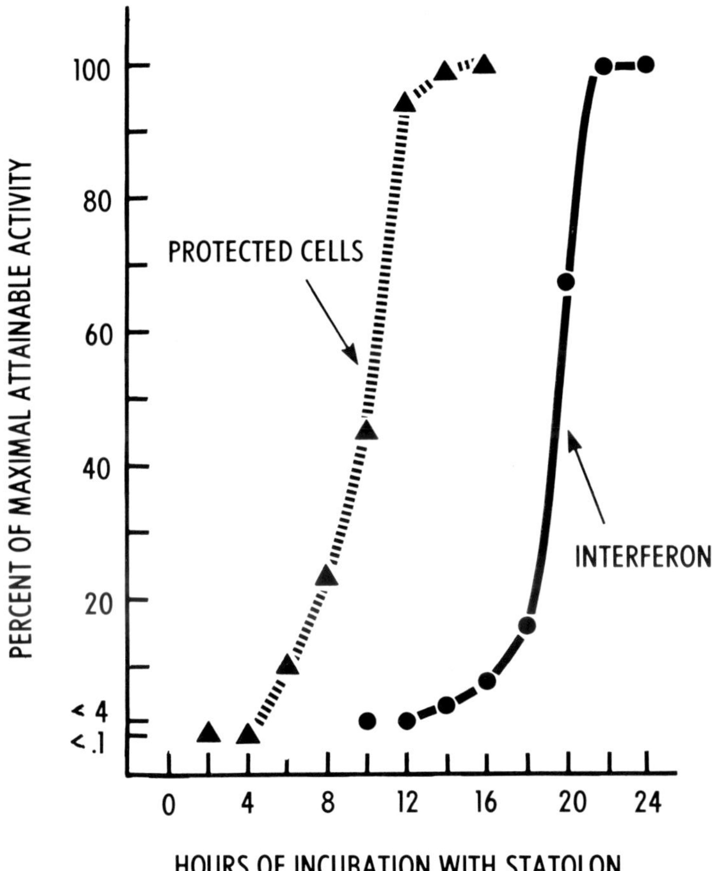

Figure 3. Development of protection and interferon production in L-cells treated with statolon.

ENHANCED RESPONSE TO INDUCTION OF INTERFERON

The difference in the speed with which protection and interferon release materialize in virus- and statolon-treated cultures provides a convenient tool to study the effect of multiple exposures of cell populations to interfering agents, unencumbered by complications of receptor destruction and regeneration.

In an initial experiment, an attempt was made to see whether a secondary contact with the same interfering virus, applied at different times following primary stimulation, might not affect the eventual yield of interferon produced. Accordingly, series of suspension cultures of L-cells were exposed to NDVuv and at given time intervals thereafter duplicate tubes were centrifuged and resuspended in fresh medium. One of each pair of cultures was then exposed a second time to NDVuv whereas the other was not handled further. Media were collected after a 2-hour interval, and the interferon content was determined by conventional plaque-inhibition assay. The data presented in Figure 4 show that the accumulation of interferon was identical in both groups and that neither the appearance nor the cessation of interferon production was influenced by the second dose of NDVuv.

It could not be decided, however, whether at least in the early part of the experiment destruction of cell receptors may have

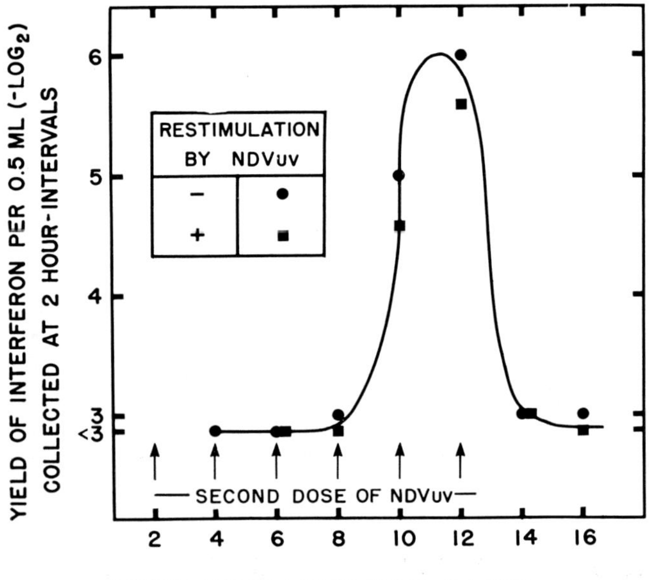

Figure 4. The effect of restimulation by NDVuv on the production of interferon in L-cells. Each square symbol represents the accumulation of interferon in cultures which sustained a second dose of NDVuv 2 hours earlier.

rendered the second dose ineffectual. The experiment was, therefore, modified by placing groups of L-cell cultures first in contact with an appropriate dose of statolon. At intervals thereafter, 2 tubes were withdrawn, centrifuged, resuspended in fresh medium, and one tube of each pair was inoculated with NDV. Twenty-four hours from the time of restimulation, media were collected and titers of interferon were determined. The results, presented in Figure 5, show that during the first 6 hours after contact with statolon the quantities of interferon produced by NDV were of the order of those obtained in controls. However, as the time interval between the application of statolon and NDV was extended, increasingly higher titers of interferon were stimulated by

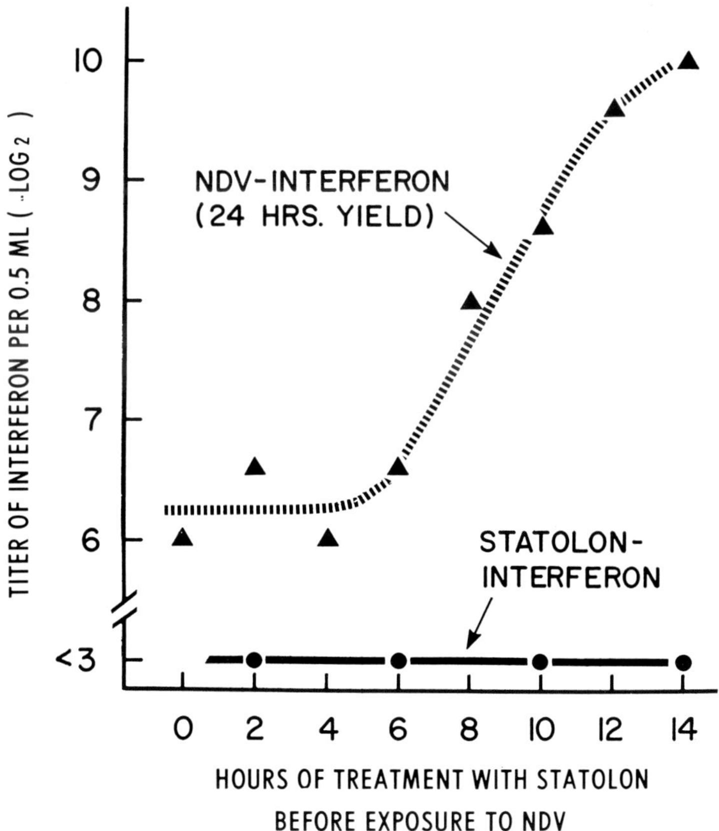

Figure 5. The enhanced response of statolon-treated L-cells to induction of interferon by Newcastle disease virus (NDV).

the virus. Thus, a pretreatment period of 14 hours with statolon caused the cells to produce 16-fold higher yields of interferon in response to a second inoculation of NDV than the amount elicited in untreated controls. The level of interferon which accumulated in cultures which received statolon alone never exceeded 5 logs$_2$ after 24 hours of incubation (not shown in the figure) and, therefore, did not interfere with the interpretation of the results.

The demonstration of enhanced yields of interferon from cultures pretreated with statolon prompted further experiments to determine whether the interferon thus obtained appeared also earlier than in control cultures. Inasmuch as contact with statolon for 14 hours gave rise to a considerably increased yield of interferon, this interval was chosen for treatment of the cultures in the next experiment. After removal of residual statolon by centrifuga-

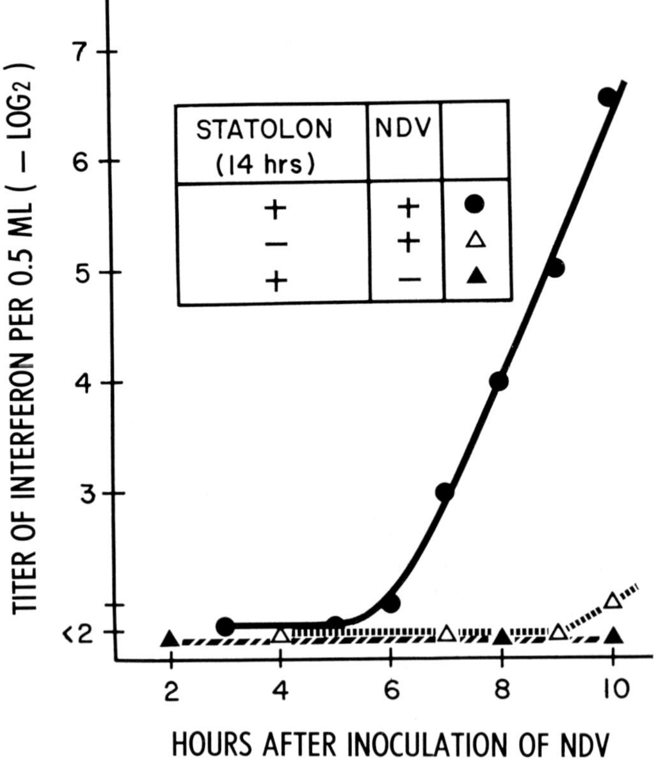

Figure 6. Early appearance of Newcastle disease virus (NDV)-induced interferon in L-cells pretreated with statolon.

tion and resuspension of the cells in fresh medium, one portion of the cultures was inoculated with NDV and the other was retained as control. A third group of tubes had received no statolon and was inoculated with NDV at the same time as pretreated cultures. At hourly intervals during the next 10 hours, media were sampled and assayed for interferon. The results are given in Figure 6. Because the cell density of the suspension culture was low in this experiment, NDV-induced interferon became detectable in the control group only after 10 hours of incubation and statolon-produced interferon remained below detectable concentrations throughout. However, in cells pretreated with statolon and subsequently exposed to NDV, interferon appeared after only 6 hours and its titer rose rapidly until by the 10th hour it was more than 20 times higher than in the corresponding control group.

It was next determined whether the enhanced response to induction of interferon observed in statolon-treated cultures could be separated from the cells which exhibit this behaviour. Accordingly, media from such cultures after residual statolon had been eliminated, as well as cell extracts, were examined for the transferability of this property. In the experiment presented in Figure 7, suspension cultures were exposed for varying lengths of time to media and extracts derived from cultures which had sustained contact with statolon during 14 hours previously. At the intervals indicated in the figure, all groups as well as controls which had not been subjected to any treatment, were inoculated with NDV and interferon production was measured after an additional incubation period of 24 hours. The results of the titrations were expressed as ratios of NDV-induced interferon in treated cells to the amount found in controls. The horizontal broken line (ratio of 1), representing the level of interferon in untreated cells, has been included for comparison. The most pronounced enhancement effect was noted when the exposure to media and extracts was restricted to 2 hours. On further incubation, the differences in titer became gradually less and were eventually absent altogether. While there was no ready explanation for this seemingly contradictory behaviour, it appeared possible that the materials used for treatment of the cell cultures harbored also some blocking activity which might counteract the enhancing effect. Moreover, it was already known from earlier studies (Paucker and Boxaca, personal communication) that a prolonged contact with cells is required for the manifestation of such blocking action. To test for this possibility, groups of L-cell cultures were pretreated for 24 hours with progressive dilutions of enhancing medium derived from cells which had been incubated with statolon for 14 hours. Upon termination of the period of treatment, the cells were exposed to NDV and after an additional incubation time of 24 hours the interferon content was determined. The yields were compared with those elicited by NDV in normal controls, and the data, shown in Figure 8, were computed as in the preceeding

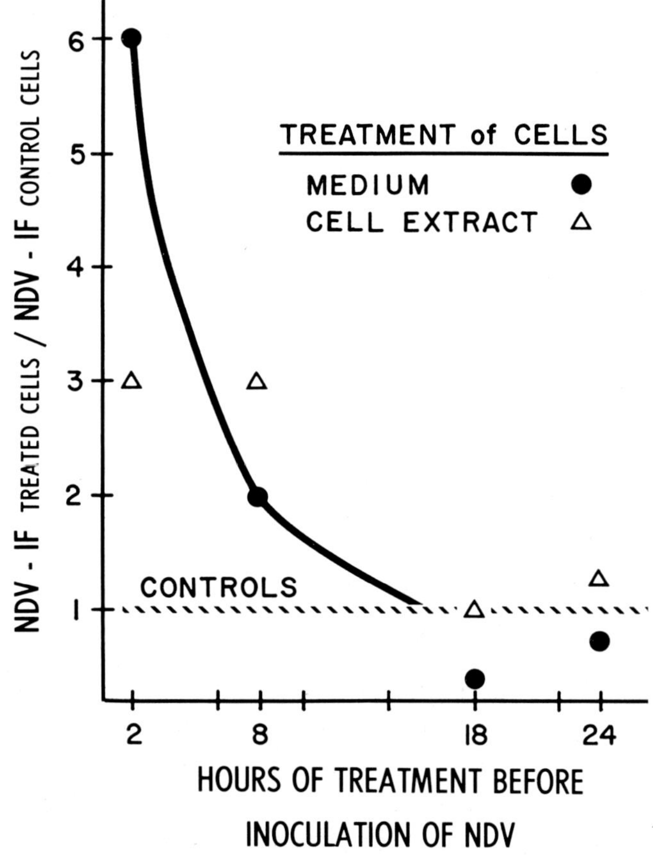

Figure 7. Transfer of enhancing activity by media and extracts from statolon-treated L-cells.

experiment. It can be seen that no increased yields of interferon were obtained on pretreatment of the cells with the undiluted medium. However, a 4-fold increase in interferon titer was obtained on exposure of cells to the 30-times diluted medium, but on further dilution the enhancing effect gradually diminished.

There is little additional information available on the factors responsible for the enhanced response of cells to the induction of interferon. Media and extracts from statolon-treated cultures which exhibit this property are, when collected at the proper time interval, devoid of interferon detectable in the plaque-inhibition assay. However, some interfering activity can usually be demonstrated in the fluorescent test but it is of low magnitude. Thus, it has not been possible so far to separate the enhancement property entirely from any remnant of interferon activity. It re-

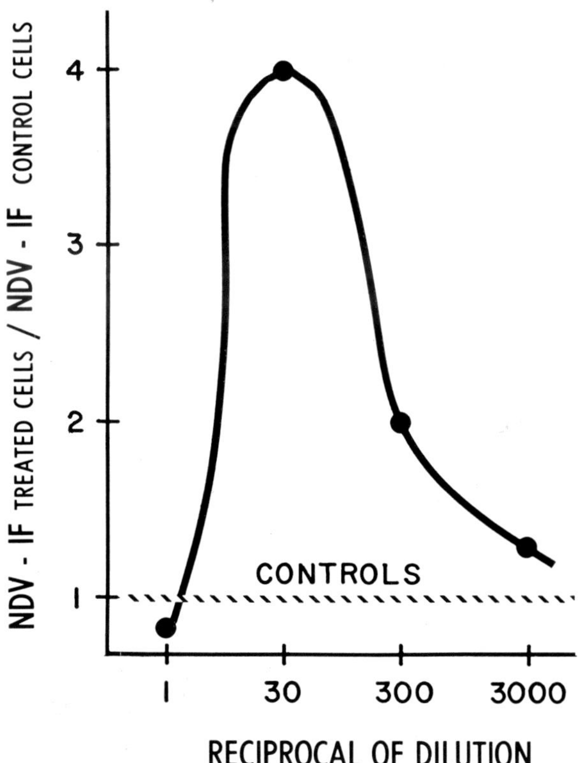

Figure 8. Titration of enhancing activity in media from statolon-treated L-cells.

sists treatment at low pH, RNase, DNase, and it is not sedimented for 2 hours at 100,000 times gravity. Like interferon, it is digested by proteolytic enzymes.

RESISTANCE TO INDUCTION OF INTERFERON

The production of interferon is in itself a process which continues during a limited period of time the extent of which depends on the type of stimulus applied as well as on the system used for study. In L-cells exposed to irradiated NDV, interferon titers reach a peak after about 18 to 20 hours in culture. Subsequently, formation of interferon gradually diminishes and ceases entirely by the time 30 hours have elapsed (17). The abatement of synthesis of interferon is associated with a declining response of the cells to restimulation of additional interferon production and leads eventually to a totally refractory condition of the cultures (17).

A similar state of resistance to secondary stimulation of interferon production by NDVuv is found in cultures initially exposed to statolon. The development of this condition was studied in the following experiment. Statolon was incorporated into a medium of 2 groups of cultures whereas a third group was maintained without addition of the compound. At the intervals indicated in Figure 9, one of the statolon-treated and the control group were inoculated with NDV and titers of interferon were measured after an additional 24 hours of incubation. The ratios between the titers attained in the 2 groups have been recorded in the figure. The second series of statolon-exposed tubes was used to determine the accumulation of interferon in the medium in the absence of further restimulation. The titers in this group were negative up to the 14th hour and had attained a maximal level after 24 hours of incubation. In the doubly treated group, increasingly higher yields of interferon were found when exposure of the cells to statolon was extended up to 14 hours. After 24 hours of treatment with statolon, the enhancement effect was less pronounced although there was at that time still a 4-fold higher titer of interferon stimulated in this than in the control group. However, when incubation in the presence of statolon was permitted to proceed for 48 hours, the cultures were totally refractory and no longer responded to restimulation of interferon production. It is apparent from these results that the enhancement of interferon synthesis represents an early manifestation of the effect of statolon on L-cells, whereas the nonresponsiveness to interferon induction develops late, at a time when primary stimulation has exercized its full effect.

As was shown previously for the enhancement phenomenon, a factor capable of blocking interferon synthesis eventually accum-

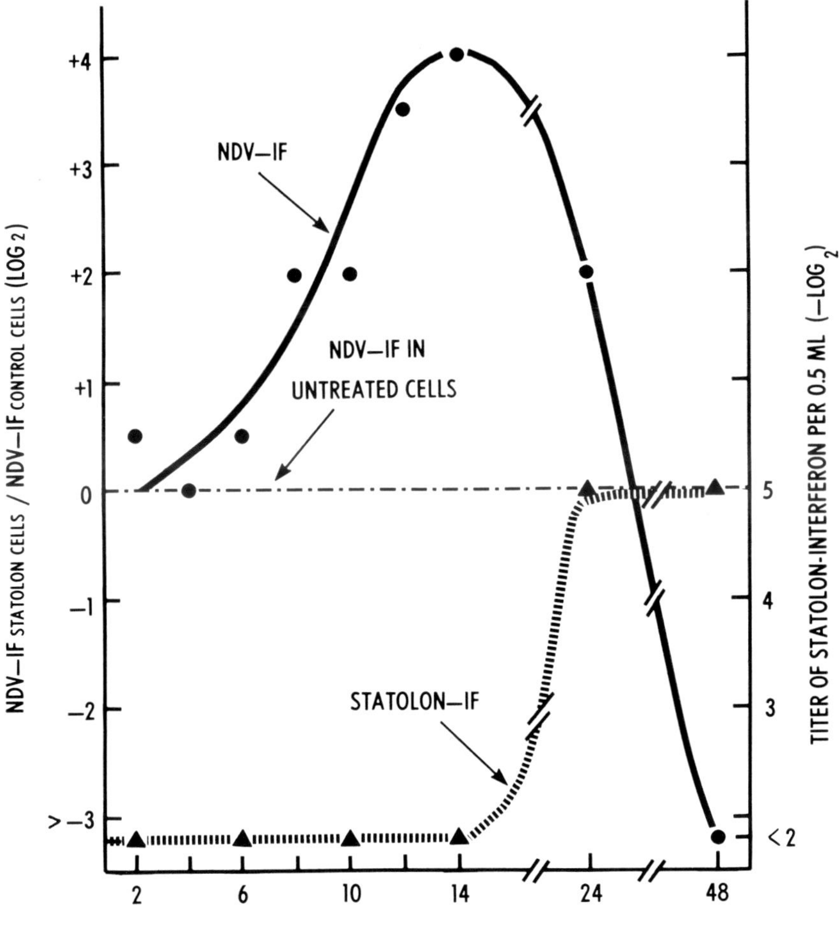

Figure 9. Restimulation of interferon production by NDV in L-cells treated with statolon.

ulates, together with interferon activity, in the medium of cell cultures which had been subjected to interfering viruses (3, 14, 21, 22). Figure 10 illustrates an experiment where cultures of L-cells were exposed to varying concentrations of an interferon preparation which had been obtained in L-cells by the action of NDV. At the end of a 24-hour incubation period, all cultures as well as controls were inoculated with NDVuv and sampled at hourly intervals for the first 6 hours and again at 24 hours. The results show that treatment of the cells with the 2 largest doses of interferon prevented completely any further synthesis of interferon on attempted restimulation. The blocking effect of the

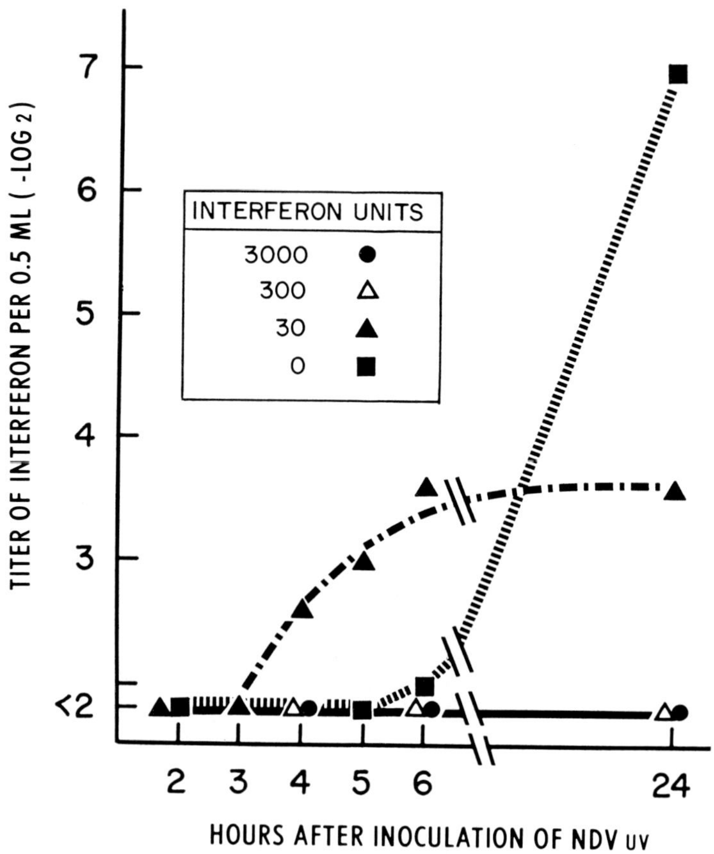

Figure 10. Suppression of interferon production in L-cells pretreated with varying doses of interferon for 24 hours.

smallest dose was partial and permitted some interferon to be produced which, in fact, appeared earlier than in the corresponding controls.

The blocking activity contained in interferon preparations was at first associated with the interferon moiety itself, but a number of observations suggested that these properties may reside in different components of the crude materials studied (16, Paucker and Boxaca, personal communication). Thus, it was observed that a considerably larger dose of interferon was required for the blocking of subsequently induced interferon production than for protection. Furthermore, resistance to viral superinfection was acquired earlier than nonresponsiveness to interferon induction. Removal of interferon from the culture medium at appropriately selected time intervals permitted the development of protection against challenge virus to proceed but in the absence of interferon in the medium, such cells failed to become refractory. Moreover, interferon preparations which had been subjected to substantial purification were largely devoid of blocking action, but attempts at locating the lost property have not yet met with unqualified success. On the other hand, the blocking action, like interferon itself, was readily neutralized by anti-interferon serum, and it operated only in the homologous species.

According to a recent report (10), allantoic fluids of embryonated eggs and media from cultures of chick embryo fibroblasts

TABLE I

ATTEMPTS AT SELECTIVE INACTIVATION OF ANTIVIRAL AND BLOCKING ACTIVITIES
IN CRUDE L-CELL INTERFERON (IF)

| | Treatment of IF[a] | Activities in Cells Exposed to IF for 24 Hrs. | | |
| | *Enzyme conc.* | *Protection*[b] | *NDVuv-IF*[c] | |
Type	*mg/ml*	*%*	*per 0.5 ml*	*Blocking*[d]
None (pH 2)	—	99.2	48	+
Trypsin	1.0	0	256	—
Pepsin	1.0	9.0	192	—
RNase	0.01	90.6	48	+
DNase	0.01	91.2	64	+
Lipase	10.0	2.0	256	—
10^5x g; 2 Hrs.	—	91.0	48	+
Control		0	256	

[a]Enzyme treatments at pH 7 except for pepsin (pH 2); 1 hour at 37°C.
[b]Based on enumeration of fluorescent cells.
[c]Twenty-four hour-yield.
[d]Reduction of IF yield in controls by at least 75%.

infected with certain viruses contained a blocking substance which while sharing many of the attributes of interferon was stable in the presence of proteolytic enzymes. Prompted by these findings, a crude interferon material prepared in L-cells was subjected to various treatments as shown in Table I. However, none of the conditions applied resulted in a clear-cut separation between the interferon and blocking activities residing in this preparation. Both were lost on contact with trypsin, pepsin and a crude lipase, whereas neither low pH, RNase, DNase or sedimentation for 2 hours at 100.000 times g had any effect whatsoever. Thus, the evidence that interferon and blocking properties belong to separate entities is still rather tenuous and rests mostly on indirect evidence.

EFFECT OF INTERFERING AGENTS ON CELLULAR GROWTH

In the course of earlier studies, it was noted that interfering preparations of inactivated NDV exerted a profound effect on the rate of division of suspended L-cells (20). On continued incubation in the presence of irradiated virus, cell growth ceased entirely and the cultures were eventually lost. This property could not be dissociated from the virus particles purified by repeated cycles of absorption and elution with chicken red cells. However, it was not certain whether the reduced rate of cellular division was an inevitable outcome of the establishment of viral interference or whether a peculiarity of the interfering virus was involved. Similar observations have now been made in cultures treated with statolon. For as long as the compound remained present in the medium, the cells exhibited a marked tendency toward clumping and failed to divide. However, as shown earlier for irradiated NDV, on removal of statolon, L-cells were found to display a considerably reduced rate of division for a number of days. In the experiment shown in Figure 11, cells which had been incubated in the presence of statolon for 24 hours were resuspended in fresh medium, and the increment in cell numbers was determined, usually at daily intervals. At the times indicated, aliquots of these as well as of cultures which had not sustained contact with statolon, were challenged with VSV to ascertain their status of protection. It can be seen that throughout the period of slow growth lasting here for about 9 days during which approximately 4 cell divisions had intervened, the cells were fully protected against VSV. After that time, the cultures gradually reverted toward a normal rate of growth but a significant degree of antiviral resistance continued to persist throughout subsequent

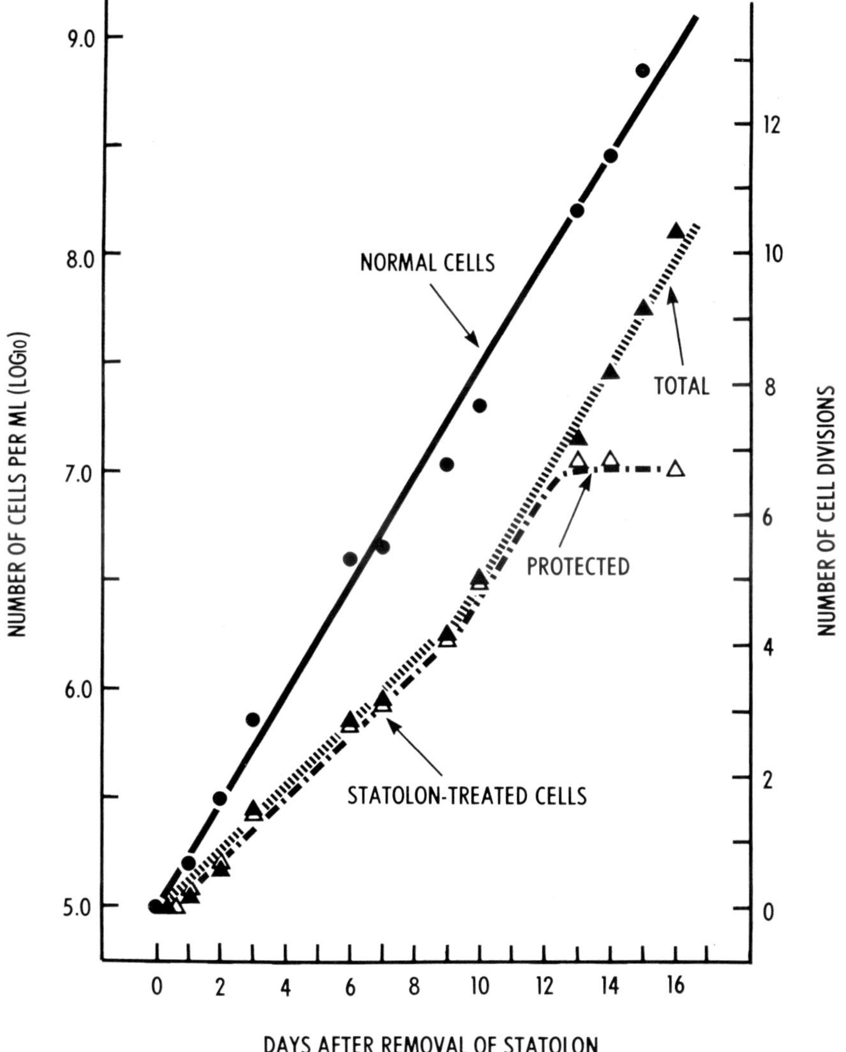

Figure 11. Persistence of protection in statolon-treated cultures of L-cells.

cell generations until the limitations of the assay made further determinations impossible. In the course of this period, the number of protected cells increased more than 100-fold over that initially exposed to statolon. The mode in which resistance to viral infection is transmitted under these conditions is unknown. Previous attempts to detect interferon in the media or extracts from cell concentrates were unsuccessful (18). However, the fact that 2 widely differing interfering agents exhibit similar growth-inhibiting propensities suggests that the observed reduction in the rate of division may involve a general pattern of behaviour in cultures subjected to interferon-inducing stimuli.

CONCLUSIONS

The development of resistance to viral infection in L-cell cultures exposed to viral and nonviral interfering agents is associated with certain changes in cellular behaviour. Concomitant with the establishment of protection, cells incubated with statolon exhibited a gradually enhanced reactivity to secondary stimulation of interferon production. Media and extracts from cultures which possessed this tendency were capable of transferring this property to fresh cells, in the absence of significant levels of interferon. The enhancement effect shared the attributes of interferon with respect to enzyme digestion, pH-stability and behaviour in the ultracentrifuge. It may represent a more sensitive criterion of interferon action than the antiviral effect or belong to an entity separable from interferon. The enhancement property residing in crude virus-induced interferon preparations has been well documented in several reports (5, 13, 14).

Upon completion of interferon release following primary stimulation by inactivated virus (3) as well as by statolon (17), cell cultures are temporarily refractory to attempts at generating additional interferon. This condition as well can be transferred to fresh cultures by means of interferon-containing media (3, 14, 21, 22). Evidence that these 2 activities found in NDV-L cell interferon belong to separate factors is as yet incomplete, but in a different host-virus system blocking activity was dissociated from interferon by digestion with proteolytic enzymes (10).

Earlier reports testified to an association between the inhibition of cell-division and exposure to inactivated interfering virus in the embryonated egg (7) as well as in suspended L-cells (20). These observations have been extended to L-cell cultures exposed to statolon. Division was found to be totally inhibited in the presence of the compound, but on removal from the medium, as previously found in cultures treated with NDVuv (20), the cells resumed a growth rate slower than normal which only after several generations reverted to that encountered in controls. During the period of reduced growth, the cultures were fully protected against challenge virus. Antiviral resistance waned gradually with the resumption of normal growth by which time the number of cells initially exposed to statolon had increased approximately 100-fold. The mode in which protection is transmitted to descendant cells in the absence of detectable interferon (18) is unknown, as is the implication of these findings in relation to the perpetuation of the state of antiviral resistance in the functioning intact organism.

ACKNOWLEDGMENTS

This investigation was supported by grant number AI-02405 from the National Institutes of Health, United States Public Health Service.

Martha Boxaca is the recipient of a fellowship from the Consejo Nacional de Investigaciones Científicas y Técnicas of the Argentine Republic.

The gift of statolon from Dr. W. Kleinschmidt, Lilly Research Laboratories, Indianapolis, Indiana, is gratefully acknowledged.

LITERATURE CITED

1. BURKE, D. C., and BUCHAN, A.: Interferon production in chick embryo cells. I. Production by ultraviolet-inactivated virus. *Virology, 26:*28-35, 1965.
2. BURKE, D. C., and ISAACS, A.: Some factors affecting the production of interferon. *Brit. J. Exper. Path., 39:*452-458, 1958.
3. CANTELL, K., and PAUCKER, K.: Quantitative studies on viral interference in suspended L cells. IV. Production and assay of interferon. *Virology, 21:* 11-21, 1963.
4. CANTELL, K., SKURSKA, Z., PAUCKER, K., and HENLE, W.: Quantitative studies on viral interference in suspended L cells. II. Factors affecting interference by uv-irradiated Newcastle disease virus against vesicular stomatitis virus. *Virology, 17:*312-323, 1962.

5. FRIEDMAN, R. M.: Effect of interferon treatment on interferon production. *J. Immunol., 96:*872-877, 1966.

6. HENLE, G., DEINHARDT, F., BERGS, V. V., and HENLE, W.: Studies on persistent infections of tissue cultures. I. General aspects of the system. *J. Exper. Med., 108:*537-560, 1958.

7. HENLE, W., HENLE, G., and WIENER KIRBER, M.: Interference between inactive and active viruses of influenza. V. Effect of irradiated virus on the host cells. *Am. J. Med. Sci., 214:*529-541, 1947.

8. ISAACS, A.: Viral interference. *Symp. Soc. Gen. Microbiol., 9:*102-121, 1959.

9. ISAACS, A., and WESTWOOD, M. A.: Duration of protective action of interferon against infection with West Nile virus. *Nature, 184:*1232-1233, 1959.

10. ISAACS, A., ROTEM, Z., and FANTES, K. H.: An inhibitor of the production of interferon ("Blocker"). *Virology, 29:*248-254, 1966.

11. KLEINSCHMIDT, W. J., and MURPHY, E. B.: Investigations on interferon induced by statolon. *Virology, 27:*484-489, 1965.

12. KLEINSCHMIDT, W. J., CLINE, J. C., and MURPHY, E. B.: Interferon production induced by statolon. *Proc. Natl. Acad. Sci., 52:*741-744, 1964.

13. LEVY, H. B., BUCKLER, C. E., and BARON, S.: Effect of interferon on early interferon production. *Science, 152:*1274-1276, 1966.

14. LOCKART, R. Z.: Production of an interferon by L cells infected with Western equine encephalomyelitis virus. *J. Bacteriol., 85:*556-566, 1963.

15. PAUCKER, K.: Interference and cell division. *Viruses, Nucleic Acids and Cancer,* Baltimore, Williams and Wilkins, 1963, pp. 430-446.

16. PAUCKER, K.: Cellular resistance to induction of interferon. *Bact. Proc., 66:* 119, 1966.

17. PAUCKER, K., and BOXACA, M.: Nonresponsiveness of L cells to the induction of interferon. *Ann. Med. exp. Fenn., 44:*274-281, 1966.

18. PAUCKER, K., and CANTELL, K.: Quantitative studies on viral inference in suspended L cells. V. Persistence of protection in growing cultures. *Virology, 21:*22-29, 1963.

19. PAUCKER, K., SKURSKA, Z., and HENLE, W.: Quantitative studies on viral interference in suspended L cells. I. Growth characteristics and interfering activities of vesicular stomatitis, Newcastle disease, and influenza A viruses. *Virology, 17:*301-311, 1962.

20. PAUCKER, K., CANTELL, K., and HENLE, W.: Quantitative studies on viral interference in suspended L cells. III. Effect of interfering viruses and interferon on the growth rate of cells. *Virology, 17:*324-334, 1962.

21. VILČEK, J.: Studies on an interferon from tick-borne encephalitis virus-infected cells (IF). IV. Comparison of IF with interferon from influenza virus-infected cells. *Acta Virol., 6:*144-150, 1962.

22. VILČEK, J., and RADA, B.: Studies on an interferon from tickborne encephalitis virus-infected cells (IF). III. Antiviral action of IF. *Acta Virol., 6:*9-16, 1962.

STUDIES ON THE MODE OF ACTION OF TWO ANTIVIRAL AGENTS: ISATIN-β-THIOSEMICARBAZONE AND INTERFERON

WOLFGANG K. JOKLIK, D.Phil. (Oxon)

IF WE KNEW THE MODE of action at the molecular level of even 1% of the drugs in use today for the modification of biological functions or malfunctions, our knowledge of cell biology and thus of life processes would be immeasurably further advanced than it is. Indeed it may be safe to predict that major advances in molecular biology will come through the elucidation of the mode of action of chemical compounds now used as "drugs." A moment's reflection will show the importance already attained by work along these lines. Among natural products, the identification of actinomycin D as a highly specific inhibitor of the transcription of RNA and its subsequent universal use for this purpose has far outweighed its importance as an antibiotic; work on the mode of action of penicillin has led to great advances in our knowledge concerning bacterial cell wall synthesis; and the gradual definition of the mode of action of streptomycin is leading to a greater understanding of ribosome structure and function. Definition of the molecular mode of action of the tetracyclines, cycloheximide and other antibiotics may confidently be expected to aid the elucidation of the mechanism of peptide bond formation. Further, synthetic compounds such as dFU and CAR have turned out to be extremely specific and useful inhibitors of DNA synthesis.

The reasons for our investigation of the mode of action at the molecular level of IBT and interferon in inhibiting the replication of vaccinia virus are therefore clear. I will attempt in this

---

The following abbreviations are used: IBT, isatin-β-thiosemicarbazone; dFU, 5'-fluorodeoxyuridine; CAR, cytosine arabinoside; UR, uridine; SDS, sodium dodecyl sulfate; mRNA, messenger RNA; RSB, reticulocyte standard buffer (10 mM Tris, 10 mM NaCl, 1.5 mM $MgSO_4$, pH 7.4); EB, elementary body; pfu, plaque forming unit.

presentation not only to outline the experimental results which were obtained, but also to indicate the wider implications of the findings as they appear at this time.

THE MODE OF ACTION OF IBT

The anti-poxvirus activity of IBT (Fig. 1) was first reported by Thompson *et al.* in 1953. Subsequent work established the usefulness of IBT in the whole animal and demonstrated that modification of the structure of the parent compound by substituents in various positions modified the antiviral spectrum (Bauer, 1963; O'Sullivan and Sadler, 1961). The N^1-methyl derivative of IBT is currently being used extensively in clinical smallpox control trials in India and Africa under the name of Marboran (Bauer, 1963).

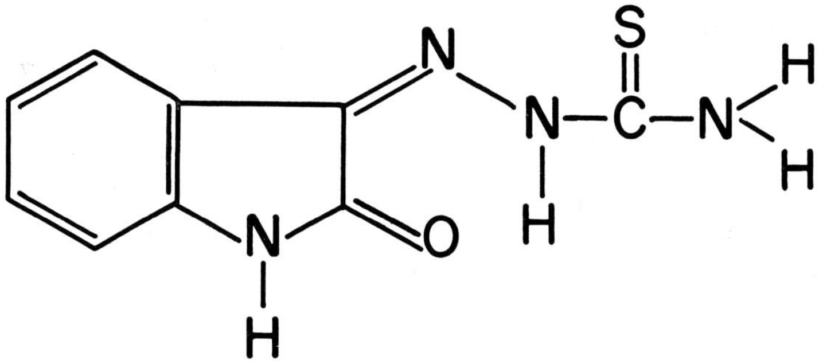

Figure 1. Configuration of isatin — β — thiosemicarbazone.

Magee and Bach (1962) showed that viral DNA is synthesized in the presence of IBT. Easterbrook (1962) demonstrated that immature viral particles are present in cells infected in the presence of IBT. It was therefore suggested that the compound interferes with the maturation of viral progeny, perhaps by preventing the correct assembly of viral structural subunits. Since there is no indication that IBT acts as an amino acid or nucleic acid base analogue or that it is incorporated by covalent linkage into either protein or nucleic acid, and since it is by no means clear what "interference with maturation" means in molecular terms, we decided to undertake an investigation of the mechanism of action of this relatively small molecule.

The system which we used was HeLa cells grown in suspension culture and infected with the WR strain of vaccinia virus (Woodson and Joklik, 1965). Figure 2 illustrates the degree of inhibition of virus yield achieved in this system with increasing concentrations of IBT. At a concentration of IBT of 15 μmolar (about 3 μg/ml) the virus yield at 24 hours after infection is reduced to 3%; at a concentration of 60 μmolar it is reduced to 1%. In most of the experiments which will be discussed, the concentration of IBT used was 15 μmolar. This concentration has no effect whatever on the multiplication of uninfected cells; in fact, cells

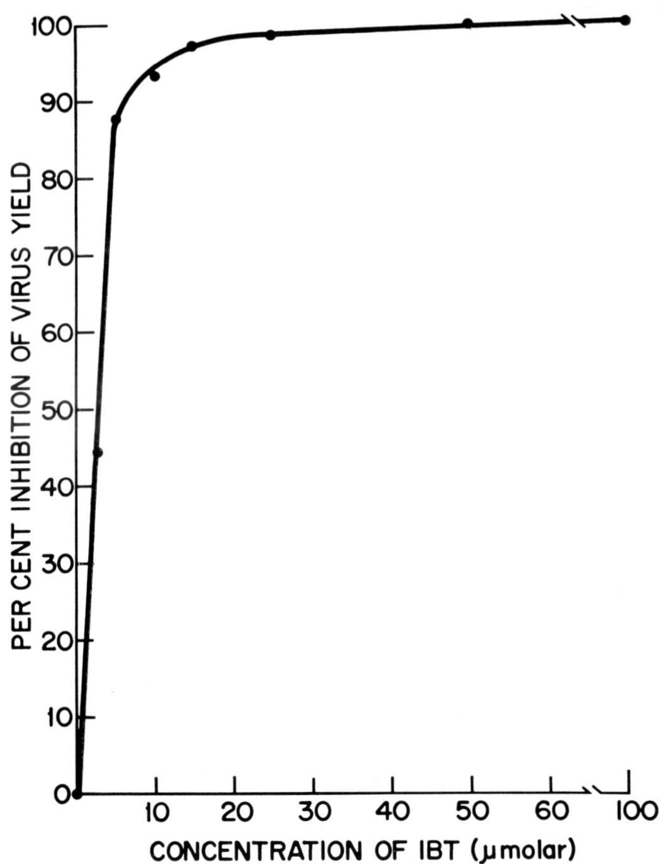

Figure 2. Effect of concentration of IBT on 24 hour virus yields. Adsorbed multiplicity of infection: 4 PFU per cell. (Reproduced from Woodson and Joklik (1956) by courtesy of *Proc. Natl. Acad. Sci.*)

can be grown perfectly well in 30 μmolar IBT. In order for IBT to inhibit viral replication maximally, it must be added at least 1 hour before the appearance of the first progeny virions. This time is between 4 and 5 hours after infection. Preexposure of cells to IBT for periods as long as 30 hours does not further decrease the yield. The later IBT is added, the smaller is the reduction of the final yield.

Uncoating of infecting virus is not inhibited by IBT, nor is the induction of the synthesis of early enzymes, in particular DNA polymerase (Woodson and Joklik, 1965). As a result, viral DNA replicates normally in the presence of IBT: both in the

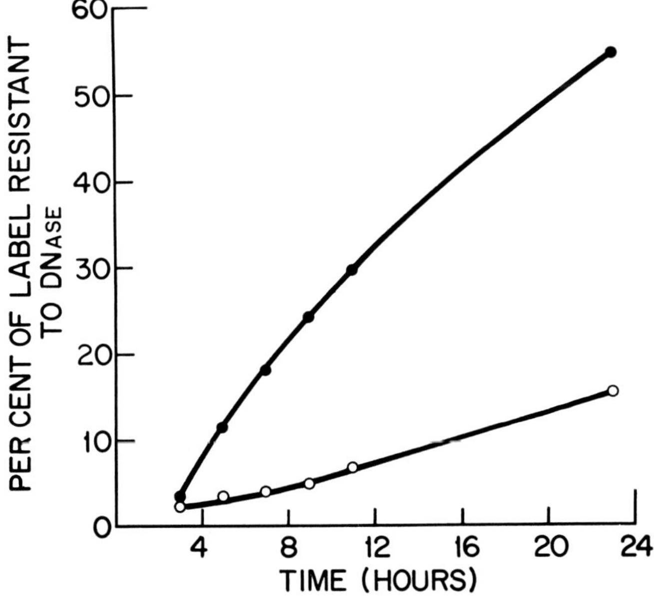

Figure 3. Coating of vaccinia progeny DNA in the absence (*filled symbols*) and presence (*open symbols*) of 15 μmolar IBT. 1.5 x 10^8 HeLa cells were infected and one half made 15 μmolar with respect to IBT at the end of the adsorption period. At 2 hours, both cultures were pulse-labeled with 5μC C^{14}-dT for 10 minutes and then centrifuged and resuspended in fresh medium (with and without IBT respectively). At intervals, samples of 1 x 10^7 cells were removed, the cytoplasmic fractions prepared and the amount of acid-insoluble label before and after incubation with 100 μg/ml DNase for 30 minutes at 37° was determined. (Reproduced from Woodson and Joklik (1965) by courtesy of *Proc. Natl. Acad. Sci.*)

presence and absence of IBT viral DNA replication commences at about $1\frac{1}{2}$ hours after infection, reaches a maximum at 2 to $2\frac{1}{2}$ hours and is over 90% complete at between 4 and $4\frac{1}{2}$ hours after infection (Joklik and Becker, 1964). Furthermore, both viral mRNA transcription and the time course of overall protein synthesis are normal during the first three hours of the infection cycle. In fact, during this early period, IBT has no detectable effect on the course of viral infection. However, after this early period, drastic changes occur. The first clue was the finding that viral DNA which had replicated was not coated (Fig. 3) and did not appear in mature virions (Fig 4) (Woodson and Joklik, 1965). We could show by means of sucrose density gradient analysis that in normal cells more than ten times more DNA replicated during the period from 120 to 130 minutes after infection is incorporated into mature virions by 11 hours after infection than in cells continuously exposed to 15 μmolar IBT. No accumulation of DNA-containing immature viral particles was detected. This finding rendered unlikely the hypothesis that IBT acted by somehow inhibiting addition of structural viral protein subunits to immature particles and thus preventing formation of mature virions.

Failure of progeny DNA to become coated suggested instead interference with *synthesis* of structural viral proteins. This had seemed unlikely at first sight, since during the early phases of our investigation we had found that the pattern of viral mRNA synthesis was the same in normal cells and cells infected in the presence of IBT; indications were that this was so not only in the "early" period (by which I mean the period before the rate of viral DNA replication has reached its peak; in this system this time corresponds to about 3 hours) but also in the subsequent "late" period.

In view of this unchanged pattern of viral mRNA synthesis, it was surprising to find that although cells infected in the presence of 15 μM IBT synthesized protein at the same rate as untreated infected cells for about $2\frac{1}{2}$ to 3 hours, there then occurred a drastic loss in this ability: at 2 hours it was normal, at 4 hours 20% and at 6 hours 10% of controls (Fig. 5). This drastic inhibition commencing at about 3 hours after infection explains why progeny

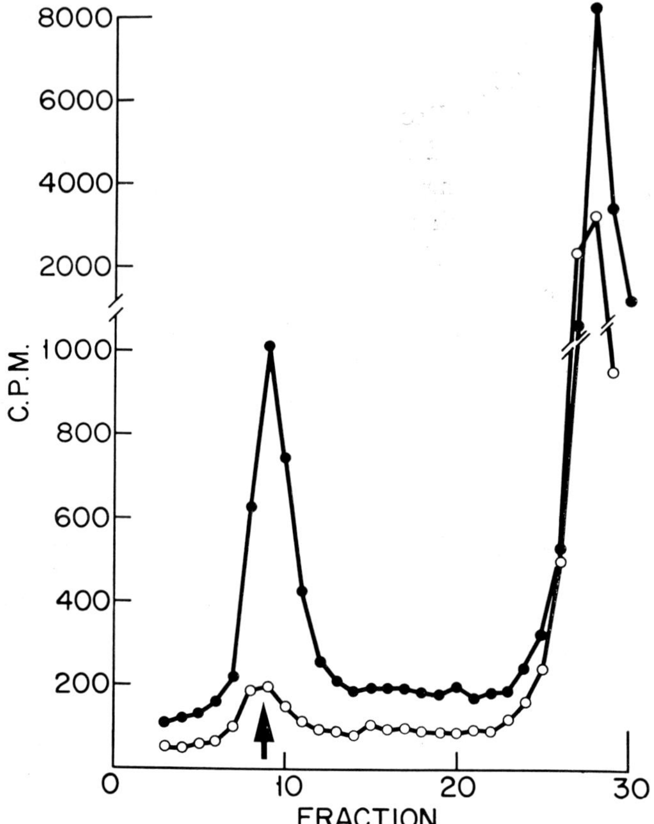

Figure 4. Distribution of progeny vaccinia DNA in sucrose density gradients 11 hours after infection in the absence (*filled symbols*) and presence (*open symbols*) of 15 μmolar IBT. Cells were infected and labeled as for Figure 3. At 11 hours after infection, the cytoplasmic fractions were layered onto 25-40% sucrose-RSB density gradients and centrifuged for 60 minutes at 15,000 RPM. The arrow indicates the position of mature virions.

viral DNA is not coated, and thus the inhibition of the formation of mature virions. However, it appears now that IBT does not interfere with maturation by inhibiting assembly of structural viral subunits the formation of which proceeds normally, but by inhibiting the formation of these viral subunits. In the presence of IBT, therefore, maturation is inhibited because of an acute shortage of material necessary for virus maturation.

The inhibitory effect of IBT on protein synthesis after 3 hours after infection was further investigated by studying polyribosome patterns. Figure 6 shows the polyribosome distribution in cytoplasmic fractions of cells infected in the presence and absence of 15 μM IBT at 2 and 4 hours after infection, as determined both by measurement of optical density and by the presence of labeled viral mRNA. At 2 hours, there is no difference between the profiles; however, at 4 hours there are very few polyribosomes present

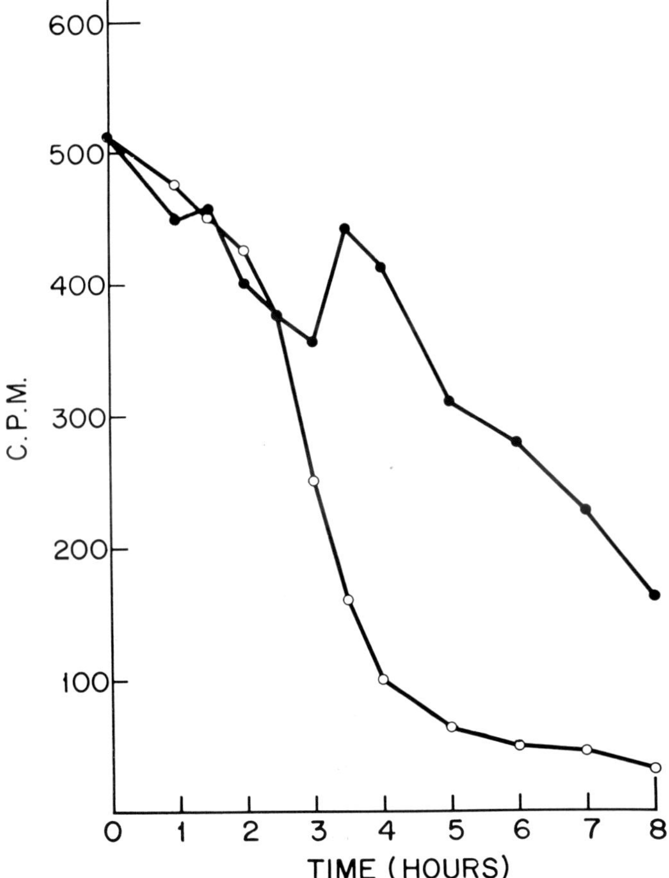

Figure 5. Effect of IBT on the incorporation of C^{14}-amino acids into cells infected in the absence *(filled symbols)* and presence of 15 μmolar IBT *(open symbols)*. Cells were pulsed for 5 minutes at each time point and incorporation into total cellular protein was determined. (Reproduced from Woodson and Joklik (1965) by courtesy of *Proc. Natl. Acad. Sci.*)

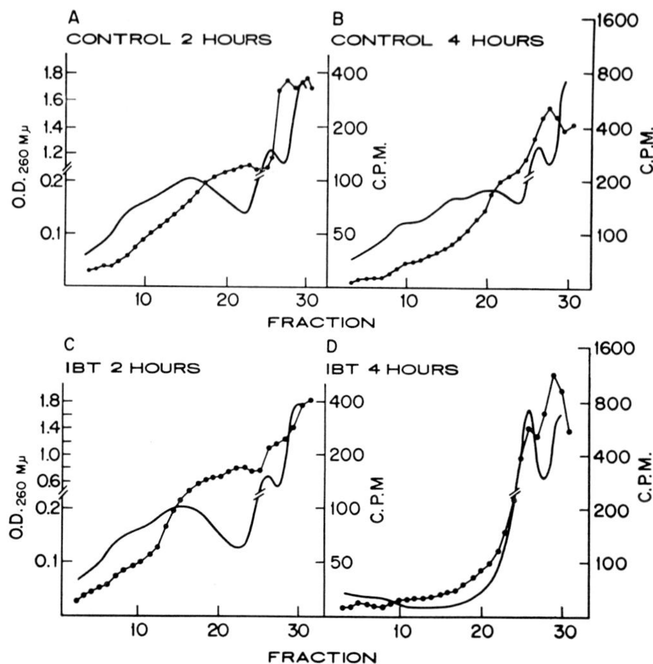

Figure 6. Optical density and radioactivity profiles of polyribosomes in the cytoplasmic fractions of HeLa cells infected for 2 and 4 hours in the absence (A and B) and presence (C and D) of 15 μM IBT. For each gradient, the cytoplasmic fraction of 4 x 10[7] cells pulse-labeled with C[14] -uridine for 10 minutes was used. Centrifugation: 15-30% sucrose density gradients, 2 hours at 25,000 rpm. (Reproduced from Woodson and Joklik (1965) by courtesy of *Proc. Nat. Acad. Sci.*)

in cells infected in the presence of IBT, and there is practically no viral mRNA present in the polyribosome region. Viral mRNA synthesized during the first 2 to 3 hours after infection is evidently capable of forming stable polyribosomes in the presence of IBT, but mRNA synthesized after this time is not. Further, it is possible to measure the rate of IBT-induced breakdown of polyribosomes. Cells infected for 4 hours were exposed to 15, 30, 60 and 120 μM IBT, and 30 and 60 minutes later the number of 74S ribosomal monomers (single ribosomes) was measured by density gradient analysis (Fig. 7). Measurement of the increase of the number of single ribosomes is the most sensitive measure of the breakdown of polyribosomes. Fifteen μM IBT has relatively little effect within one hour, while 60 and 120 μM IBT caused all the 74S ribosomes that could be released from polyribosomes by IBT (about 80% of the total) to be released within 1 hour. It appears, therefore, that not only does IBT prevent the formation of stable polyribosomes by mRNA synthesized after 3 hours after infection, but that in the presence of IBT preexistent polyribosomes breakdown. When similar concentrations of IBT were added to infected cells before 3 hours after infection there was no release of ribo-

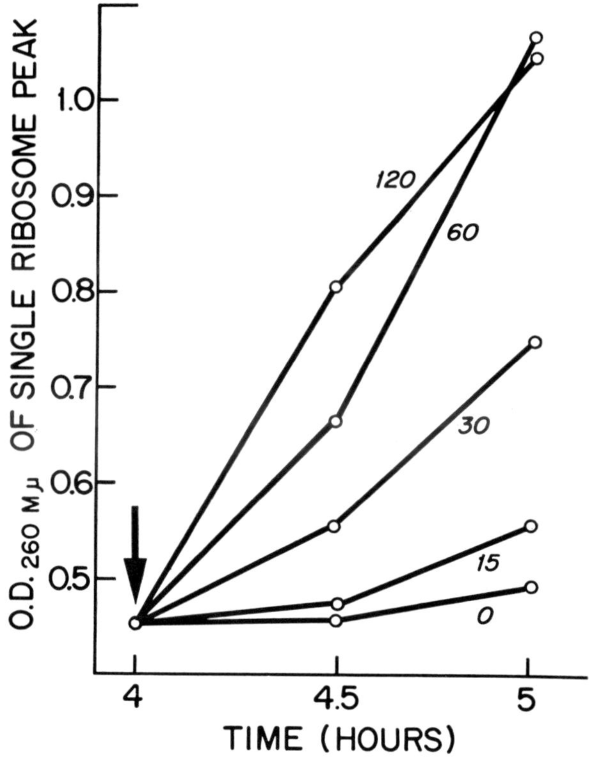

Figure 7. Effect of IBT concentration on the release of single 74S ribosomes from polyribosomes. At 4 hours after infection a cell culture was divided and IBT was added to aliquots to give final concentrations of 0, 15, 30, 60 and 120 μM. Samples were removed for analysis 30 and 60 minutes later. The cytoplasmic fractions were placed onto 15-30% sucrose density gradients, centrifuged for 2 hours at 25,000 rpm, and the optical density of the 74S single ribosome peaks was measured. (Reproduced from Woodson and Joklik (1965) by courtesy of *Proc. Nat. Acad. Sci.*)

somes. Polyribosomes may thus be said to be resistant to IBT during the first 3 hours of the infection cycle, and sensitive thereafter.

We examined next both the size and fate immediately after transcription of viral mRNA transcribed after 3 hours after infection in cells treated with IBT. Let us take size first. The method used was to administer brief pulses (from 4 to 13 minutes) of C^{14}-UR to infected cells and to analyze by means of sucrose density gradient centrifugation the labeled RNA species, liberated from combination with protein by SDS, in the cytoplasmic fractions. It has already been shown (Becker and Joklik, 1964) that the only labeled RNA species larger than 4S in such cytoplasmic cell fractions is vaccinia mRNA. This is so because all host cell RNA is transcribed in the nucleus and is not transported to the cytoplasm before about 15-20 minutes after formation. By contrast, vaccinia mRNA transcription proceeds in the cytoplasm.

The experimental plan was: we examined normally infected cells and cells infected in the presence of 15μm IBT, each at 2 hours after infection (when protein synthesis was normal and polyribosomes were resistant to IBT) and at 4 hours after infection (when protein synthesis was greatly inhibited and polyribosomes were sensitive to IBT) ; and we used 4 minute pulses of C^{14}-UR, which allowed examination of the size of mRNA almost immediately after transcription, and 9 and 13 minute pulses which allowed in addition examination of mRNA molecules which were up to 7 or 8 and 11 or 12 minutes old respectively. The results were: at 2 hours after infection the size of mRNA after pulses of 4, 9 and 13 minutes was about 10-12S no matter whether IBT was present or not. At 4 hours after infection in the absence of IBT, the S value of mRNA was 16-20S, irrespective of the duration of the pulse (Fig. 8A). In the presence of IBT, however, the median S was 16-20 after a pulse of 4 minutes, but only about 8 after pulses of 9 or 13 minutes (Fig. 8B). Clearly, there has been a pronounced change in the time interval between 4 and 9 minutes after transcription: during this time interval the sedimentation coefficient has decreased from 16-20 to 8S, which corresponds on the average to the introduction of 2 breaks into each mRNA molecule.

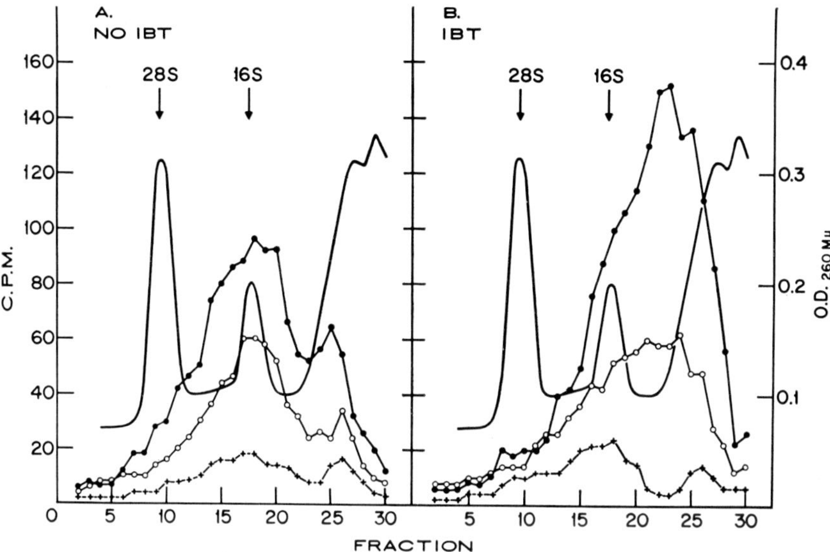

Figure 8. Effect of IBT on the size of viral mRNA. Cells were pulse-labeled with H³-uridine for 4 (x), 9 (o), and 13 (●) minutes at 4 hours after infection. The cytoplasmic fractions derived from 10^7 cells were treated with SDS and centrifuged for 16 hours at 24,000 rpm in 15-30% sucrose-SDS density gradients. (Reproduced from Woodson and Joklik (1965) by courtesy of *Proc. Nat. Acad. Sci.*)

A similar conclusion regarding the occurrence of an event which profoundly affects mRNA when IBT is present later than 3 hours after infection was reached from studies of the fate of viral mRNA. The experimental plan was similar to the previous one but instead of determining the sedimentation coefficient of mRNA, sucrose density gradient analyses were carried out in order to determine the cytoplasmic structures with which the mRNA was associated. In cells infected for 2 or 4 hours in the absence of IBT or for 2 hours in the presence of IBT, and pulse-labeled for 9 minutes, about 80% of the viral mRNA is found in polyribosome form, and the rest in a region which overlies the 40S sub-ribosomal particle peak (Fig. 9A). With a pulse of 4 minutes, the fraction of mRNA in polyribosome form is somewhat less, with a pulse of 13 minutes it is slightly more. In cells infected in the presence of IBT for 4 hours and pulsed for 9 or 13 minutes, only

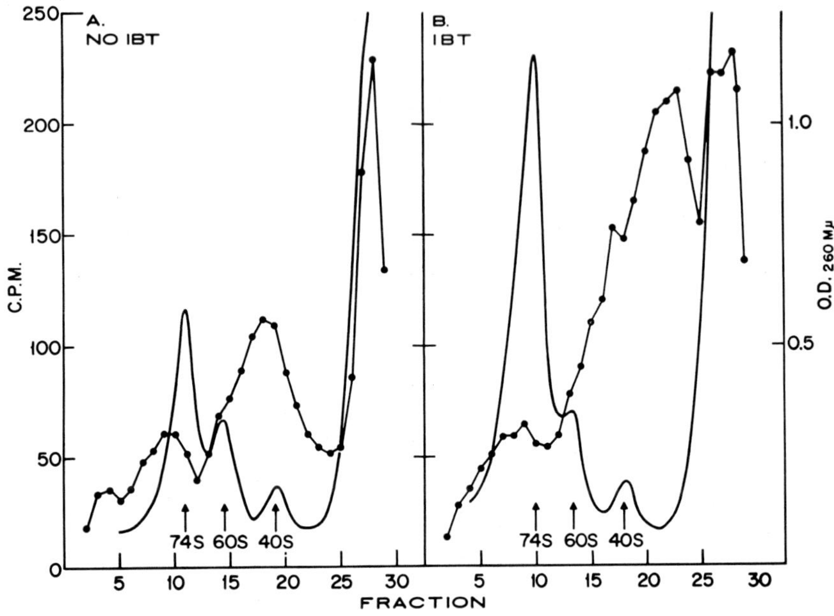

Figure 9. Effect of IBT on the fate of vaccinia mRNA. Cells were labeled with C14-uridine for 9 minutes at 4 hours after infection. The cytoplasmic fractions derived from 2 x 10⁷ cells were centrifuged for 16 hours at 20,000 rpm in 15-30% sucrose-RSB density gradients at 5°C.

10 to 20% of the total mRNA is in polyribosomal form, there is practically no accumulation of mRNA in the 40S region, but most of the labeled RNA is found in the region of about 8 to 12S (Fig. 9B) . By contrast, the corresponding 4 minute pulse gave a profile identical with that derived from cells infected in the absence of IBT for 4 hours and also pulsed for 4 minutes. Again, there occurs a drastic change in the fate of viral mRNA transcribed in cells infected in the presence of IBT for 4 hours after the first 4 minutes of transcription: whereas immediately after transcription the mRNA is incorporated into polyribosomes, most of the mRNA leaves this fraction between 4 and 9 minutes.

The conclusions from the experiments described so far can be summarized thus: in cells infected in the presence of IBT for 4 hours, the viral mRNA which is being transcribed is normal in size and fully functional as judged by its ability to be incorporated

polyribosomes formed by late viral mRNA (transcribed from progeny genomes). In the presence of dFU, polyribosomes also persist in the cell for 6 to 7 hours although they are fewer in number than in cells infected under normal conditions. For most of this time, these are polyribosomes held together by early viral mRNA molecules (since polyribosomes held together by host cell mRNA breakdown within the first 1 to 2 hours after infection, and no mRNA molecules transcribed from progeny viral genomes are formed in such cells). In the presence of dFU plus IBT, the polyribosome patterns are exactly the same as in cells infected in the presence of dFU alone; whereas of course, as we have already seen, in cells infected in the presence of IBT alone there are practically no polyribosomes present after 3 hours. Thus again, IBT is without effect in cells in which the formation of progeny viral genomes has been prevented.

Why is the presence of progeny viral genomes necessary in order for IBT to exert its effect? We know that there are transcribed from progeny viral genomes messenger RNA sequences which are not transcribed from parental genomes (Oda and Joklik, in press). All the facts which I have just outlined are explicable on the hypothesis that a messenger RNA transcribed from progeny viral genomes codes for the synthesis of a protein through which IBT exerts its effect. This fits in with the fact that if protein synthesis is prevented by the addition of puromycin, IBT likewise is without effect.

This state of affairs is reminiscent of another effect due to the presence of progeny viral genomes in the infected cell. It has already been mentioned above that infection of HeLa cells with vaccinia virus induces the formation of certain enzymes, in particular a new DNA polymerase (Jungwirth and Joklik, 1965). The synthesis of this enzyme proceeds for a period of about 3 hours and is "switched off" at 3 to 4 hours after infection. If the formation of progeny viral genomes is prevented by the presence of dFU, the synthesis of this enzyme proceeds in an uncontrolled manner, and is not switched off: the level of DNA polymerase continues to increase until at least 8 to 9 hours after infection. Once again the presence of progeny viral genomes in the cell profoundly affects the patterns of macromolecular synthesis. McAuslan (1963) has shown that switch-off of the synthesis of "early" virus-coded proteins involves both the transcription of messenger RNA and its translation into polypeptide chains. This switchoff of the synthesis of early viral enzymes must involve a selective mechanism, since the synthesis of certain proteins is switched off at a time when the synthesis of other proteins, for instance certain viral structural proteins, is proceeding at a rapid rate. The protein causing switch-off must be able to differentiate between viral messenger RNAs coding for early enzymes and those coding for viral structural proteins. An attractive hypothesis is that IBT destroys the selectivity of the switch-off protein by an allosteric mechanism so that it then prevents the functioning of not only some, but of all viral messenger RNAs. However, there is at yet no evidence for this suggestion, and other explanations are no less likely.

The most important implications of the effect of IBT are those concerned with the control of translation of viral messenger RNAs. IBT provides a set of conditions under which all viral messenger RNAs are prevented from functioning normally. The elucidation of the mechanism by which this is achieved could well be of great significance.

STUDIES ON THE MECHANISM OF ACTION OF INTERFERON

Interferons are formed in response to infection with both RNA and DNA viruses. Interferons fail to affect virus multiplication if host cell messenger RNA and protein synthesis are inhibited (Taylor, 1964; Friedman and Sonnabend, 1964; Levine, 1964); it has therefore been postulated that they induce the synthesis of protein (s) which is the actual inhibitor of virus multiplication. The mechanism by which this postulated inhibitory protein acts is unknown. Neither infectious progeny RNA nor replicative intermediate are synthesized in interferon-treated cells (De Somer *et al.*, 1962; Friedman and Sonnabend, 1965; Gordon *et al.*, 1966) ; and there is some autoradiographic evidence that viral DNA replication is also inhibited (Friedman *et al.*, 1965) . Further, an early function of RNA genomes, namely the shutting-off of host cell RNA synthesis, has been reported to be prevented in interferon-treated cells (Levy, 1964; Levy *et al.*, 1966) . These observations are in line with an earlier suggestion (Joklik, 1965) that since interferons prevent the multiplication of both RNA and DNA viruses, the most likely locus of action would be the earliest common event, namely the exercise of messenger RNA function. In the case of RNA viruses this function is exercised by parental viral RNA itself, in the case of DNA viruses it is exercised by messenger RNA transcribed from parental viral genomes.

We have recently studied the mechanism of action of interferon in L cells infected with vaccinia virus (Joklik and Merigan, 1966). The interferon used in these investigations was partially purified by Dr. T. M. Merigan of Stanford University. In all experiments which I will describe, interferon was used as follows: suspension cultures of L cells were exposed to interferon for 16 hours. The cells were then centrifuged, resuspended at a concentration of 10^7 cells/ml in adsorption medium and infected with vaccinia virus at an added multiplicity of 1,000 EB (20-30 pfu) per cell. After 15 minutes, during which 50% of the virus adsorbed, the cells were diluted twenty-fold into normal growth medium and used for experiments. Under these conditions interferon inhibited virus multiplication by about 95%.

We quickly found that the mechanism of action of interferon was quite different from that of IBT (Joklik and Merigan, 1966). In interferon-treated infected cells, there is no induction of the synthesis of "early" enzymes, and therefore no replication of viral DNA. No progeny viral DNA genomes therefore ever arise in such cells. These results suggested inhibition either of the transcription of viral messenger RNA or of its translation. Figure 11 shows the kinetics of transcription, as calculated from summation of the appropriate areas on sucrose density gradient profiles. Far from being inhibited in interferon-treated cells, the transcription of viral messenger RNA is actually enhanced. This effect is greater the lower the multiplicity; at a multiplicity of 175 EB/cell the rate of transcription of viral mRNA in cells pretreated with interferon is about 5 times that in normal infected cells. However, the rate of transcription falls off more rapidly in interferon-treated than in normal cells. Figure 12 shows a typical sucrose density gradient sedimentation profile for messenger RNA transcribed in interferon-treated and normal cells at $1\frac{1}{2}$ hours after infection: the size

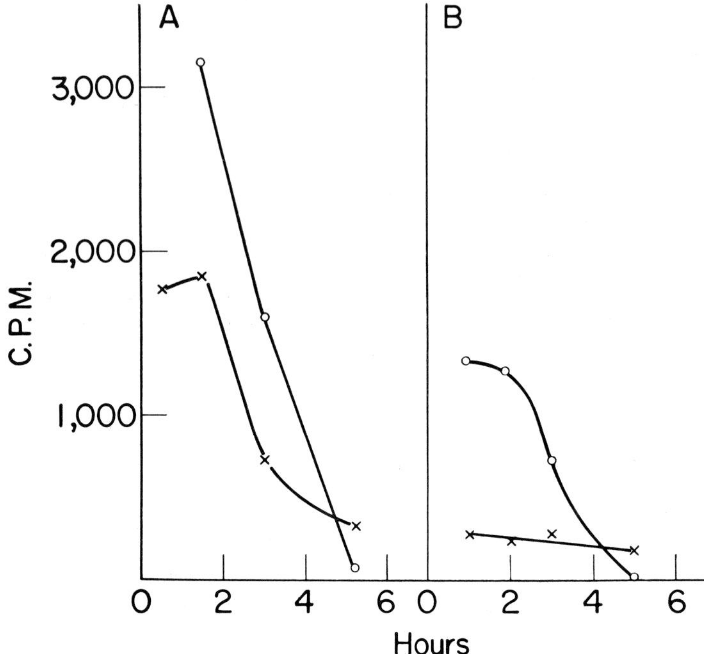

Figure 11. Kinetics of formation of vaccinia messenger RNA in normal and interferon-treated cells. Each point was derived from summation of the appropriate area of an SDS-sucrose density gradient profile. A) Multiplicity of infection 750; B) Mutiplicity of infection 175. *Crosses*: no interferon; *circles*: interferon-treated. (Reproduced from Joklik and Merigan (1966) by courtesy of *Proc. Nat. Acad. Sci.*)

of the messenger RNA being transcribed in the two kinds of cells is the same. We conclude that viral messenger RNA of normal size is being transcribed in interferon-treated cells from parental viral genomes (since no progeny viral genomes arise in such cells) at a rate which is certainly not less, but rather more, than that in untreated cells. The reason for the accelerated rate of viral messenger RNA formation in interferon-treated cells is not clear at the moment. Two possible explanations are: 1) that there is a control mechanism which regulates the rate of transcription of viral messenger RNA in untreated cells and which fails to operate in interferon-treated cells; and 2) that the messenger RNA transcribed in interferon-treated cells represents not only "early" viral messenger RNA, but viral messenger RNA transcribed from the whole of the parental genome. These possibilities are being investigated in our laboratory at this time.

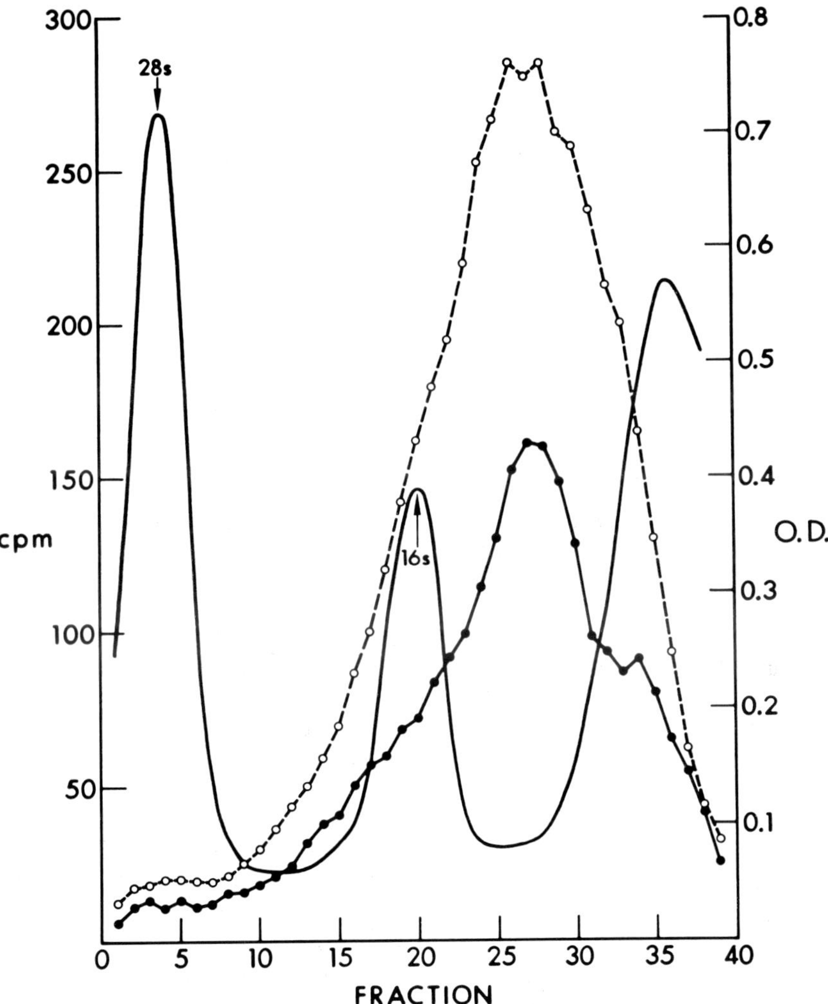

Figure 12. Size of vaccinia messenger RNA transcribed in normal cells and cells pretreated with interferon, both infected with vaccinia virus for 1½ hours at a multiplicity of 500 EB per cell. Conditions of centrifugation as for Figure 8. *Filled symbols*: normal cells; *open symbols*: cells pretreated with interferon.

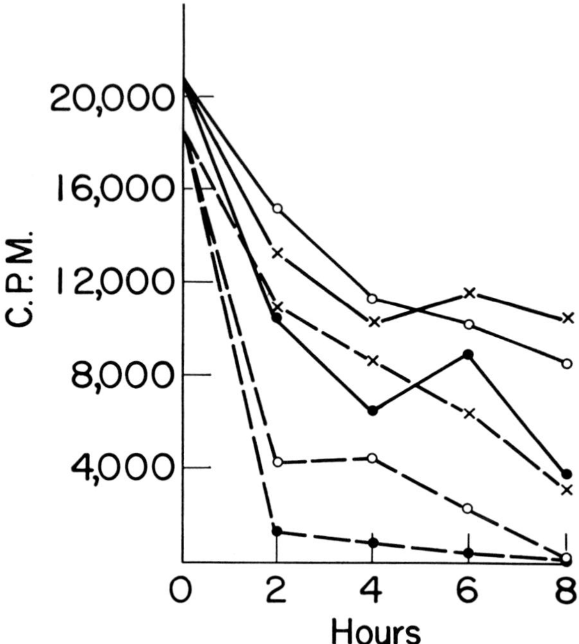

Figure 13. Rate of protein synthesis in normal and interferon-treated L cells infected with vaccinia virus. Each point represents the amount of radioactivity incorporated into the cytoplasmic fraction of 1.35 x 10⁷ cells in Eagle's medium containing one tenth of the usual concentration of amino acids, pulsed for 4 minutes with 15 μC of a uniformly C¹⁴-labeled amino acid mixture (1.5 mC/mg). *Continuous curves*: no interferon; *broken curves*: interferon-treated. *Crosses*: multiplicity of infection 50; *open circles*: 150; *closed circles*, 500. (Reproduced from Joklik and Merigan (1966) by courtesy of *Proc. Nat Acad. Sci.*)

Since interferon pretreatment did not inhibit the transcription of vaccinia messenger RNA, we investigated next its translation. Figure 13 shows the rate of protein synthesis in infected pretreated cells as compared with infected untreated cells. In the latter, the rate of protein synthesis falls to about half of that in uninfected cells within the first 4 hours, and this is then followed by an increase when virus-coded protein synthesis is at its peak. In infected cells pretreated with interferon, however, there occurs a dramatic decrease in the incorporation of amino acids into protein, which is almost complete by about 4 hours after infection. As the multiplicity of infection is decreased, these effects are correspondingly diminished, but even at a multiplicity as low at 50 EB/cell, corresponding to only about 1 PFU/cell, a multiplicity at which it is almost impossible to detect any synthesis of vaccinia messenger RNA, this effect is clearly detectable.

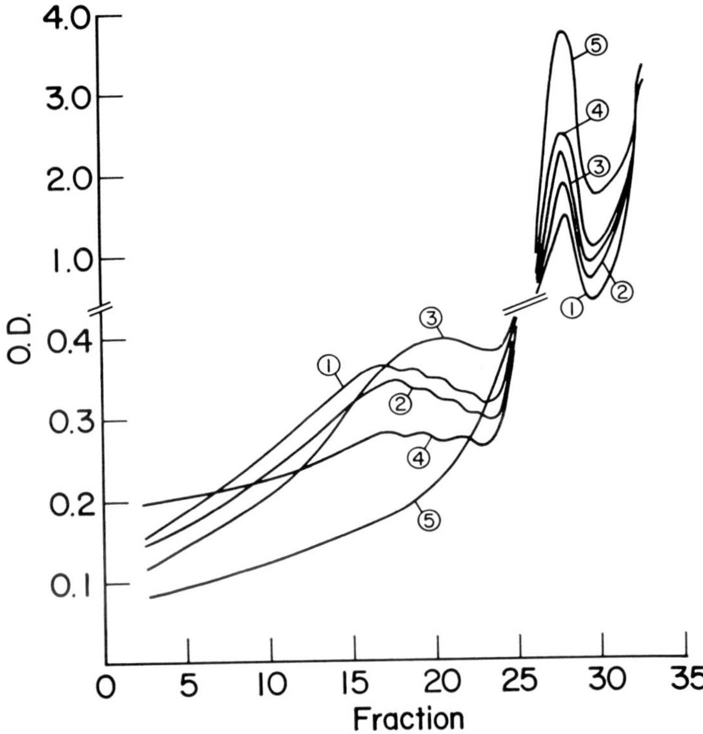

Figure 14. Polyribosomes in normal and interferon-treated L cells infected with vaccinia virus. Each profile was derived from a 15 to 30% sucrose density gradient charged with the cytoplasmic fraction of 5×10^7 L cells; 2 hours centrifugation at 25,000 rpm at 2°. Multiplicity of infection, 500. *Curve 1:* uninfected cells, normal or interferon-treated; *curves 2, 3 and 4:* normal cells infected for 1, 2 and 4 hours respectively; *curve 5:* interferon-treated cells infected for 1, 2 or 4 hours. (Reproduced from Joklik and Merigan (1966) by courtesy of *Proc. Nat. Acad. Sci.*)

This rapid inhibition in the rate of protein synthesis in infected cells pretreated with interferon is due to a rapid disaggregation of polyribosomes. Figure 14 shows the polyribosome patterns obtained from L cells infected at a multiplicity of 500 EB/ cell at 1, 2 and 4 hours after infection, compared with the pattern obtained from uninfected cells. In normal infected cells, there occurs only a relatively small decrease in the total number of polyribosomes during the first 4 hours after infection. In cells pretreated with interferon, however, there are very few polyribosomes by the end of the first hour after infection and none are detectable at the end of the second hour. This finding was confirmed in experiments in which cells were pulsed for 3 minutes with a uniformly C^{14}-labeled mixture of amino acids and the radioactivity profiles of the polyribosomes determined. The labeling technique is very sensitive for detecting small amounts of polyribosomes, but even using it, no functioning polyribosomes were detected in cells pretreated with interferon and infected for 2 hours.

The question thus arises as to the fate of the viral messenger RNA which is being transcribed at a rapid rate from the invading parental viral genomes. Apparently, these messenger RNA molecules cannot attach to ribosomes to form polyribosomes, since no polyribosomes are detectable at a time when transcription of this messenger RNA occurs at a rapid rate (for instance, at 2 hours after infection). Direct experiments showed that this was indeed the case. Just as in the investigation concerning the mechanism of action of IBT, infected cells were pulse-labeled with C^{14}-UR for short periods of time (4, 9 and 13 minutes), and the cytoplasmic fractions were then centrifuged into sucrose density gradients in such a manner as to separate messenger RNA molecules in polyribosome form from free messenger RNA molecules. The results are shown in Table I. As the duration of the pulse increased, a gradually increasing large proportion of messenger RNA molecules synthesized in cells not pretreated with interferon was found in the polyribosome fraction. By contrast, only a small constant proportion of such molecules was associated with polyribosomes

TABLE I

PERCENTAGE OF VACCINIA MESSENGER RNA ASSOCIATED WITH POLYRIBOSOMES
IN NORMAL AND INTERFERON-TREATED CELLS

Duration of Pulse	No Interferon	Interferon-treated
4 min	41	16
9 min	57	17
13 min	66	18

in infected cells which had been pretreated with interferon. However, even this slight association could well have been due to some nonspecific adsorption effect, or to some mechanism not directly concerned with polyribosome formation. One may conclude that in interferon-treated cells viral messenger RNA does not form polyribosomes.

We may now reexamine the experiments concerning the inhibition of protein synthesis and the disaggregation of polyribosomes in infected cells pretreated with interferon. Since viral messenger RNA does not form polyribosomes in interferon-treated cells, the polyribosomes which are being rapidly disaggregated in such cells following infection are clearly host cell polyribosomes. It would, however, be incorrect to conclude that pretreatment with interferon promotes disaggregation of host cell polyribosomes. In fact, in L cells infected with mengovirus disaggregation of host cell polyribosomes can be clearly observed before reformation of polyribosomes by progeny viral RNA occurs, and pretreatment with interferon does not accelerate this disaggregation. Similarly, disaggregation of host cell polyribosomes is most probably rapid in L cells infected with vaccinia virus no matter whether the cells have been pretreated with interferon or not. However, since such pretreatment prevents vaccinia messenger RNA from reforming polyribosomes, a process which normally results in the smooth exchange of ribosomes from combination with host cell messenger RNA to combination with viral messenger RNA, the rapid disaggregation of host cell polyribosomes is unmasked in such cells. Further, we may conclude that the disaggregation of host cell polyribosomes on infection with vaccinia virus is not due to genetic functioning of the invading viral genomes, since the viral messenger RNA which is being transcribed evidently cannot express itself. It must be concluded that this disaggregation is the result of the introduction into the cell of viral structural protein subunits, and presumably occurs without the necessity for protein synthesis. Since we would like to believe that what is true for one viral messenger RNA is also true for others, we would expect the situation to be the same in cells infected with poliovirus and mengovirus also: that is, that disaggregation of host cell polyribosomes is due to the introduction into the cell of viral protein, rather than to the formation of a special protein coded for by the incoming viral genome.

There is one further clearly discernible effect of pretreatment with interferon. Such cells rapidly disintegrate beginning at about 3 hours after infection with vaccinia virus. Figure 15 illustrates this phenomenon. The number of cells that can be classified visually as "intact" decreases somewhat when normal L cells are infected with high multiplicities of vaccinia virus. This effect is greatly magnified for cells pretreated with interferon. Even at multiplicities of infection as low as 50 EB/cell, practically all cells

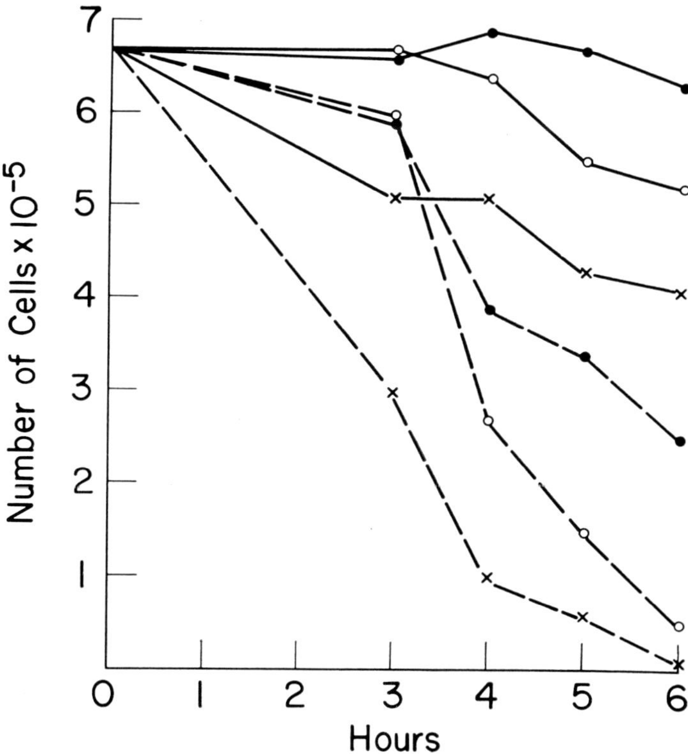

Figure 15. Number of "intact" cells in cultures of normal and interferon-treated L cells infected with vaccinia virus. *Continuous curves*: normal cells; *broken curves*: interferon-treated cells. *Crosses*: multiplicity of infection 500; *open circles*, 200; *closed circles*, 100. Cell destruction was somewhat less when the initial cell concentration was 2.5 x 10⁵ cells/ml. (Reproduced from Joklik and Merigan (1966) by courtesy of *Proc. Nat. Acad. Sci.*)

have been converted to "ghosts" by 6 hours after infection. The reason for this is not clear. It is difficult to attribute this rapid cellular disintegration to a specific protein synthesized after infection, whether coded by the host cell or the viral genome (since protein synthesis is severely inhibited very soon after infection of such cells). It does not seem likely that the mere fact that protein synthesis is inhibited is responsible for this cellular decay, since cells in which protein synthesis has been inhibited by puromycin do not disintegrate nearly as rapidly. It is possible that infection with vaccinia virus causes severe damage to the cell membrane of cells pretreated with interferon, and that cellular decay is a consequence of this damage. On the other hand, it is possible that rupture of lysosomes occurs in cells pretreated with interferon when they are infected.

It is clear, then, that interferon does not protect cells against invasion by virus; on the contrary, cells pretreated with interferon are killed as the result of an abortive infection cycle and disintegrate within a relatively short period of time. Pretreatment with interferon converts cells into virus scavenging structures, in which invading virus particles are uncoated, but in which no mature viral progeny formation is possible. It might be said that cells which have the misfortune to be invaded by virus sacrifice themselves while neutralizing virus, and so save the uninfected members of the population. There is every likelihood that this situation demonstrated *in vitro* also holds for the intact organism.

Finally, we may assess what insight into the essential processes occurring during the vaccinia virus multiplication cycle has been provided by this investigation of the mechanism of action of interferon. In the first place, there is clearly a difference between normal host cell messenger RNA and early vaccinia messenger RNA. Treatment with interferon does not cause disaggregation of normal host cell polyribosomes, nor does it have any effect on the rate of protein synthesis in uninfected cells. Clearly, host cell messenger RNA can combine with ribosomes while vaccinia messenger RNA cannot. The reason for this difference is not clear. On the one hand, it is conceivable that ribosomes in cells pretreated with interferon differ in some way from ribosomes in normal cells. On the other hand, it is possible that there is produced in cells pretreated with interferon some factor which so alters vaccinia messenger RNA that it is no longer able to combine with ribosomes. There is no clue as yet as to which of these mechanisms is operative. Secondly, the work with interferon has brought to light the existence of a mechanism for breaking down host cell polyribosomes on infection. The rapidity of this process has previously not been appreciated, since the ribosomes liberated by this breakdown apparently quickly attach to viral messenger RNA and reform polyribosomes which now translate viral message. Interferon may thus be used as a reagent for specifically preventing polyribosome reformation, and should be of value in studying the process of host cell polyribosome disaggregation. As of now, it appears that this disaggregation of host cell polyribosomes is due to the introduction into the cell of foreign protein, that is, of structural viral protein. So far, the evidence has been that information stored in the viral genome must be expressed in the infected cell in order to effect this disaggregation; but in view of the finding that viral message is prevented from being translated in cells pretreated with interferon, it is difficult to see how a virus-coded protein causing disaggregation could arise. Thirdly, interferon should prove valuable in the study of cellular decay following virus infection. This process occurs rapidly, within 4 to 6 hours, and at relatively low multiplicities of infection. It is truly remarkable that introduction of as few as 20 to 50 virus particles, and thus of relatively few protein molecules, into a relatively huge cell, can cause such profound changes.

Summary. IBT inhibits the formation of vaccinia virus progeny. It does not affect viral DNA replication or viral messenger RNA synthesis, or the functioning of "early" viral messenger RNA. However, in the presence of IBT, the functional half-life of vaccinia messenger RNA is reduced from about 30-60 minutes to rather less than 5 minutes. The effect is conditional upon the transcription of late vaccinia messenger RNA. It is likely that a protein coded by late vaccinia messenger RNA, that is messenger RNA transcribed from progeny viral genomes, is responsible for this effect.

Interferon acts by a different mechanism. In cells pretreated with interferon, transcription of vaccinia messenger RNA proceeds normally, but such cell messenger RNA cannot associate with ribosomes to form polyribosomes. As a result, virus-coded DNA polymerase is not synthesized and no progeny viral DNA is formed. The net result is an abortive viral multiplication cycle which results in complete cell destruction by 4 to 6 hours after infection.

These studies have provided further insight into the essential processes occurring during the proliferative phase of the vaccinia virus multiplication cycle.

REFERENCES

1. BAUER, D. J.: The chemotherapy of ectromelia infection with isatin-β-dialkyl thiosemicarbazones. *Brit. J. Exper. Pathol., 44:*233-242, 1963.
2. BAUER, D. J.: Clinical experience with the antiviral drug marboran (1-methylisatin 3-thiosemicarbazone). *Ann. New York Acad. Sci., 130:*110-117, 1965.
3. BECKER, Y., and JOKLIK, W. K.: Messenger RNA in cells infected with vaccinia virus. *Proc. Nat. Acad. Sci. (U.S.), 51:*577-585, 1964.
4. DE SOMER, P., PRINZIE, A., DENYS, P., JR., and SCHONNE, E.: Mechanism of action of interferon. 1. Relationship with viral ribonucleic acid. *Virology, 16:* 63-70, 1962.
5. EASTERBROOK, K. B.: Interference with the maturation of vaccinia virus by isatin-β-thiosemicarbazone. *Virology, 17:*245-251, 1962.
6. FRIEDMAN, R. M., and SONNABEND, J. A.: Inhibition of interferon by p-fluorophenylalanine. *Nature, 203:*366-367, 1964.
7. FRIEDMAN, R. M., and SONNABEND, J. A.: Inhibition by interferon of production of double-stranded Semliki Forest Virus ribonucleic acid. *Nature, 206:*532, 1965.
8. FRIEDMAN, R. M., SONNABEND, J. A., and McDEVITT, H.: Interferon inhibition of cytoplasmic DNA accumulation in vaccinia virus infection. A radioautographic study. *Proc. Soc. Exper. Biol. & Med., 119:*551-553, 1965.
9. GORDON, I., CHENAULT, S. S., STEVENSON, D., and ACTON, J. D.: Effect of interferon on polymerization of single-stranded and double-stranded mengovirus ribonucleic acid. *J. Bact., 91:*1230-1238, 1966.
10. JOKLIK, W. K.: The molecular basis of the viral eclipse phase. *Prog. Med. Virology, 7:*44-96, 1965.
11. JOKLIK, W. K., and BECKER, Y.: The replication and coating of vaccinia DNA. *J. Mol. Biol., 10:*452-474, 1964.
12. JOKLIK, W. K., and MERIGAN, T. C.: Concerning the mechanism of action of interferon. *Proc. Nat. Acad. Sci., Wash., 56:*558-565, 1966.
13. JUNGWIRTH, C., and JOKLIK, W. K.: Studies on "early" enzymes in HeLa cells infected with vaccinia virus. *Virology, 27:*80-93, 1965.
14. LEVINE, S.: Effect of actinomycin D and puromycin dihydrochloride on action of interferon. *Virology, 24:*586-588, 1964.
15. LEVY, H. B.: Studies on the mechanism of interferon action. 2. The effect of interferon on some early events on Mengo Virus infection in L cells. *Virology, 22:*575-579, 1964.

16. Levy, H. B., Snellbaker, L. F., and Baron, S.: Effect of interferon on RNA synthesis in Sindbis virus infected cells. *Proc. Soc. Exper. Biol. & Med., 121:* 630-632, 1966.
17. Magee, W. E., and Bach, M. K.: Biochemical studies on the antiviral activities of the isatin-β-thiosemicarbazones. *Ann. New York Acad. Sci., 130:*80-91, 1965.
18. McAuslan, B. R.: The induction and repression of thymidine kinase in the poxvirus-infected HeLa cell. *Virology, 21:*383-389, 1963.
19. O'Sullivan, D. G., and Sadler, P. W.: Agents with high activity against Type 2 poliovirus. *Nature, 192:*341-343, 1961.
20. Taylor, J.: Inhibition of interferon action by actinomycin. *Biochem. and Biophys. Res. Comm., 14:*447-451, 1964.
21. Thompson, R. L., Davis, J., Russell, P. B., and Hitchings, G. H.: Effect of aliphatic oxime and isatin thiosemicarbazones on vaccinia infection in the mouse and in the rabbit. *Proc. Soc. Exper. Biol. & Med., 84:*496-499, 1953.
22. Woodson, B., and Joklik, W. K.: The inhibition of vaccinia virus multiplication by isatin-β-thiosemicarbazone. *Proc. Nat. Acad. Sci., Wash., 54:*946-953, 1965.

ACTION OF ACTINOMYCIN AND CYCLOHEXIMIDE ON REOVIRUS REPLICATION

A. J. SHATKIN AND B. RADA

THE REOVIRUSES ARE A GROUP of respiratory and enteric viruses comprising three distinct serotypes which share a common complement-fixing antigen (1). Although widely distributed in nature, they lack a clearly-established relationship to disease states. They differ from RNA-containing enteroviruses in several ways. For example, they are 70 mμ in diameter and contain RNA of molecular weight ten million (2) as compared to the values of 27 mμ and two million for poliovirus (3). In tissue culture cells, a single cycle of reovirus replication requires 16 hours or longer and includes a latent period of 6-7 hours (4), an interval corresponding to a complete infectious cycle of poliovirus (5).

Another characteristic feature of these viruses is their helical, double-stranded RNA (6). The biochemical events associated with the replication of a virus containing double-stranded RNA are of considerable interest. Analysis of these events, particularly virus-specific RNA synthesis, is difficult because cell RNA synthesis continues in reovirus- infected cells and masks the virus-directed processes (7). Similar difficulties in other virus-cell systems have been circumvented by using actinomycin to suppress cell RNA formation selectively (8, 9). Although reovirus replication is markedly inhibited by actinomycin at a concentration of 2 μg/ml (4), it is unaffected at lower concentrations of the antibiotic which reduce cellular RNA synthesis by more than 90% (10, 11, 12). Under these conditions, it is possible to study the formation of virus-induced RNA.

In Table I are shown the specific activities of the RNA in uninfected L cells which had been incubated with the indicated concentrations of actinomycin for 4 hours followed by a 1-hour exposure to uridine-2-C^{14}. The synthesis of cell RNA was progressively inhibited with increasing drug levels. At 0.5 μg/ml, the rate of cell RNA synthesis was 11% that of growing cells, and the RNA consisted mainly of 4S species. Included for comparison are the 24-hour yields of infectious virus obtained from cells treated with the antibiotic beginning at $2\frac{1}{2}$ hours after infection. In contrast to cell RNA synthesis, at a drug concentration of 0.5 μg/ml or less, no reduction in virus titer was observed. As shown in Figure 1, the kinetics of virus formation were unaltered by the presence of 0.5 μg actinomycin/ml. In both treated and untreated cultures, there was a 6-7 hour eclipse period followed by a logarithmic increase in infectious virus which reached 2000 PFU/ cell in this experiment.

TABLE I

EFFECT OF ACTINOMYCIN ON REOVIRUS REPLICATION
AND L CELL RNA SYNTHESIS

Actinomycin $\mu g/ml$	*RNA* $cpm/\mu g$	*% RNA Inhibition*	*Virus Titer PFU/ml*	*% Virus Inhibition*
0.0	103	—	2.5×10^7	—
0.1	42	58	2.8×10^7	—
0.2	23	77	2.4×10^7	—
0.3	21	79	4.9×10^7	—
0.4	15	85	4.7×10^7	—
0.5	11	89	2.3×10^7	—
1.0	8	92	3.0×10^6	88
2.0	6	94	6.6×10^5	97
5.0	2	98	1.5×10^6	94

L-929 mouse fibroblasts were concentrated to 5×10^6/ml in Eagle's medium (23) containing 2% bovine fetal serum and infected with type 3 reovirus at an input multiplicity of about 20 PFU/cell. After adsorption for 2 hours, unadsorbed virus was removed by washing with the above medium, and the cells were resuspended at a density of 2-2.5×10^5/ml. Virus titers were determined by plaque assay (4) after incubation at 37°C for 24 hours. The titer at $5\frac{1}{2}$ hours after infection was 1.7×10^5 PFU/ml.

Under conditions of actinomycin suppression of cell RNA synthesis, virus-induced RNA formation was detected beginning at about 7 hours after infection (Fig. 2). The specific activity of newly formed RNA in uninfected cells incubated for one hour with uridine-2-C^{14} was reduced from 180 cpm/μg to 4 cpm/μg by prior exposure for 2 hours or longer to 0.5 μg actinomycin/ml. In infected cells, there was a marked increase in RNA synthesis, and at $12\frac{1}{2}$ to $13\frac{1}{2}$ hours after infection the rate was 14-fold greater than in uninfected, actinomycin-treated cells.

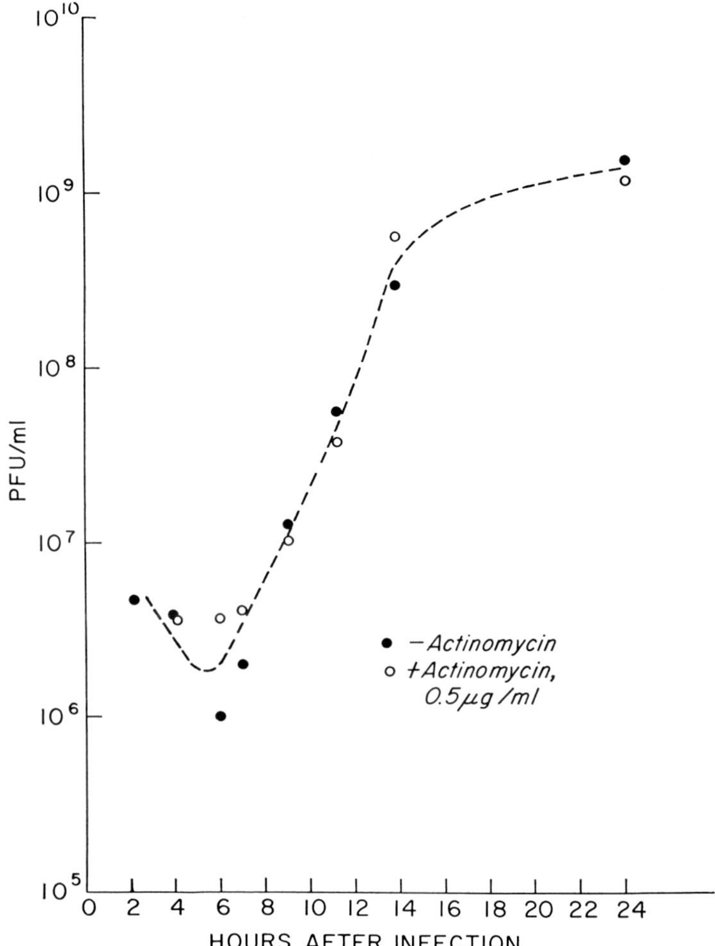

Figure 1. Time course of reovirus formation. In this and subsequent experiments, cells were infected with an input multiplicity of about 100 PFU/cell and, after adsorption, resuspended at a concentration of 4-5 x 10^5/ml. Actinomycin was added at 2½ hours after virus inoculation.

Since it is known that helical ribonucleotide polymers function poorly as templates for protein synthesis *in vitro* (13, 14, 15), it seemed possible that, in addition to double-stranded virus progeny RNA, reovirus-infected cells might contain virus-induced, single-stranded RNA which could function as a messenger or template for virus protein synthesis. To examine this possibility,

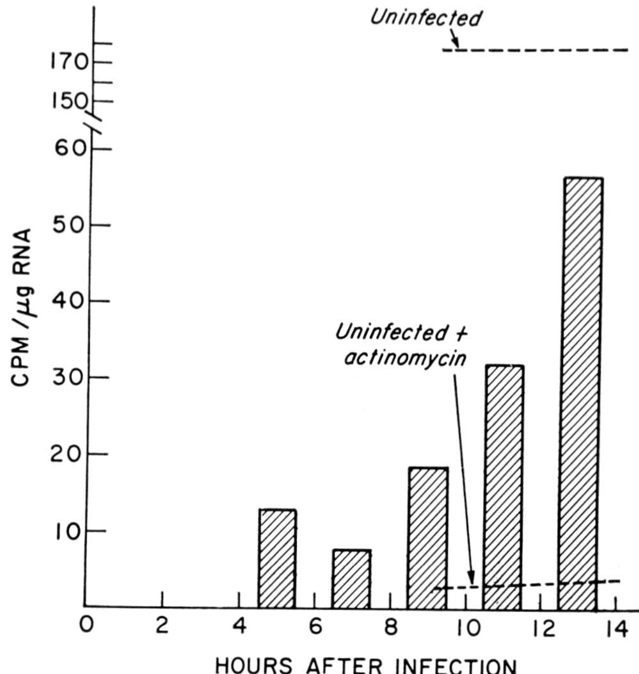

Figure 2. Stimulation of RNA synthesis in actinomycin-treated cells following reovirus infection. Infected cultures containing 0.5 μg actinomycin/ml were exposed to uridine-2-C[14] (2 x 10-6M, specific activity = 24.4 μc/μmole) for 1 hour at the indicated times. The cells were then chilled, centrifuged, extracted two times with 5% perchloric acid (PCA) at 4°C, and hydrolysed with 0.3N KOH at 37°C for 16 hours. After neutralization at 4°C with PCA and removal of the resulting precipitate by centrifugation, aliquots were plated for counting, and RNA was measured by the orcinol method (24). Uninfected cells were treated with actinomycin for 2 hours or longer before exposure to uridine-2-C[14] for 1 hour.

actinomycin-treated cells were exposed to uridine-2-C[14] at different times after infection, and the newly formed RNA was analyzed by centrifugation in linear sucrose density gradients. The sedimentation pattern of RNA extracted with phenol from either uninfected or infected cells includes the optical density peaks corresponding to 28 and 16S ribosomal RNA and 4S soluble RNA (Fig. 3). The distribution of newly formed, acid-precipitable RNA was determined both before and after incubation with

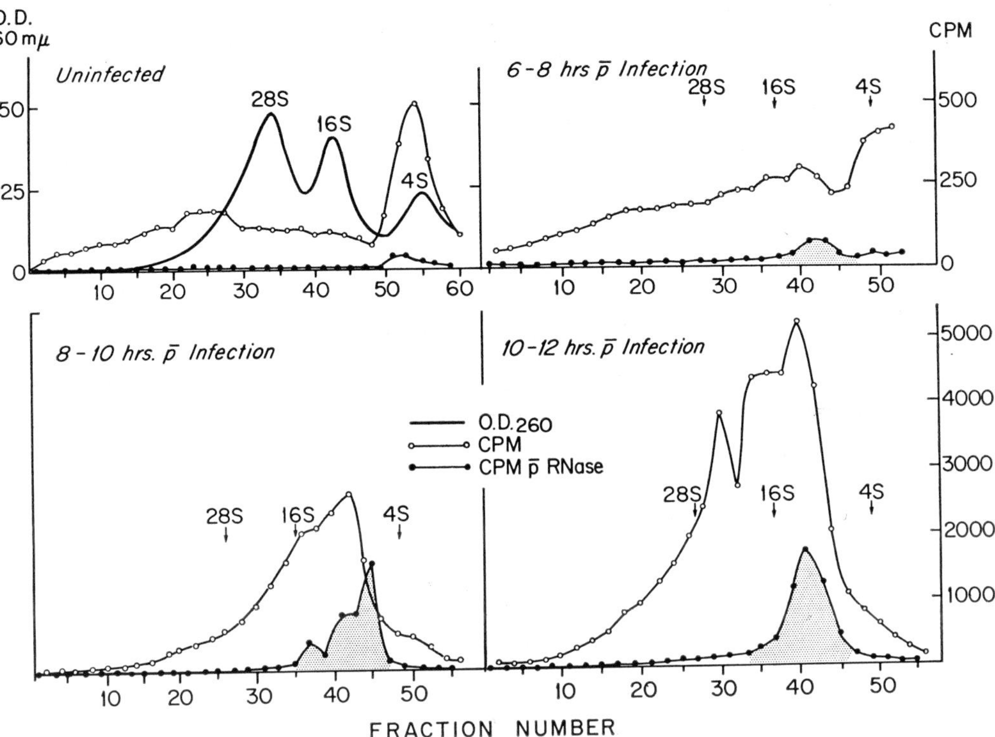

Figure 3. Sedimentation of newly formed RNA in sucrose density gradients. Antibiotic-treated cell cultures were incubated with uridine-2-C^{14} for the indicated time intervals, chilled, centrifuged, and washed with Eagle's medium at 4°C. The cells were resuspended at a density of 10^7/ml in 0.01M acetate buffer, pH 5.1 containing 0.1M NaCl and 0.001M MgCl$_2$. Sodium dodecyl sulfate (final concentration 0.34%) and an equal volume of H$_2$O-saturated phenol were added. The cell suspension was then shaken for 5 minutes at 60°C and rapidly cooled at 4°C. After centrifugation to separate the phases, the aqueous portion was re-extracted with phenol, and the RNA was precipitated from the aqueous extract at −20°C by adding two volumes of ethanol. The RNA was dissolved in acetate buffer (0.1M NaCl, no MgCl$_2$) and overlaid onto 15-30% linear sucrose gradients in the same solution. After centrifugation for 16 hours at 24,000 rpm in the Spinco SW-25.1 rotor, samples were collected through a hole punctured in the bottom of the tube. The optical density of the effluent was continuously monitored. Even-numbered, 0.5 ml fractions (O) were precipitated with 5% PCA and counted. Odd-numbered fractions (●) were incubated with 2 μg RNase/ml in 0.01M phosphate buffer, pH 7.4, 0.15M NaCl for 30 minutes at 37°C and then precipitated for counting.

TABLE II

BASE COMPOSITION ANALYSIS
Moles/100 Moles Nucleotides

	Cytidylic	Adenylic	Guanylic	Uridylic		$\dfrac{A+G}{C+U}$
	Acid	*Acid*	*Acid*	*Acid*	$G+C$	
RNA from Purified Reovirus	23.5	26.2	24.0	26.3	47.5	1.01
RNA Synthesized 9-12 Hrs. p Infection						
Double-stranded	24.2	26.1	23.2	26.5	47.4	0.97
Single-stranded	22.2	25.6	25.2	27.0	47.4	1.03

RNA was extracted from antibiotic-treated cells incubated 9-12 hours after infection with phosphate-P^{32} (10^{-5}M phosphate, 7.5 μc/ml). Single- and double-stranded RNA were separated on an MAK column, and their base compositions were compared with RNA-P^{32} which had been extracted from reovirus and further purified by equilibrium density gradient centrifugation in Cs_2SO_4 (16). Base composition analysis was performed by high voltage, paper electrophoresis of alkaline hydrolysates of the P^{32}-labelled RNA in pH 3.5 pyridine-acetate buffer as described previously (25).

ribonuclease (RNase). Alternate fractions collected from the gradients were either precipitated directly with 5% perchloric acid, or before precipitation, incubated for 30 minutes at 37°C with 2 μg RNase/ml in pH 7.4, 0.01M phosphate buffer and 0.15M NaCl, conditions which degrade single-stranded RNA but not double-stranded reovirus RNA (16). In uninfected cells treated with actinomycin, some incorporation of RNase-sensitive radioactivity into 4S RNA and a larger, heterogeneous component persisted. In infected cells, there was a progressive increase in the rate of RNA synthesis. The major fraction of the newly formed RNA was heterogeneous and nuclease-sensitive. In addition, 10-25% was resistant to enzyme digestion and sedimented at 10-12S, the position of double-stranded RNA extracted with phenol from purified reovirus.

When chromatographed on methyl-esterified albumin-Kieselguhr (MAK) columns, the double-stranded RNA eluted at a NaCl concentration of 0.7-0.8M and was well-separated from single-stranded RNA which eluted at 1.0-1.2M NaCl (17). The base composition of the RNA extracted from infected cells and purified by MAK chromatography is shown in Table II. The double-stranded RNA had a guanylic plus cytidylic acid content of 47.4% and a purine/pyrimidine ratio close to unity. These values are nearly identical to those obtained for RNA extracted from purified reovirus and banded in Cs_2SO_4. On the basis of its size, secondary structure and base composition, it is likely that the RNase-resistant RNA is progeny virus RNA.

The single-stranded, virus-induced RNA is made on a reovirus RNA template since it formed RNase-resistant complexes when annealed with denatured reovirus RNA (Table III). This interaction was concentration dependent (17) and highly specific. It did not occur when the RNA was annealed with native reovirus RNA, denatured rice dwarf virus double-stranded RNA, L cell RNA, or denatured L cell DNA. Furthermore, the single-stranded RNA did not self-anneal indicating that it represents copies of only one strand of the double-stranded reovirus RNA.

In a wide variety of organisms including poliovirus-infected tissue culture cells (18), protein synthesis occurs on polyribosomes, clusters of ribosomes linked together by single-stranded

TABLE III
ANNEALING OF REOVIRUS-DIRECT SINGLE-STRANDED RNA

Additions to RNA	Annealed	RNase	Acid-precipitable cpm/0.6 ml	% RNase Resistant
None	—	—	521	....
None	—	+	34	7
None	+	+	64	12
Denatured reovirus RNA (27 μg)	+	+	451	87
Native reovirus RNA (27 μg)	+	+	45	9
L-cell RNA (16 μg)	+	+	53	10
Denatured L-cell DNA (38 μg)	+	+	26	5
Denatured rice dwarf virus RNA (20 μg)	+	+	33	6

RNA was extracted from cells labeled with uridine-2-C^{14} 10-12 hours after infection, and single-stranded RNA was separated by MAK chromatography. An aliquot was dissolved in 0.5 ml 0.3M NaCl-0.005M phosphate buffer, pH 7 and mixed with 0.1 ml 0.1XSSC (standard saline citrate = 0.15M NaCl, 0.015M sodium citrate) containing denatured reovirus RNA. The virus RNA was denatured in 0.1XSSC by heating for 10 minutes at 100°C followed by quick cooling. The mixture was then placed in a water bath at 90°C and cooled to room temperature over a period of several hours. Samples were digested with 2 μg RNase/ml for 30 minutes at 37°C, precipitated with PCA and counted.

messenger RNA (19, 20, 21). The single-stranded, reovirus-directed RNA is also present in polyribosomes in infected cells (Fig. 4). Actinomycin-treated cells which had been incubated with uridine-2-C^{14} for 1 hour beginning at $10\frac{1}{4}$ hours after infection were homogenized, and the cytoplasmic fraction was centrifuged in a sucrose density gradient. The acid-precipitable radioactivity sedimented in the polyribosome region of the gradient, moving as a broad peak below the 74S single ribosomes. When the extract was digested before centrifugation with 2 μg RNase/ml for 10 minutes at 0-4°C, conditions which partially digest single-stranded RNA, the polyribosomes were degraded to single ribosomes. The partially digested virus-specific RNA remained attached to the single ribosomes and sedimented as a peak of radioactivity at 74S. The newly formed RNA was also purified from polyribosomes by phenol extraction. It was 92% nuclease-sensitive in the presence of 0.25M NaCl, but when annealed with denatured reovirus RNA became more than 80% RNase-resistant, indicating that it is single-stranded, virus-directed RNA. Its location in polyribosomes suggests that one of its functions is to serve as a messenger for virus protein synthesis.

The newly formed single-stranded and double-stranded RNAs are both directed by a virus RNA template, but the enzymes involved in their synthesis are unknown. Purified, double-stranded reovirus RNA does not function *in vitro* as a template for DNA-dependent RNA polymerase or DNA polymerase of *E. coli* (16). Although the possibility remains that the polymerases of L cells in tissue culture can use intact reovirus RNA as a template, the results of the following experiment suggest that the induction of new enzymes is required for reovirus production. Replicate cultures of actinomycin-treated, infected cells were set up, and at hourly intervals during the infectious cycle protein synthesis was inhibited by adding 2 μg cycloheximide/ml, a concentration which reduces protein synthesis by 90% in less than 1 hour. All

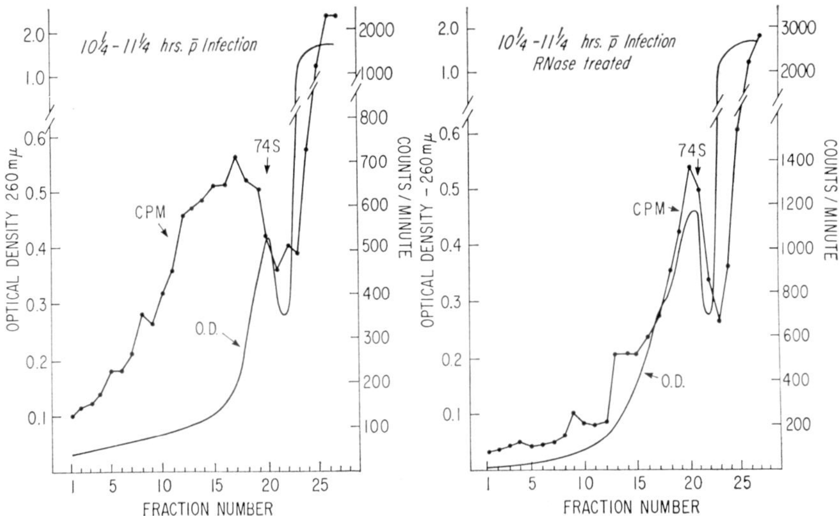

Figure 4. Sedimentation of cytoplasmic fractions from reovirus-infected cells. Actinomycin-treated cultures were exposed to uridine-2-C[14] 10¼-11¼ hours after infection. The cells were chilled, concentrated to 4 x 10[7]/ml in hypotonic media (0.01M tris buffer, pH 7.8, 0.01M NaCl, 0.0015M MgCl$_2$), and ruptured in a glass tissue homogenizer. After removing the unbroken cells and nuclei by low-speed centrifugation (800g - 2 minutes), one-half of the extract was made 0.25% with respect to sodium desoxycholate. The second half was incubated for 10 minutes at 4°C with 2 μg RNase/ml followed by the addition of detergent. The extracts were then centrifuged through 5-30% sucrose density gradients (2½ hours - 24,000 rpm in Spinco SW 25.1 rotor) and the fractions collected were assayed for acid-precipitable radioactivity. The gradients contained 19,480 cpm (left) and 19,960 cpm (right) of which 2,850 cpm and 1,050 cpm respectively were present in pellets.

cultures were then incubated with uridine-2-C[14] from 11-12 hours after infection, and the specific activity of the RNA was determined. As shown in Figure 5, there was no virus-specific RNA synthesized in cultures inhibited by cycloheximide at 6 hours or earlier in the infectious cycle. From the 6th to 9th hour, virus-specific RNA synthesis became progressively resistant to inhibition by cycloheximide.

Total virus-induced RNA synthesis was not detectably reduced by treatment with cycloheximide beginning 9-10 hours after infection. However, analysis of the RNA formed under these

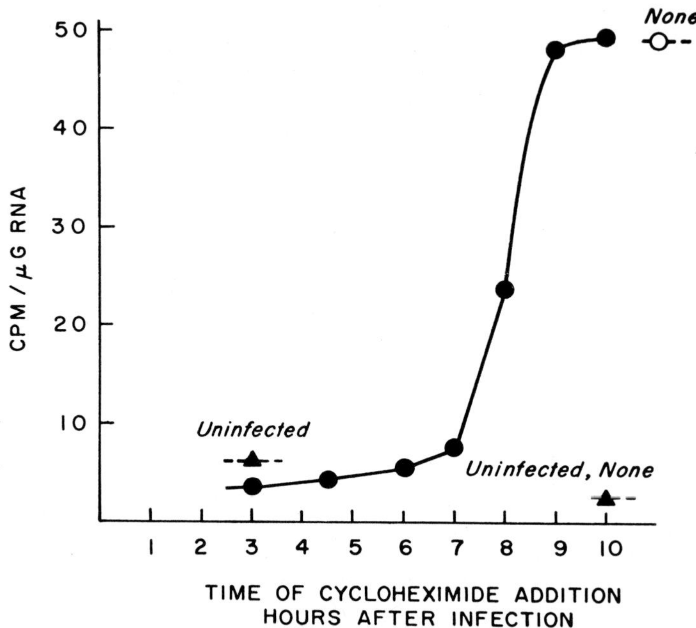

Figure 5. Effect of cycloheximide on reovirus-directed RNA synthesis. A series of actinomycin-treated, infected cell cultures was set up and incubated at 37°C. At indicated times (●), one in the series received 2 µg cycloheximide/ml to inhibit protein synthesis. All cultures were incubated with uridine-2-C[14] 11-12 hours after infection. The final specific activity of the newly synthesized RNA in each culture is plotted at the time of cycloheximide addition and is compared to an infected culture receiving no cycloheximide (O). Uninfected cells (▲) were similarly treated but without the addition of virus. Values are shown for an uninfected culture which received no cycloheximide and one which was incubated for 8 hours prior to the addition of uridine-2-C[14].

conditions revealed that there was a highly selective inhibition in the synthesis of double-stranded RNA. An actinomycin-treated culture was divided at 10 hours after infection, and in one of the resulting cultures protein synthesis was inhibited by adding cycloheximide. Both were then incubated with uridine-2-C^{14} from 11-12 hours after infection, the cells were extracted with phenol, and the RNA was sedimented in sucrose density gradients. In this experiment, 10% of the total RNA synthesized in the absence of cycloheximide was double-stranded and nuclease-resistant (Fig. 6A). In the culture which had received cycloheximide at 10 hours after infection, double-stranded RNA synthesis was reduced by more than 50%, whereas single-stranded RNA formation was not inhibited (Fig. 6B). These findings suggest that the two types of RNA are synthesized by different enzymes, and that the protein responsible for the formation of double-stranded RNA turns over more rapidly than that involved in single-stranded RNA synthesis. Similar findings have been reported using puromycin to inhibit protein synthesis (22).

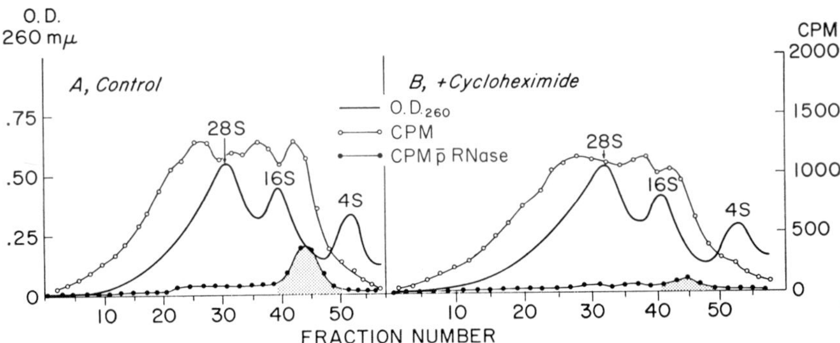

Figure 6. Inhibition of double-stranded RNA synthesis by cycloheximide. At 10 hours after infection, an actinomycin-treated culture was divided and 2 μg cycloheximide/ml was added to one of the resulting cultures. Both were incubated with uridine-2-C^{14} 11-12 hours after infection. The RNA was extracted and centrifuged in 15-30% linear sucrose density gradients as described in the legend to Figure 3.

In summary, reovirus-induced RNA synthesis was studied in cells treated with 0.5 μg actinomycin/ml to suppress cell RNA formation but not virus replication. Both single-stranded and double-stranded virus-specific RNAs were detected 6-7 hours after infection. The double-stranded RNA has a base composition, secondary structure, and size similar to RNA extracted from purified reovirus, and probably is the progeny virus RNA. The single-stranded material is heterogeneous in size and apparently represents copies of only one strand of the duplex RNA. Its presence in polyribosomes suggests that it functions as a template for virus protein synthesis.

The production of both types of RNA requires protein synthesis 6-9 hours after infection. At later times in the infectious cycle, the synthesis of only double-stranded RNA depends on continued protein formation, suggesting that two (or more) enzymes of different stability are involved in reovirus-induced RNA synthesis. If new enzymes whose synthesis is directed by single-stranded messenger RNA are required for reovirus replication, then it seems possible that the cell genome codes for the enzyme which synthesizes the reovirus-specific single-stranded RNA. The single-stranded RNA could then direct the synthesis of both virus coat proteins and an enzyme for the formation of double-stranded RNA. Alternatively, the parental double-stranded RNA may be reduced to a functional single-stranded messenger RNA inside the cell, although attempts to demonstrate such a conversion have been unsuccessful. Another possibility is that the virus particle, in addition to double-stranded RNA, contains an amount of non-helical RNA adequate to code for a new enzyme.

REFERENCES

1. SABIN, A. B.: Reoviruses. A new group of respiratory and enteric viruses formerly classified as ECHO type 10 is described. *Science, 130:*1387-1389, 1959.
2. GOMATOS, P. J., and TAMM, I.: The secondary structure of reovirus RNA. *Proc. Natl. Acad. Sci. (U. S.), 49:*707-714, 1963.
3. SCHAFFER, F. L., and SCHWERDT, C. E.: Purification and properties of poliovirus. *Advances in Virus Research, 6:*159-204, 1959.
4. GOMATOS, P. J., TAMM, I., DALES, S., and FRANKLIN, R. M.: Reovirus type 3: Physical characteristics and interaction with L cells. *Virology, 17:*441-454, 1962.
5. DARNELL, J. E., and LEVINTOW, L.: Poliovirus protein: Source of amino acids and time course of synthesis. *J. Biol. Chem., 235:*74-77, 1960.
6. LANGRIDGE, R., and GOMATOS, P. J.: The structure of RNA. Reovirus RNA and transfer RNA have similar three-dimensional structures, which differ from DNA. *Science, 141:*694-698, 1963.

7. GOMATOS, P. J., and TAMM, I.: Macromolecular synthesis in reovirus-infected L cells. *Biochem. Biophys. Acta, 72*:651-653, 1963.

8. REICH, E., FRANKLIN, R. M., SHATKIN, A. J., and TATUM, E. L.: Action of actinomycin D on animal cells and viruses. *Proc. Natl. Acad. Sci. (U. S.), 48*:1238-1245, 1962.

9. SHATKIN, A. J.: Actinomycin inhibition of ribonucleic acid synthesis and poliovirus infection of HeLa cells. *Biochem. Biophys. Acta, 61*:310-313, 1962.

10. SHATKIN, A. J.: Actinomycin and the differential synthesis of reovirus and L cell RNA. *Biochem. Biophys. Res. Commun., 19*:506-510, 1965.

11. KUDO, H., and GRAHAM, A. F.: Synthesis of reovirus ribonucleic acid in L cells. *J. Bacteriol., 90*:936-945, 1965.

12. LOH, P. C., and SOERGEL, M.: Growth characteristics of reovirus type 2: Actinomycin D and the preferential synthesis of viral RNA. *Proc. Soc. Exper. Biol. & Med., 122*:1248-1250, 1966.

13. NIRENBERG, M. W., and MATTHAEI, J. H.: The dependence of cell-free protein synthesis in *E. coli* upon naturally occurring or synthetic polyribonucleotides. *Proc. Natl. Acad. Sci. (U. S.), 47*:1588-1602, 1961.

14. SINGER, M. F., JONES, O. W., and NIRENBERG, M. W.: The effect of secondary structure on the template activity of polyribonucleotides. *Proc. Natl. Acad. Sci. (U. S.), 49*:392-399, 1963.

15. MIURA, K. I., and MUTO, A.: Lack of messenger RNA activity of a double-stranded RNA. *Biochem. Biophys. Acta, 108*:707-709, 1965.

16. SHATKIN, A. J.: Inactivity of purified reovirus RNA as a template for *E. coli* polymerases *in vitro*. *Proc. Natl. Acad. Sci. (U. S.), 54*:1721-1728, 1965.

17. SHATKIN, A. J., and RADA, B.: Reovirus-directed RNA synthesis in infected L cells. *J. Virology, 1*:24-35, 1967.

18. SCHARFF, M. D., SHATKIN, A. J., and LEVINTOW, L.: Association of newly formed viral protein with specific polyribosomes. *Proc. Natl. Acad. Sci. (U. S.), 50*:686-694, 1963.

19. GIERER, A.: Function of aggregated reticulocyte ribosomes in protein synthesis. *J. Mol. Biol., 6*:148-157, 1963.

20. WARNER, J. R., KNOPF, P. M., and RICH, A.: A multiple ribosomal structure in protein synthesis. *Proc. Natl. Acad. Sci. (U. S.), 49*:122-129, 1963.

21. WETTSTEIN, F. O., STAEHELIN, T., and NOLL, H.: Ribosomal aggregate engaged in protein synthesis: Characterization of the ergosome. *Nature, 197*:430-435, 1963.

22. KUDO, H., and GRAHAM, A. F.: Selective inhibition of reovirus-induced RNA in L cells. *Biochem. Biophys. Res. Commun., 24*:150-155, 1966.

23. EAGLE, H.: Amino acid metabolism in mammalian cell cultures. *Science, 130*:432-437, 1959.

24. MEJBAUM, W.: Estimation of small amounts of pentose especially in derivatives of adenylic acid. *Hoppe-Seyler's Z., 258*:117-120, 1939.

25. SEBRING, E. D., and SALZMAN, N. P.: An improved procedure for measuring the distribution of $P^{32}O_4$ among the nucleotides of ribonucleic acid. *Anal. Biochem., 8*:126-129, 1964.

INHIBITION OF POLIOVIRUS
REPLICATION BY GUANIDINE

DAVID BALTIMORE

Guanidine is one of the few chemical compounds which interferes with intracellular virus-specific processes but not with host cell metabolism. This strongly basic substance inhibits the growth of many enteroviruses, but not that of other classes of viruses (see Tamm and Eggers, 1963, for early references and a general discussion of inhibition by guanidine). In the course of studies on the multiplication of poliovirus, we have been investigating the mechanism of action of guanidine. Recent advances in the understanding of poliovirus replication have led to a new suggestion about the mode of action of guanidine. As background for this hypothesis, the following is a brief summary of our current knowledge about poliovirus growth.

Three types of RNA molecules are known to be involved in the intracellular multiplication of poliovirus. One is the viral RNA, which is a single molecule composed of about 6,000 nucleotides (Schaffer, 1962). This RNA, which has a sedimentation coefficient of about 355 (Darnell, 1962), is both the genetic material of the virus and the messenger RNA for the synthesis of viral proteins (Penman, Becker and Darnell, 1964). The second virus-specific molecule is double-stranded poliovirus RNA which contains both a strand of viral RNA and a strand of RNA with a complementary base sequence (Baltimore, 1966). This second RNA has no known function in replication and seems to be a by-product of infection (Baltimore, unpublished results). The third type of molecule is known as the replicative intermediate (Erickson, Fenwick and Franklin, 1964). This is a complex of double- and single-stranded RNA which is thought to be an intermediate in the synthesis of viral RNA (Baltimore and Girard, 1966). The mechanism by which this molecule plus the viral RNA polymerase can synthesize new chains of viral RNA has been discussed in some detail (Fenwick, Erickson, and Franklin, 1964; Ochoa, Weissmann, Borst, Burdon and Billeter, 1964).

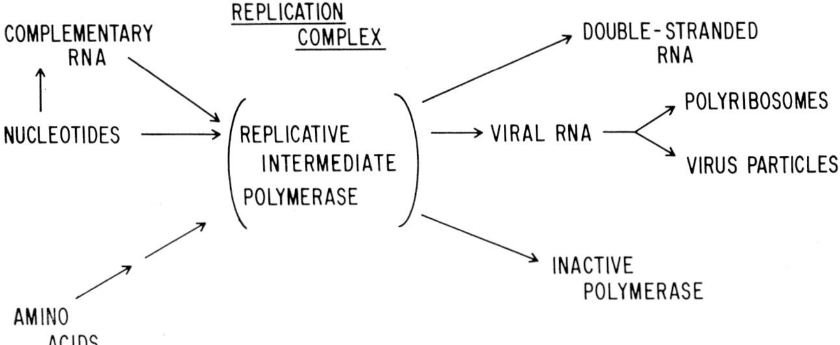

Figure 1. Scheme of the synthesis and processing of poliovirus RNA.

Our present understanding of the interrelation of these three molecules in the replication of RNA is schematically summarized in Figure 1. The central unit in this scheme is a recently identified structure which is called the "replication complex" (Girard, Baltimore and Darnell, 1967). This membrane-bound particle contains viral RNA polymerase, replicative intermediate and possibly other material. It sediments in a sucrose gradient at an average rate of 250 S. The replication complex is the site of viral RNA synthesis; soon after the completion of a molecule, the molecule is released from this complex to be incorporated into polyribosomes and virus particles (Girard *et al.,* 1967).

It is important to know whether a given molecule of RNA must pass through a polyribosomal stage before entering a virus particle. Although this question is not settled, the RNA in polyribosomes appears not to be the source of RNA for particles, on the basis of studies using cycloheximide. This drug is known to stop the movement of ribosomes along the messenger RNA in polyribosomes (Wettstein, Noll and Penman, 1964; Colombo, Felicetti and Baglioni, 1966), and would therefore be expected to stop the transfer of RNA from polyribosomes to virus particles if such a process occurred. Since it fails to interfere with the incorporation of RNA into particles (Baltimore, Girard and Darnell, 1966), it is unlikely that RNA passes through polyribosomes into virus particles. Therefore, as shown in Figure 1, soon after a molecule of RNA is released from the replication complex, a decision must be made about its ultimate fate. The per cent of RNA entering virus particles depends on the time after infection. Before 2.5 hours after infection, apparently due to a lack of functional coat protein, all newly made RNA goes to polyribosomes (Baltimore, Girard and Darnell, 1966).

Besides viral RNA, there are two other "products" of the replication complex shown in Figure 1. The first is double-stranded RNA. Numerous lines of evidence indicate that as molecules of replicative intermediate cease to function as templates, they are released from the replication complex as double-stranded RNA (Baltimore and Girard, 1966; Baltimore, unpublished). This process is occurring throughout infection and is balanced by a synthesis of new strands of complementary RNA which form new molecules of replicative intermediate.

The second material released from the replication complex is "inactive viral RNA polymerase." Although there is no direct evidence for turnover of the viral RNA polymerase, studies on the inhibition of RNA synthesis by inhibitors of protein synthesis lead to the conclusion that it occurs. These experiments will be discussed below. As with the replicative intermediate, this loss of polymerase must be balanced by a synthesis of new polymerase. Since both of the known components of the replication complex appear to be unstable, the structure as a whole may only have a transient existence.

With this background, we can return to a consideration of guanidine's mechanism of action. A number of years ago we suggested that guanidine might interfere with the synthesis of the viral RNA polymerase (Baltimore, Eggers, Franklin and Tamm, 1963). This hypothesis arose because it was felt that the drug must interfere with RNA synthesis in some manner and yet no direct *in vitro* action could be shown. Also, addition of guanidine at three hours after infection caused a reduction in polymerase activity measured at $3\frac{1}{2}$ hours. This would be consistent with a blockade in polymerase formation accompanied by the normal breakdown of polymerase and would make inhibition of RNA synthesis by guanidine analogous to inhibition of RNA synthesis by compounds which block protein synthesis (Eggers, Baltimore and Tamm, 1963) .

In order to further test this idea, we have compared inhibition of RNA synthesis by cycloheximide, the above mentioned inhibitor of protein synthesis, with inhibition by guanidine. The effect of cycloheximide on the amount of labeled RNA in the replication complex is shown in Figure 2. For this experiment, tritiated uridine was added to infected cells at three hours after infection and the amount of acid-precipitable radioactivity in the replication complex was determined (Girard *et al.*, 1967) . In a control culture, the radioactivity reached an equilibrium value in about 15 minutes (Curve A, Figure 2 and Girard *et al.*, 1967), but if cycloheximide was added at 5 minutes after the uridine, the

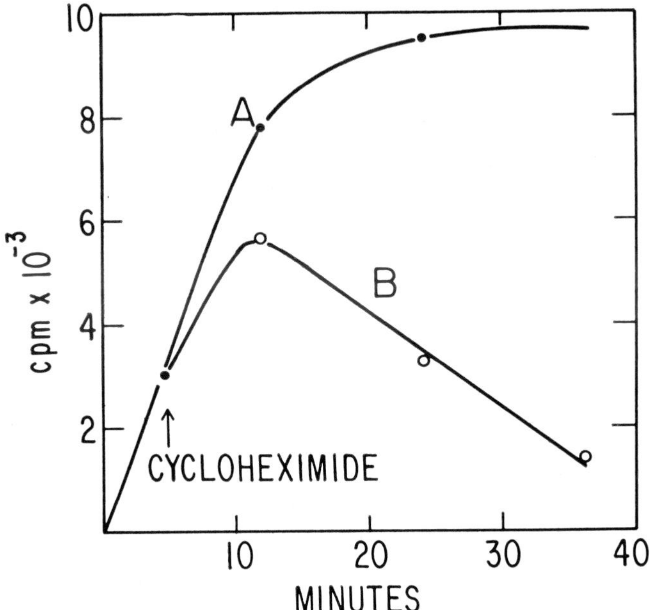

Figure 2. Effect of guanidine on the production of labeled poliovirus and the synthesis of viral RNA. Cells were treated with guanidine at about three hours after infection and then pulsed with tritiated uridine for 10 or 15 minutes. The amounts of labeled virus and labeled 35 S viral RNA (determined on sucrose gradients) were ascertained and expressed as per cent of virus or RNA found in untreated cells pulsed for the same length of time. Composite of three different experiments.

amount of label in the replication complex rose for a short while and then declined (Curve B, Figure 2). This result substantiates the notion previously put forth that the viral RNA polymerase is unstable and must be constantly renewed by *de novo* protein synthesis (Baltimore and Franklin, 1963). Although cycloheximide must affect the synthesis of viral coat proteins, there appears to be a large enough pool of protein so that the RNA which is made in the presence of the drug is not prevented from entering virus particles (Baltimore *et al.*, 1966). Also, the ratio of production of double- to single-stranded viral RNAs is not affected by cycloheximide treatment (Baltimore, unpublished).

The effects of guanidine differ completely from those of cycloheximide, which rules out the possibility that guanidine might prevent polymerase synthesis. Labeled RNA in the replication complex does not decrease, virus formation is impaired and the ratio of double- to single-stranded RNA is changed.

The first effect to be noted was that virus particle synthesis is inhibited by guanidine more rapidly than viral RNA synthesis (Fig. 3). In this experiment, the rates of RNA and virus synthesis were determined at various times after the addition of guanidine. Cultures of cells which had been infected for three hours were treated with the drug and then pulse-labeled with uridine for 10 or 15 minutes. The amounts of labeled 35 S viral RNA and labeled virus were then determined as described previously (Baltimore *et al.*, 1966), and expressed as per cent of a control, untreated culture. It is clear from Figure 3 that viral RNA synthesis is inhibited much more gradually than is the formation of virus. At a time when 50% of the normal rate of RNA synthesis is still occurring, only 5% of the normal amount of labeled virus is found. Furthermore, 20 minutes after guanidine treatment, the rate of RNA synthesis stabilizes at about 10% of normal while virus synthesis is undetectable. This result suggests that it may not be the formation of RNA which is affected directly by guanidine but rather the handling of finished RNA molecules.

This extremely rapid blocking of virus formation led us to investigate the location of the viral RNA made in the presence of guanidine. Under conditions which show that in untreated infected cells RNA is made in the 250 S replication complex and

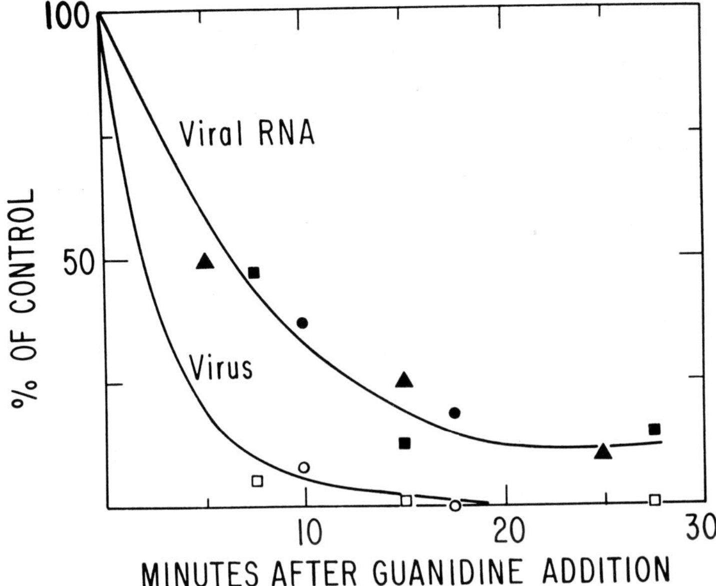

Figure 3. Effect of cycloheximide on the amount of RNA in the replication complex. Each point was determined by sucrose gradient analysis in the presence of EDTA as described in Girard *et al.*, 1967. Cultures received tritiated uridine at three hours after infection. A) Untreated. B) Treated with 133 μg/ml of cycloheximide five minutes after the addition of the tritiated uridine.

then released (Girard *et al.*, 1967), RNA made in the presence of guanidine is found to be trapped in a new structure which we have called the "guanidon." The guanidon sediments heterogeneously with an average rate of about 750 S. It is both the site of RNA synthesis and the site of accumulation of finished molecules. It therefore functions as the replication complex in the presence of guanidine but differs from the normal complex in that RNA is unable to leave the guanidon to form polyribosomes or virus particles. As quickly as 5 minutes after the addition of guanidine, almost all of the RNA being made is located in the guanidon.

The existence of the guanidon suggests that guanidine is acting on a heretofore unrecognized viral function. This function is the release of completed chains of viral RNA from the replication complex. Our present hypothesis is that guanidine interferes with such a function. Since guanidine-resistant and guanidine-dependent mutants can be isolated (Ledinko, 1962), this function must be carried out by a virus-coded protein which might even be the viral RNA polymerase.

This hypothesis does not explain why guanidine inhibits RNA synthesis and thus we must assume that the block in RNA processing causes an inhibition of RNA synthesis. To investigate this question, we have examined the RNA which is synthesized in the presence of guanidine and have found that it is very aberrant. There is approximately an equal amount of double-stranded and single-stranded viral RNA made and kinetic analysis is consistent with the idea that both RNAs are products of the replicative intermediate. What single-stranded RNA is made is not complementary RNA. One interpretation of these results is that the pathways shown in Figure 1 are normally operative in the presence of guanidine except that the lack of release of completed RNA strands from the replication complex causes a 10-fold reduction in the amount of single-stranded RNA being synthesized. This suggestion is being investigated at present.

In summary, the results on guanidine inhibition of poliovirus synthesis suggest that guanidine does not directly prevent RNA synthesis. The data is consistent with the idea that the drug interferes with the processing of finished chains and that these cannot leave the site of synthesis. This generates a structure which we have called the "guanidon." RNA synthesis is indirectly inhibited in such a manner that single-stranded RNA synthesis is much more drastically affected than double-stranded RNA synthesis.

This work was carried out under Research Grant No. CA-07592 from the National Cancer Institute.

REFERENCES

1. BALTIMORE, D.: *J. Mol. Biol., 18:*421, 1966.
2. BALTIMORE, D., EGGERS, H. J., FRANKLIN, R. M., and TAMM, I.: *Proc. Nat. Acad. Sci. (U. S.), 49:*843, 1963.
3. BALTIMORE, D., and FRANKLIN, R. M.: *Cold Spring Harbor Symp. Quant. Biol., 28:*105, 1963.
4. BALTIMORE, D., and GIRARD, M.: *Proc. Natl. Acad. Sci., 56:*741, 1966.
5. BALTIMORE, D., GIRARD, M., and DARNELL, J. E.: *Virology, 29:*179, 1966.

6. COLOMBO, B., FELICETTI, L., and BAGLIONI, C.: *Biochim. Biophys. Acta, 119:* 109, 1966.
7. DARNELL, J. E.: *Cold Spring Harbor Symp. Quant. Biol., 27:*149, 1962.
8. EGGERS, H. J., BALTIMORE, D., and TAMM, I.: *Virology, 21:*281, 1963.
9. ERIKSON, R. L., FENWICK, M. L., and FRANKLIN, R. M.: *J. Mol. Biol., 10:*519, 1964.
10. FENWICK, M. L., ERIKSON, R. L., and FRANKLIN, R. M.: *Science, 146:*527, 1964.
11. GIRARD, M., BALTIMORE, D., and DARNELL, J. E.: *J. Mol. Biol.,* In press.
12. LEDINKO, N.: *Cold Spring Harbor Symp. Quart. Biol., 27:*309, 1962.
13. OCHOA, S., WEISSMANN, C., BORST, P., BURDON, R., and BILLETER, M.: *Fed. Proc., 23:*319, 1964.
14. PENMAN, S., BECKER, Y., and DARNELL, J. E.: *J. Mol. Biol., 8:*541, 1964.
15. SCHAFFER, F. L.: *Cold Spring Harbor Symp. Quant. Biol., 27:*89, 1962.
16. TAMM, I., and EGGERS, H. J.: *Science, 142:*24, 1963.
17. WETTLSTEIN, F. O., NOLL, H., and PENMAN, S.: *Biochim. Biophys. Acta, 87:*525, 1964.

STUDIES ON THE REPLICATION OF LARGE RNA VIRUSES

WILLIAM S. ROBINSON

INTRODUCTION

ACTINOMYCIN D HAS BEEN A useful tool in studies on the replication of certain RNA containing viruses because it effectively blocks cellular RNA synthesis without interrupting viral RNA synthesis. In this way, viral RNA synthesis can be observed in cells in the absence of all other RNA synthesis. This method has been particularly useful in studies with the picorna viruses (1) and the RNA phages (2).

We have attempted to use actinomycin D in the same way to study viral specific RNA synthesis in chick embryo fibroblast cultures infected with Rous sarcoma virus and its helper Rous associated virus (RSV + RAV) and with two different myxoviruses: Newcastle disease virus (NDV) and influenza virus. Using the same experimental methods, we have observed what appears to be fundamental differences in the replication of these three viruses. The myxoviruses were chosen for comparison with RSV + RAV because of certain structural and biological similarities. All possess single stranded RNA as their nucleic acid (3-5), and they are among the largest and most complex in structure of the RNA viruses (6). Mature virus particles are not seen to accumulate in infected cells as is the case in cells infected with smaller RNA viruses such as poliovirus (7). Instead the avian tumor viruses and the myxoviruses probably mature by a process of "budding" from cell membranes (8) from which they acquire cellular lipid and protein in the formation of their outer envelope. This process results in a mature virion with a high content of lipid. All three viruses can be grown in chick embryo fibroblast cultures under conditions where all the cells can be infected synchronously. Furthermore, the rates of NDV and flu production in such cultures are probably not too different from the rate of RSV + RAV production. The total amount of virus produced by NDV and flu infected cells when measured by H³-uridine incorporation into the RNA of mature virus recovered from the culture medium at the end of one growth cycle is somewhat less than half that found in RSV + RAV labeled in the same way for a comparable period of time.

I will briefly review some of our experiments concerned with viral specific RNA synthesis in chick embryo fibroblast cultures infected with each of the three viruses in order to illustrate certain interesting differences. Actinomycin D was used in each case to inhibit cellular RNA synthesis with the expectation that viral specific RNA synthesis might then be observed.

I. Viral Specific RNA Synthesis in NDV Infected Cells (9)

The L-Kansas strain of NDV was used for infection of cultures under conditions which infect all cells. To label only viral specific RNA actinomycin D was added to the culture medium in a concentration of 2 μgm/ml to inhibit cell RNA synthesis and 30 minutes later H³-uridine was added. At the appropriate time, total RNA was then isolated from the cells by phenol extraction in the presence of sodium dodecyl sulfate and EDTA followed by alcohol precipitation.

Figure 1 shows the results of sucrose gradient fractionation of the RNA from a culture of uninfected cells (A) and NDV infected cells (B), after incubation with uridine-³H for 2 hours (6 to 8 hours after infection in the case of B) in the presence of actinomycin D. It can be seen that actinomycin D almost completely inhibited incorporation of uridine-³H into RNA of the uninfected culture (A). In contrast, at least three tritium labeled RNA components are seen in the RNA from NDV infected cells (B). The fastest sedimenting labeled RNA (peak in fraction 7) is in the position (57-S) expected for intact viral RNA. A second peak of labeled RNA (peak in fractions 13 and 14) sediments just ahead of 28-S cell RNA and is designated 35-S. The third peak (fraction 19) can be seen in the position of 18-S cell RNA.

The 18-S and 35-S labeled RNA components first appear in the cell 3 to 4 hours after infection and increase in amount up to about 10 hours when cell death occurs. The 57-S component is

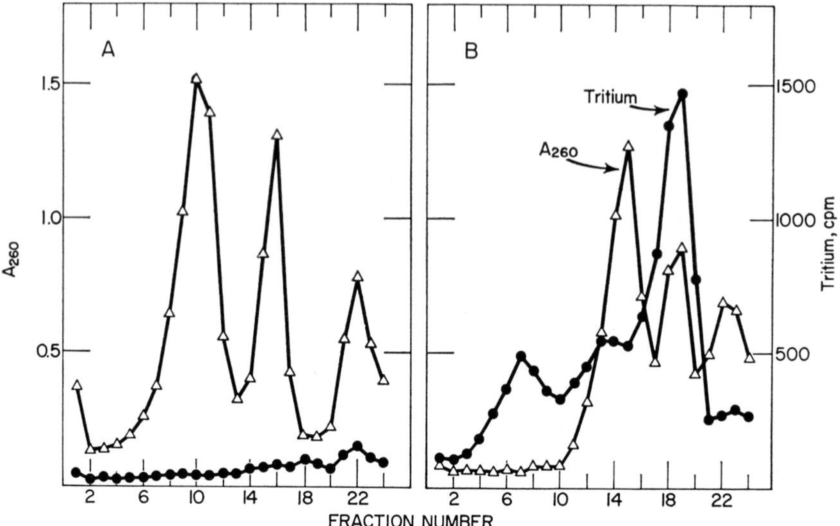

Figure 1. RNA synthesis in uninfected cells (A) and NDV infected cells (B) in the presence of actinomycin D. Actinomycin D (2 μgm/ml) was added to a culture of uninfected cells (A) and to cells 6 hr after infection with NDV (B). Thirty min later uridine-^{3}H (5 μc/ml, 5 mc/μmole) was added to both. After 2 more hr, the cell RNA was recovered from both and fractionated by sucrose gradient centrifugation. Tritium ($\bullet-\bullet$) and A$_{260}$ ($\triangle-\triangle$) were determined on each fraction of the sucrose gradient. Data are from Bratt and Robinson (9).

usually seen in a smaller amount than shown in Figure 2 and frequently is not seen at all. This could be due to the rapid exit from the cell of 57-S RNA into virus.

A fourth viral specific RNA component designated 22-S is always found when the RNA is sedimented for a longer time for better separation of RNA in the 22-S region (9).

These findings indicate that significant amounts of viral specific RNA are synthesized in the presence of actinomycin D. This is consistent with studies of the effects of actinomycin D on overall virus production which indicate that moderate inhibition occurs only during the early part of the NDV growth cycle in chick embryo fibroblasts (10). When NDV (11) and Sendai virus (12), another parainfluenza virus, are grown in other kinds of cells such as chorioallantoic membrane cells they are uneffected by high concentrations of actinomycin D at any stage in the growth cycle.

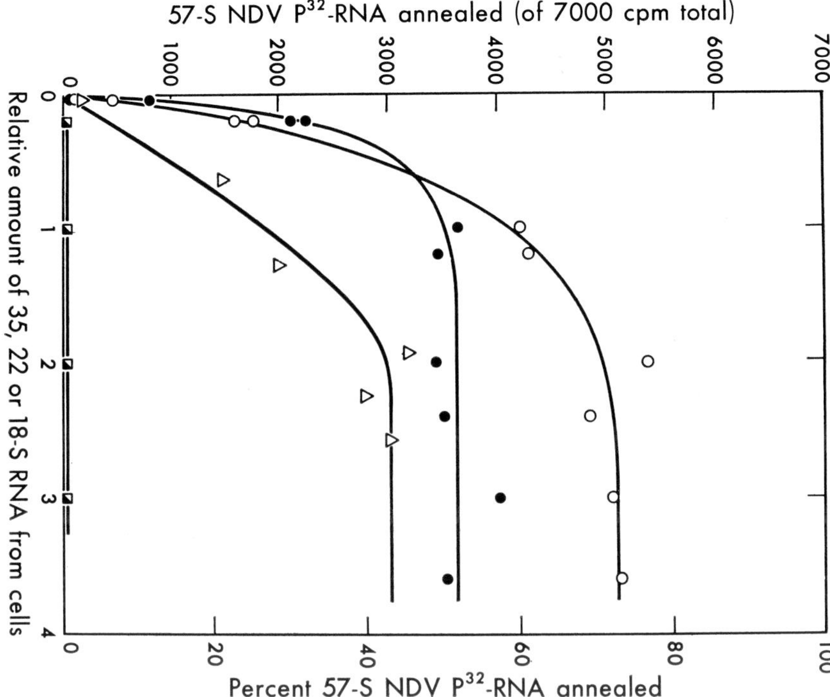

Figure 2. Annealing ³²P labeled 57-S NDV RNA with varying amounts of 35 (o—o), 22 (Δ—Δ) and 18-S (o—o) viral specific RNA's from NDV infected cells and with 18 (□—□) and 28-S (□—□) RNA's from normal cells. Annealing was carried out as described in Table 1. One unit on the abscissa represents the amount of RNA recovered from a single culture of cells. Data are from Bratt and Robinson (9).

The labeled RNA components shown in Figure 1 were almost completely digested to acid soluble products by pancreatic RNase in 0.2 M salt indicating that they were predominantly single stranded. However, a small amount of RNase resistance was found predominantly in the 35-S RNA component in different experiments. The proportion of RNase resistant RNA is much greater after a short period of labeling (e.g., 6 minutes) than after a long period such as 2 hours (Bratt and Robinson, unpublished observations). Such RNase resistant RNA probably represents base paired RNA analogous to that found in cells infected with other RNA viruses (13-16) which is thought to be an intermediate in viral RNA synthesis.

The base compositions of the 18-S and 35-S RNA components were found to be nearly complementary to that of the intact viral RNA (9). In order to test for complementarity of base sequence, tritium labeled 18-S, 22-S and 35-S viral specific RNAs were incubated under annealing conditions with unlabeled 57-S RNA from NDV and then tested for RNase resistance. The results are shown in Table I. The labeled 18, and 22-S RNA's were completely resistant and the 35-S RNA about 97% resistant to RNase after incubating with a large excess of 57-S viral RNA but not with TMV RNA or with ribosomal RNA from normal cells. In addition, there was no annealing when 18-S ribosomal RNA, labeled in normal cells in the absence of actinomycin D, was incubated with NDV RNA. Thus, in this experiment, the three RNA

TABLE I

ANNEALING ³H LABELED RNA

Unlabeled RNA (0.73 μgm) Added to the Annealing Mixture	TCA Insoluble cpm after Annealing and Digestion with RNase			
	18-S ³H-RNA from Infected Cells	22-S ³H-RNA from Infected Cells	35-S ³H-RNA from Infected Cells	18-S ³H Ribosomal RNA from Uninfected Cells
57-S NDV	7007	2200	2794	190
TMV	100	36	76	204
Uninfected cell 18-S	88	33	—	—
Uninfected cell 28-S	—	—	65	—
Total cpm added to the annealing mixture (i.e. no RNase digestion)	6736	2160	2707	3400

Tritium labeled RNA from infected cells was prepared as described in Figure 1. Tritium labeled 18-S cell RNA was prepared from normal cells in the same way after incubation of the cells with uridine-³H for 12 hours in the absence of actinomycin D. Unlabeled viral RNA was prepared from purified virus grown in eggs and was fractionated to obtain uncontaminated 57-S RNA as previously described (Duesberg and Robinson, 1965). TMV was a gift of Dr. C. A. Knight and the RNA was prepared by phenol extraction. Annealing was done by placing solutions containing the appropriate nucleic acids in 0.25 *M* salt at 90°C and cooling slowly to 37° over 6 hr. To test for annealing, each RNA sample in 0.1 *M* salt and 0.002 *M* MgCl₂ was then incubated with pancreatic RNase (10 μgm/ml) at 37° for 1 hr following which TCA precipitable radioactivity was determined. The data are from Robinson and Bratt, 1966.

components from infected cells were entirely complementary to viral RNA. The finding of unusually large amounts of RNA complementary to viral RNA in NDV infected cells has also been described by Kingsbury (47). In all experiments, the 18-S RNA component annealed completely to viral RNA. However, the 22 and 35-S RNA components did not always anneal completely to viral RNA. In different experiments, from 90 to 100% of the 35-S component and from 95 to 100% of the 22-S RNA were found to be complementary to viral RNA (9). Such findings are in accord with the observation that a small amount of 35-S RNA component in different experiments is in the form of base paired RNA.

The maximum fraction of the 57-S RNA component labeled in infected cells that could be annealed to 57-S RNA from virus was 30% (9). This suggests that about 1/3 of the labeled 57-S RNA component found in infected cells is complementary to viral RNA and 2/3 is probably identical to viral RNA.

When the 57-S RNA from NDV labeled with ^{32}p was incubated with various amounts of 18, 22 and 35-S RNA from infected cells annealing was again observed. Figure 2 shows that the maximum fraction of ^{32}p labeled viral RNA that could be annealed to the 35-S component was 70%, to the 18-S component 50% and to the 22-S component 40%. Only 70% could be annealed to a mixture of the 18, 22 and 35-S components (9) indicating that the three probably share base sequences in part and their sum does not make the entire complement of the 57-S viral RNA.

Annealing in this way with labeled viral RNA, it is possible to show that the complementary RNA components in the NDV infected cell are present in approximately the same amount and with the same sedimentation distribution in cells not treated with actinomycin D as in cells treated with actinomycin D (9). This suggests that actinomycin D does not significantly influence synthesis of these viral specific RNA components.

Fractionation of NDV infected cells into cytoplasmic and nuclear fractions after incubation with uridine-^{3}H for about 15 minutes in the presence of actinomycin D revealed that the labeled viral specific RNA was almost exclusively in the cytoplasm (9). Further fractionation of the cytoplasmic extract showed that most of the labeled RNA was attached to the polyribosomes. Figure 3

shows the results of sedimentation of the RNA isolated from each
of three parts of a sucrose gradient after centrifugation of a cytoplasmic extract from infected cells in such a way as to delineate
the polyribosomes and single ribosomes (9). Figure 3-1 shows the
RNA from the polyribosomes. A small amount of cellular 28 and
and 18-S RNA (fractions 12 to 18) and a large amount of 4-S
RNA (used as carrier during RNA isolation) are indicated by the

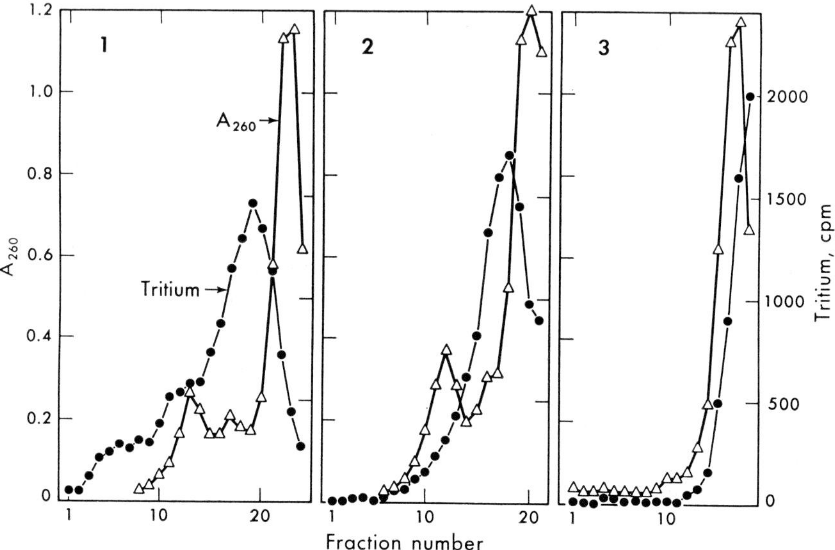

Figure 3. Viral specific RNA recovered from polysomes (1), single ribosomes
(2) and structures sedimenting more slowly than single ribosomes (3) after
sucrose gradient fractionation of the cytoplasmic extract from NDV infected
cells as described by Robinson and Bratt, 1966. Infected cells were incubated
with uridine-^{3}H in the presence of actinomycin D as described in Figure 1.
Cells were then removed from the culture dish with trypsin and fractionated
into cytoplasm and nucleus by dounce homogenization and differential
centrifugation. The cytoplasmic extract was made 0.5% with respect to
deoxycholate and fractionated in a sucrose gradient in such a way as to
delineate the polyribosome and single ribosomes. The regions of the sucrose
gradient containing the polyribosomes, the single ribosomes and the material
sedimenting more slowly than single ribosomes were recovered and the total
RNA was isolated from each by phenol extraction using S-RNA as carrier.
Each RNA preparation was then fractionated by sucrose gradient centrifugation as described in Figure 1 and 2_{260} (Δ—Δ) and tritium (●—●) were determined on each fraction of each gradient (data from Robinson and Bratt,
unpublished).

A_{260} tracing. The labeled viral specific RNA from the polysomes consist of 57-S RNA (fractions 4 to 8), 35-S RNA (fractions 10 to 12) and 18-S RNA (fractions 17 to 21). The labeled 18 and 35-S RNAs annealed to viral RNA as shown in Table 1. Figure 3-2 shows the RNA from the single ribosomes. Cellular RNA and carrier 4-S RNA are again indicated by the A_{260} tracing. Most of the labeled RNA is the 18-S viral specific RNA. Figure 3-3 shows that almost no cellular RNA and no labeled RNA sedimenting faster than 4-S RNA was recovered from the top of the polysome gradient (in structures sedimenting more slowly than single ribosomes).

Thus, it appears that in cells infected with NDV several viral specific RNA components can be labeled with uridine ^{3}H in the presence of actinomycin D. The RNA is predominantly single stranded although a small amount of base paired RNA is found. From 90 to 100% of three of the RNA components is complementary to viral RNA. One third of the RNA in the other component (57-S) is complementary to viral RNA and 2/3 is probably identical to viral RNA. Of the total viral specific RNA made, more than 90% is complementary in base sequence to the RNA of the virus.

The function of the complementary RNA pieces is unknown; however, the three occurring in greatest amount (18, 22 and 35-S) have common base sequences in part and their sum does not complete the total complement of viral RNA. In addition, they appear to be attached to the polyribosomes of the cell.

The finding of large amounts of single stranded RNA complementary in base sequence to viral RNA is so far unique for NDV. This is not the case in cells infected by small RNA viruses such as the RNA phage (17) and polio virus (18) where the predominant RNA made in the infected cell in the presence of actinomycin D is the viral strand. In addition, large amounts of RNA complementary to viral RNA are not found in cells infected with influenza virus or with RSV + RAV as will be described in the following sections.

II. Influenza Virus Infected Cells (19)

Similar experiments have been done with chick embryo fibroblast cultures infected with the PR-8 strain of influenza virus and the results indicate that viral RNA synthesis in influenza virus infected cells differs significantly from that in NDV infected cells.

Figure 4-a shows the results of sucrose gradient fractionation of the total RNA isolated from influenza virus infected cells fol-

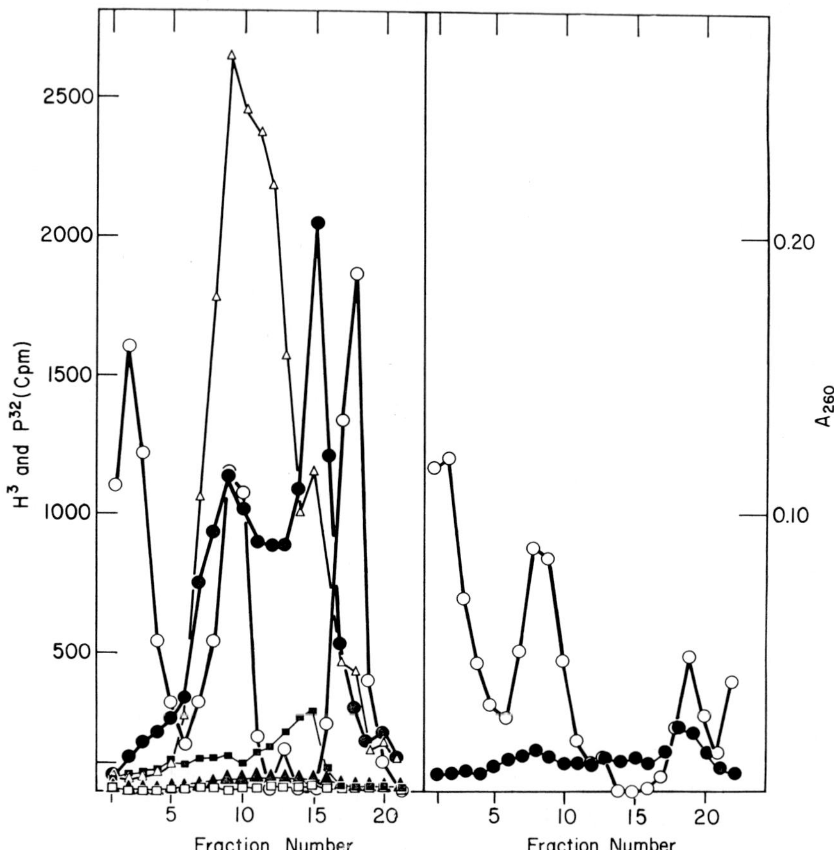

Figure 4. RNA synthesis in influenza virus infected (A) and uninfected cells (B) in the presence of actinomycin D. Actinomycin D (5 μgm/ml) was added to a culture of uninfected cells (B) and to cells 4 hr after infection with influenza virus. Thirty minutes later uridine-^{3}H (5 c/mM, 20 μc/ml) was added to each culture. After 150 minutes, more RNA was isolated from the cells of each and fractionated by sucrose gradient sedimentation. Absorbance at 260 mμ ($\circ$—$\circ$), TCA precipitable ^{32}P ($\triangle$—$\triangle$) and tritium ($\bullet$—$\bullet$) before RNase digestion, TCA precipitable ^{32}P ($\blacktriangle$—$\blacktriangle$) and tritium ($\blacksquare$—$\blacksquare$) after RNase digestion and TCA precipitable tritium after heat denaturation and RNase digestion ($\square$—$\square$) were determined on aliquots of each fraction. Data are from Duesberg and Robinson (19).

lowing incubation with ³H-uridine in the presence of actinomycin D from 4 to 6 hours after infection. ³²P labeled RNA from influenza virus grown in embryonated eggs was mixed with the cell RNA before centrifugation to serve as a sedimentation marker. Cellular 28-S (peak in fraction 2) and 18-S (peak in fraction 9) and carrier 4-S (peak in fraction 18) RNAs were detected by UV absorbancy. It can be seen that the ³²P labeled RNA from influenza virus grown in embryonated eggs consists of several components not distinctly separated. The fastest sedimenting viral RNA component sediments at about 18-S (³²P peak in fraction 9). The other viral RNA components have estimated sedimentation constants of 16, 14 and 9-S. All preparations of influenza virus RNA isolated in this way have the same four components. The significance of the four components is not certain although it is possible to recover RNA from virus in the form of an aggregate sedimenting at about 38-S when the RNA is extracted in a buffer containing divalent cation such as Mg^{++}. The 38-S component can then be dissociated into the slower sedimenting components by exposure to EDTA and low ionic strength (19). This finding suggests that the four components may come from the same virus particle. The recovery of such a 38-S component has also been reported by others (48, 49).

In the experiment shown in Figure 4-a, ³H-uridine has been incorporated into several viral specific RNA components recovered from the cells. The fastest sedimenting component appears to coincide with the 18-S component of viral RNA (³H peak in fraction 9) and another prominent component (³H peak in fraction 15) appears to coincide with the 14-S component of viral RNA (³²P peak in fraction 15). No RNA synthesis is observed in the uninfected cell in the presence of actinomycin D (Fig. 4-b).

Treatment of the RNA in each fraction shown in Figure 4-a with RNase in the presence of 0.2 M salt demonstrates that ³²P labeled viral RNA is rendered completely acid soluble. In contrast, a significant fraction of the viral specific RNA from the cell resists RNase digestion. The RNase resistant RNA sediments as heterogeneous material with a peak in fraction 15. When recovered from virus infected cells, the RNase resistant RNA can be shown to have other properties of double stranded RNA. It demonstrates a sharp thermal transition with Tm about 78°C in 0.01 M salt and its buoyant density in Cs_2SO_4 is significantly lower than that of single stranded viral RNA (19).

The labeled RNase resistant RNA represents about 15% of the total labeled viral specific RNA after a labeling period of 150 minutes (Figure 4-a) and 25% after a short 4 minute pulse (19). This suggests that the base paired structure could be an intermediate in synthesis of the larger fraction of single stranded RNA.

Annealing experiments similar to the ones described for NDV revealed that only about 5% of the total labeled viral specific RNA from infected cells could be annealed to a large excess of viral RNA (19). About one third of this was in the double stranded form and two thirds was single stranded RNA (19). Therefore, unlike NDV, most of the labeled viral specific RNA in influenza virus infected cells is probably RNA with the base sequence of the viral strands.

These results indicate that significant amounts of viral specific RNA are synthesized in the presence of actinomycin D in influenza virus infected cells. Although replication of influenza virus in chick fibroblast cultures is completely inhibited by relatively low concentrations of actinomycin D added early in infection (10), synthesis of viral components is apparently unaffected at later stages. This is consistent with the finding that actinomycin D added late in infection does not inhibit influenza virus production (10, 11) and more recent experiments suggesting that actinomycin D added early in infection interrupts the appearance of viral specific RNA replicase activity (20). Replication of fowl plague virus which is closely related to influenza virus is also inhibited by actinomycin D only during the early stages of infection (50). Thus, actinomycin D probably inhibits an early step in virus replication before the onset of viral RNA synthesis. The same may be true for NDV replication in chick fibroblast cultures.

Furthermore, it can be concluded that viral RNA synthesis in influenza virus infected cells appears to have similarities to that found in cells infected by the picorna viruses (1) and the RNA phages (2, 15-17). That is, RNA synthesis proceeds in the presence of actinomycin D, the major fraction of viral specific RNA found in infected cells is single stranded with base sequence like that of viral RNA and a smaller fraction of RNA consists of so-called "replicative form" consisting at least in part of base paired RNA.

III. RSV + RAV Infected Cells

RSV is an avian tumor virus which transforms all cells infected in chick fibroblast cultures. The Bryan strain of RSV is defective in that it can infect and transform cells but infectious virus is not produced unless the cells are infected also by a helper virus such as Rous associated virus (RAV) or one of several other leukosis viruses (21, 22). The RSV that is produced has physical, antigenic and host range properties indistinguishable from those of the helper virus. It has not been possible to separate the two physically (21, 23). Thus, studies concerned with RSV production are done with cells producing RSV and RAV and the probable ratio of RAV to RSV is about 10 (21). Such a mixture will be referred to as RSV + RAV. In the RNA from such a mixture there is no evidence that RSV RNA and RAV RNA can be distinguished by physical methods (3).

When chick embryo fibroblast cultures are infected with certain stocks of the Bryan strain of RSV + RAV, 80 to 100% of the cells can be infected with RSV as well as with RAV. The infected cells become transformed and produce large amounts of virus for many days. Using such cultures, it is possible to show that actinomycin D at concentrations above 0.1 μg/ml rapidly and almost completely inhibits incorporation of uridine-³H into the RNA of the mature virus recovered from the culture medium. This inhibition occurs at any time after infection. In the experiment shown in Figure 5, cultures were used one week after infection. Actinomycin D (0.5 μg/ml) and 30 minutes later uridine-³H were added to one culture (B) and only uridine-³H was added to the control culture (A). Culture medium was removed at different times, virus was purified from the medium, the RNA recovered and the amount of radioactivity in the 71-S viral RNA determined. It can be seen that production of virus (predominantly RAV) in the presence of actinomycin D is reduced to around 5% of that in the absence of actinomycin D. Higher concentrations of actinomycin D inhibit virus production by a greater

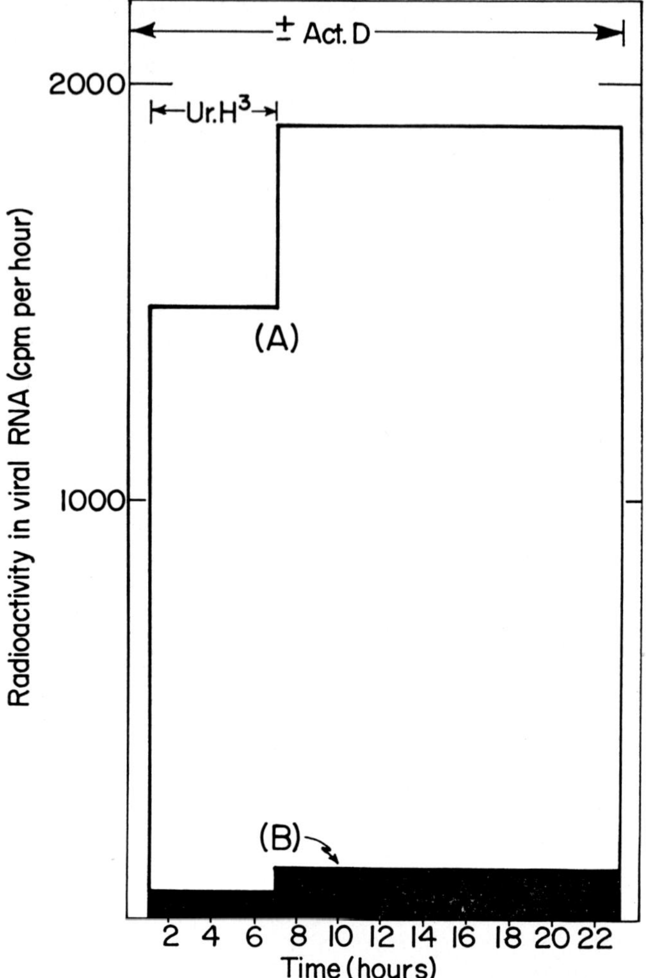

Figure 5. Effect of actinomycin D on production of tritium labeled virus by RSV + RAV infected cells. Two cultures of cells one week after infection with RSV + RAV were used. Actinomycin D (0.5 μgm/ml) was added to one culture (B) and after 30 min., uridine-³H (5 μc/ml, 1 mc/μmole) was added to that culture and to a control culture without actinomycin D (A). After 6 hr, the medium from both cultures was removed for virus purification and fresh medium was added to both. Actinomycin D was again added to culture B. After another 16 hr, the medium was again removed from both for virus recovery. Virus was purified from each sample of medium and viral RNA was extracted and the amount of radioactivity in the 71-S viral RNA determined. The shaded area represents the radioactivity expressed as tritium cpm per hour of incubation in virus produced by culture B and the unshaded area that produced by culture A. Data are from Robinson (24).

amount. Thus, it appears that actinomycin D inhibits RAV as well as RSV production at any time after infection.

Many experiments have been done in an effort to characterize viral specific RNA synthesis in RSV + RAV infected cells. The experiments have been for the most part negative. For example, when infected cells are incubated with uridine-^{3}H in the presence of different concentrations of actinomycin D, almost no labeled RNA can be found in the infected actinomycin D treated cells that is not also present in uninfected control cultures. An exception to this is the irregular finding of very small amounts of labeled RNA sedimenting with characteristics of intact 71-S RSV + RAV RNA detected at certain actinomycin D concentrations (24). The amount of this labeled RNA is never more than 5% of that found in mature virus in the culture medium of infected cells labeled for a comparable time in the absence of actinomycin D. With concentrations of actinomycin D higher than about 2 μg/ml, no such labeled RNA is observed. From these experiments, we conclude that actinomycin D under the conditions of our tissue culture not only inhibits virus production but also inhibits synthesis of RSV and RAV RNA.

Furthermore, annealing experiments with labeled viral RNA have not detected complementary RNA in the infected cell as can easily be done in influenza virus and NDV infected cells and has been done with cells infected with RNA phage (17), tobacco mosaic virus (25, 26). This suggests that complementary RNA in RSV + RAV producing cells is either not present or present in amounts too low to detect by methods used for other RNA virus infected cells.

In summary, it can be said that actinomycin inhibits RAV and RSV production at any time after infection of chick fibroblast cultures. Numerous other studies (27-31) have shown that production of infectious RSV is completely inhibited by actinomycin D. Recently, similar findings have been made with two other RNA tumor viruses, avian myeloblastosis virus (32) and in our laboratory Rauscher mouse leukemia virus (33). On the other hand, under most conditions the replication of RNA phages (34), the picorna viruses (1) as mentioned before, and the arboviruses (35, 36) is not inhibited by actinomycin D even at high concentrations. Exceptions to this have been reported for polio virus (37-39) and RNA phage (40) under special conditions. As previously discussed, inhibition of the replication of some of the myxoviruses occurs only at an early step in the growth cycle (10, 11) and this appears to distinguish them from the RNA tumor viruses. In the case of a double stranded RNA virus, Shatkin (41) has shown that Reovirus replication proceeds in the presence of moderate concentrations of actinomycin D. Thus, inhibition of replication at any time after infection may be characteristic of all RNA tumor viruses and may distinguish them from other groups of RNA viruses. It may indicate a common mechanism of replication for the RNA tumor viruses.

The mechanism of inhibition of RSV + RAV replication by actinomycin D is not known but it does appear that viral RNA synthesis is actually interrupted. Based on the inhibition of RSV replication by actinomycin D, Temin (29, 42) suggested that a DNA "provirus" was synthesized after RSV infection and that this new DNA served as a template for synthesis of RSV RNA in the cell. Temin's subsequent experiments annealing RSV RNA to cell DNA appear to be inconclusive (43). His unusual idea of a DNA provirus seems unjustified on the basis of inhibitor data alone. Actinomycin D does block cellular RNA synthesis and in this way it could interrupt a cell function essential for RSV replication. On the other hand, inhibition of DNA dependent RNA synthesis may not be the only action of actinomycin D in the cell. Recently experiments have suggested that at times actinomycin D may be involved in promoting breakdown of RNA in the cell (44, 45). In addition, there are indications that morphogenesis of phage may be interrupted by actinomycin D (46). Other effects in the cell may not yet be known. Furthermore, the observation that replication of viruses such as poliovirus and the RNA phage is inhibited by actinomycin D under special conditions [such as modification of the cells to be infected by incubation with insulin or serum-free medium (38)] indicates that inhibition of replication of the RNA tumor viruses need not necessarily be due to a different mechanism of virus replication but could be due to an altered condition of the infected cells. It seems apparent that de-duction of biochemical mechanisms from inhibitor effects alone is not always justified.

I wish to thank Dr. H. Rubin for support during part of this work and Dr. H. Robinson for critical review of the manuscript. I also am indebted to Doctors P. H. Duesberg and M. A. Bratt, who contributed many ideas and much work to experiments reported here.

This investigation was supported by U.S. Public Health Service Research Grant, CA 08557 from the National Cancer Institute.

SUMMARY

Viral specific RNA synthesis has been studied in chick embryo fibroblast cultures infected with three different RNA viruses: Newcastle disease virus (NDV), influenza virus (flu) and Rous sarcoma virus with its helper Rous associated virus (RSV + RAV). Cell cultures at different times after infection with a particular virus were incubated with uridine-H^3 in the presence of actinomycin D (1 to 5 μg/ml). At the appropriate time, cell material was then removed from the culture plate with a solution containing sodium dodecyl sulfate (SDS) and the RNA was recovered by phenol extraction and alcohol precipitation. The RNA was then fractionated by sucrose gradient sedimentation. In the case of NDV infected cells, at least four distinct components of RNA were labeled in the presence of actinomycin D. Three of these (18, 22 and 35-S) representing more than 90% of the viral specific RNA synthesized were almost completely complementary to viral RNA and were found almost exclusively in the polyribosome fraction of the cell. The fourth component sedimented in the position of viral RNA (57-S). An RNA component with properties of base paired RNA was found sedimenting in the 35-S region.

Flu infected cells on the other hand incorporated uridine-H^3 in the presence of actinomycin D predominantly into the viral strand of RNA and less than 5% of the labeled RNA was complementary to viral RNA. An RNA component with properties of base paired RNA was also found in flu infected cells.

Finally, experiments with RSV + RAV infected cells demonstrated that actinomycin D (> 0.1 μgm/ml) completely blocks the incorporation of uridine-H^3 into the RNA of whole virus appearing in the culture medium. In addition, very little uridine-H^3 was incorporated into the RNA of RSV + RAV infected cells in the presence of actinomycin D. This action of actinomycin D occurred at any time after infection and suggests that actinomycin D inhibits RSV + RAV RNA synthesis. In this way, RSV + RAV appears to differ from all other RNA viruses.

REFERENCES

1. DARNELL, J. E., PENMAN, S., and BALTIMORE, D.: In *Perspectives in Virology*, E. M. POLLARD, Vol. 4, p. 16, 1965.
2. KELLY, R. B., GOULD, J. L., and SINSHEIMER, R. L.: *J. Mol. Biol.*, *11*:562, 1965.
3. ROBINSON, W. S., PITKANEN, A., and RUBIN, H.: *Proc. Natl. Acad. Sci.*, *54*:137, 1965.
4. DUESBERG, P. H., and ROBINSON, W. S.: *Proc. Natl. Acad. Sci.*, *54*:794, 1965.
5. DUESBERG, P. H., and ROBINSON, W. S.: *J. Mol. Biol.*, In press.
6. WILDY, P., and WATSON, D. H.: *Cold Spring Harbor Symp.*, *27*:25, 1962.
7. HORNE, R. W., and NAGINGTON, J.: *J. Mol. Biol.*, *1*:333, 1959.
8. MORGAN, C., RIFKIND, R. A., and ROSE, H. M.: *Cold Spring Harbor Symp.*, *27*: 57, 1962.
9. BRATT, M. A., and ROBINSON, W. S.: *J. Mol. Biol.*, In press.
10. GRANOFF, A., and KINGSBURY, D. W.: In *Cellular Biology of the Myxoviruses*, Ed. G. E. W. WOLSTENHOLME and J. KNIGHT, p. 96, 1964.
11. BARRY, R. D., IVES, D. R., and CRUICKSHANK, F. G.: *Nature*, *194*:1139, 1962.
12. WHITE, D. O., and CHEYNE, I. M.: *Nature*, *208*:813, 1965.
13. MONTONGNIER, L., and SANDERS, F. K.: *Nature*, *199*:1178, 1963.
14. BALTIMORE, D., BECKER, Y., and DARNELL, J. E.: *Science*, *143*:1178, 1964.
15. KELLY, R. B., and SINSHEIMER, R. L.: *J. Mol. Biol.*, *8*:602, 1964.
16. FENWICK, M. L., ERICKSON, R. L., and FRANKLIN, R. M.: *Science*, *146*:527, 1964.
17. WEISSMAN, C., BORST, P., BURDON, R. H., BILLETER, M. A., and OCHOA, S.: *Proc. Natl. Acad. Sci.*, *51*:682, 1964.
18. DARNELL, J. E.: *Cold Spring Harbor Symp.*, *27*:149, 1962.
19. DUESBERG, P. H., and ROBINSON, W. S.: *J. Mol. Biol.*, In press.
20. HO, P. K., and WALTERS, C. P.: *Biochemistry*, *5*:231, 1966.
21. HANAFUSA, H., HANAFUSA, T., and RUBIN, H.: *Proc. Natl. Acad. Sci.*, *49*:572, 1963.
22. RUBIN, H. J.: *Cellular Comp. Physiol. Suppl.*, *1*:*64*:173, 1964.
23. HANAFUSA, H., HANAFUSA, T., and RUBIN, H.: *Virology*, *22*:591, 1964.
24. ROBINSON, W. S.: In *Viruses Inducing Cancer*, Ed. W. J. BURDETTE, p. 107, 1966.
25. SHIPP, W., and HASELKORN, R.: *Proc. Natl. Acad. Sci.*, *52*:401, 1964.
26. BURDON, R. H., BILLETER, M. A., WEISSMANN, C., WARNER, R. C., OCHOA, S., and KNIGHT, C. A.: *Proc. Natl. Acad. Sci.*, *52*:768, 1964.
27. BATHER, R.: *Proc. Am. Assoc. Cancer Res.*, *4*:4, 1963.
28. EIDENOFF, M. L., BATES, B., PEREZ, A., and DE LA SIERRA, A.: *Proc. Am. Assoc. Cancer Res.*, *4*:18, 1963.
29. TEMIN, H. M.: *Virology*, *20*:577, 1963.
30. BADER, J. P.: *Virology*, *22*:462, 1964.

31. VIGIER, P., and GOLDE, A.: *Virology, 23*:511, 1964.
32. ALLEN, D. W.: *Biochim. Biophys. Acta, 114*:606, 1966.
33. DUESBERG, P. H., and ROBINSON, W. S.: Submitted for publication.
34. HAYWOOD, A. M., and SINSHEIMER, R. L.: *J. Mol. Biol., 6*:247, 1963.
35. HELLER, E.: *Virology, 21*:652, 1963.
36. TAYLOR, J.: *Biochim. Biophys. Res. Commun., 14*:447, 1962.
37. GRADO, C., FISCHER, S., and CONTRERAS, G.: *Virology, 27*:623, 1965.
38. COOPER, P.: *Virology, 28*:663, 1966.
39. SCHAFFER, F. L., and GORDON, M.: *J. Bacteriol., 91*:2309, 1966.
40. HAYWOOD, A. M., and HARRIS, J. M.: *J. Mol. Biol., 18*:448, 1966.
41. SHATKIN, A. J.: *Biochim. Biophys. Res. Commun., 19*:506, 1965.
42. TEMIN, H. M.: *Virology, 23*:486, 1964.
43. TEMIN, H. M.: *Proc. Natl. Acad. Sci., 52*:323, 1964.
44. REICH, E., and GOLDBERG, I. H.: *Progress in Nucleic Acid Research, 3*:184, 1964.
45. HANDSCHACK, W., and LINDIGKEIT, R.: *Biochim. Biophys. Res. Commun., 23*: 793, 1966.
46. KORN, D., PROTASS, J. J., and LEIVE, L.: *Biochim. Biophys. Res. Commun., 19*: 473, 1965.
47. KINGSBURY, D. W.: *J. Mol. Biol., 18*:204, 1966.
48. BRUENING, G., and AGRAWAL, H.: *Proc. Natl. Acad. Sci., 55*:818, 1966.
49. PONS, M.: *Virology,* In press.
50. BARRY, R. D.: In *Cellular Biology of the Myxoviruses,* Ed. G. E. W. WOLSTEN-HOLME and J. KNIGHT, p. 51, 1964.

RIBOSOMAL, VIRAL AND REPLICATIVE VIRAL RNA SPECIES:

An Ultrastructural Comparison

NICOLE GRANBOULAN, KLAUS SCHERRER
AND RICHARD M. FRANKLIN

THE EXTENSIVE STUDIES ON THE ultrastructure of various types of DNA molecules resulted from the development of new methods for preparation of nucleic acids for electron microscopy, particularly the protein monolayer technique of Kleinschmidt *et al.* (29). Until quite recently, comparatively few ultrastructural studies had been made on purified RNA of ribosomal or viral origin (27, 3, 21, 26). These earlier studies brought attention to the difficulty of obtaining extended single-stranded RNA molecules which were then prepared for microscopy without using the protein monolayer technique.

In recent years, biochemical and biophysical studies resulted in a more precise description of the different RNA populations of the normal cell, viral RNAs, and the new species of RNA molecules appearing in infected cells during the replication of an RNA virus. These findings lent new impetus to the search for improved methods for preparing RNA for electron microscopy.

The application of the method used so successfully for DNA presented two major difficulties. In the first place, there was breakage of double-stranded viral RNA during extraction or spreading on monolayers. Despite this, length measurements made on double-stranded RNA revealed that the small molecules could be obtained as single pieces. Thus, it was possible to visualize the double-stranded replicative form of two RNA bacteriophages: M12 (1) and R17 (16). But it was almost impossible to observe entire molecules of RNA from reovirus or wound tumor virus (13, 28). This difficulty could be overcome if the RNA was released by gentle rupture of the capsid just before (18) or during

366

spreading. The second difficulty concerns the intra- or inter-molecular aggregation of single-stranded RNA. This aggregation is presumably due to intra- or inter-molecular H-bonding. Because of this phenomenon, the examination of isolated and elongated single-stranded RNA molecules was impossible (Fig. 1). The use of

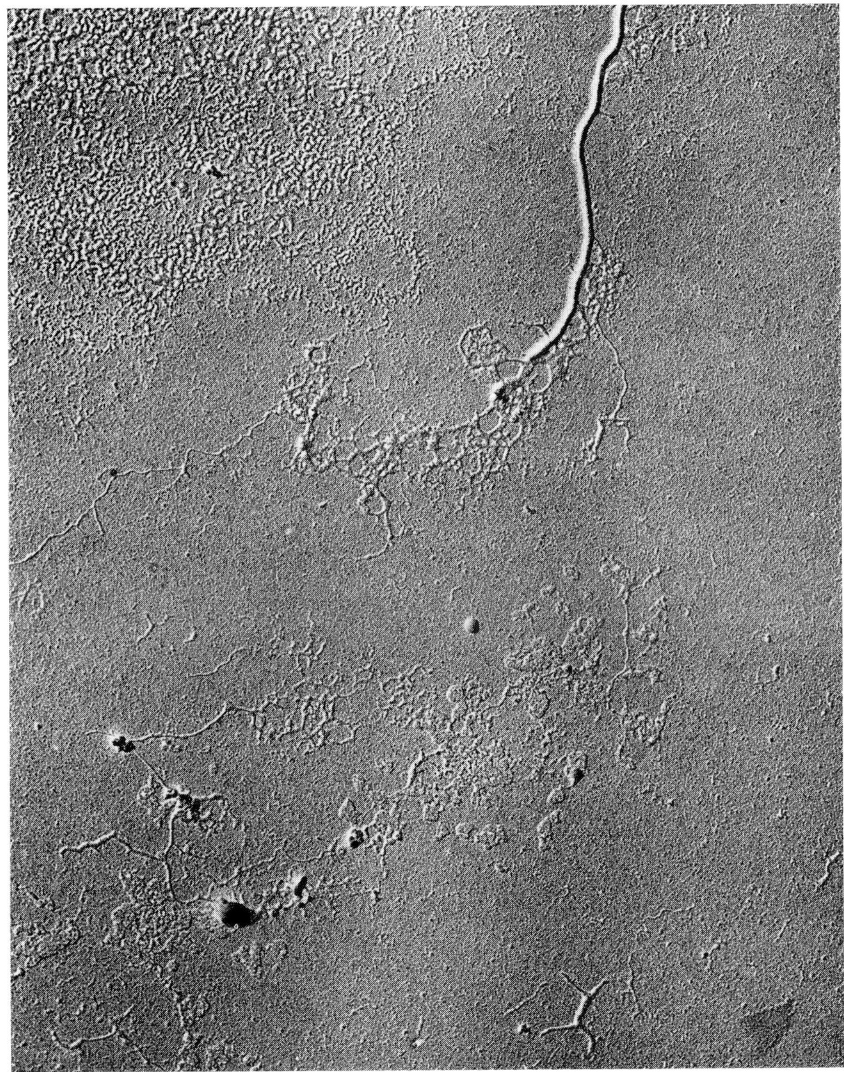

Figure 1. Single-stranded viral RNA extracted from turnip yellow mosaic virus, spread without urea. Numerous intra-or inter-molecular tanglings. M = 22,000 X.

urea during spreading, however, suppresses this aggregation (19) (3F).

Fortunately these technical difficulties were overcome at a time when very interesting problems were posed to electron microscopists due to the recent studies of biochemists and virologists. Some of these problems are the following: 1) What are the lengths of different ribosomal RNA fractions (28S and 18S RNA), their precursors (45S and 35S RNA) and of the messenger-like polydisperse 50-80S RNA, in normal cells (36, 38, 39, 41, 45)? How do these molecules compare in length with RNA from various types of viruses? 2) The single-stranded RNA extracted from oncogenic viruses have very high sedimentation coefficients (67 to 73S) which are assumed to be due to RNAs with molecular weights in the order of about 10^7 daltons. Examples are the RNAs of avian myeloblastosis virus (23, 25, 33), Rous sarcoma virus (22, 34), Rauscher leukemia virus (32, 12, 6), and mouse mammary tumor virus (4). These high sedimentation coefficients could also be due to special configurations of the single-stranded RNA and therefore ultrastructural studies were called for. 3) Replicative intermediate was found in bacteria infected with RNA bacteriophages and was believed to be the double-stranded template for viral RNA with nascent single strands (8). Replicative intermediate was recently purified (10) and therefore its ultrastructure could be compared with both single-stranded RNA extracted from the virion and with double-stranded replicative form (16).

In this paper an ultrastructural comparison of all of the above mentioned RNA species will be presented.

MATERIALS AND METHODS

I. Sources of RNA

1) **Ribosomal RNA and Heavy RNA Fractions from Normal Cells.** These RNAs were extracted from duck erythroblasts by the hot phenol method according to Scherrer *et al.* (40). The preparations were reprecipitated from 0.01 M EDTA with 0.1 M NaCl and 66% ethanol. Fractionations were carried out on linear sucrose gradients (15-30% $^W/_W$) in 0.01 M EDTA, 0.05 M NaCl at 0°C. Individual fractions were chosen according to the optical

114 molecules of lengths between 2 and 5 μ was 3.43 $\pm$ 0.80 μ (Chart 4). The distribution of lengths of molecules from a preparation of pure 45S r-RNA precursor fraction extracted from cells treated with Actinomycin D showed a peak between 3 and 3.50 μ (37). From this ultrastructural study, the different values of the sedimentation coefficients must be attributable to the different lengths of the molecules and not to differences in their configurations.

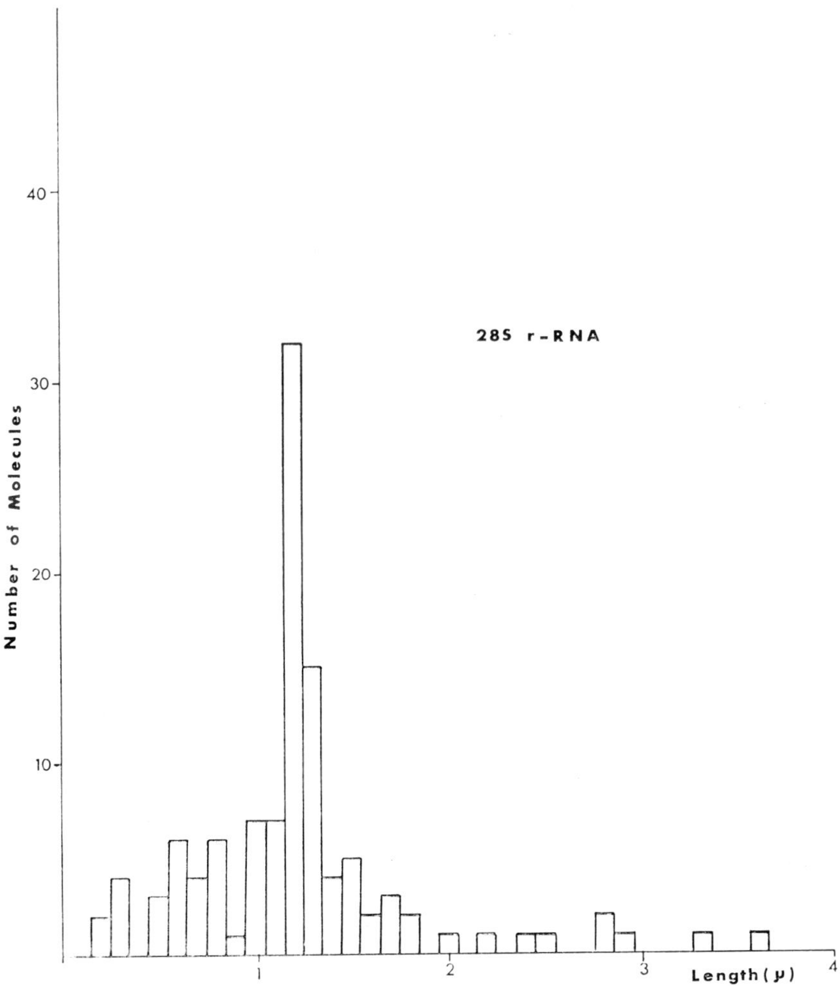

Chart 2. Distribution of the lengths of 113 molecules from the 28S ribosomal RNA fraction, spread in the presence of urea. Mean length = 1.14 $\pm$ 0.36 μ.

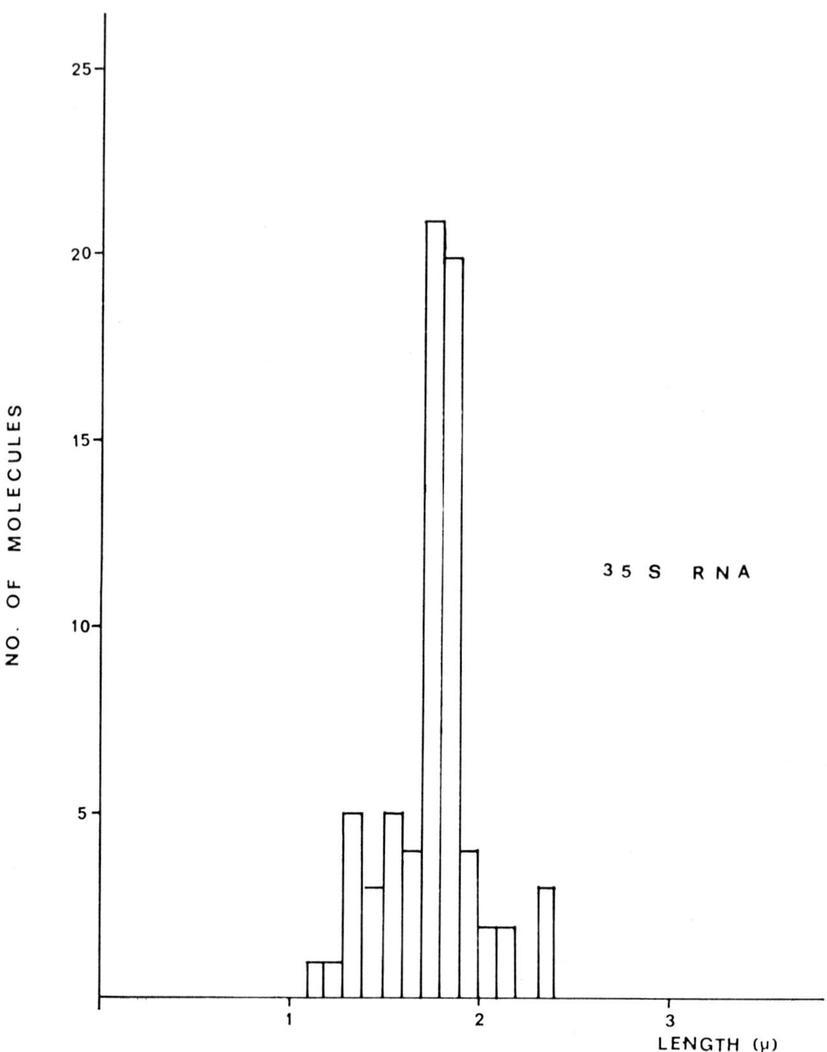

Chart 3. Distribution of the lengths of 71 molecules from the 35S precursor
fraction of ribosomal RNA, spread in the presence of urea. Mean length =
1.78 ± 0.23 μ.

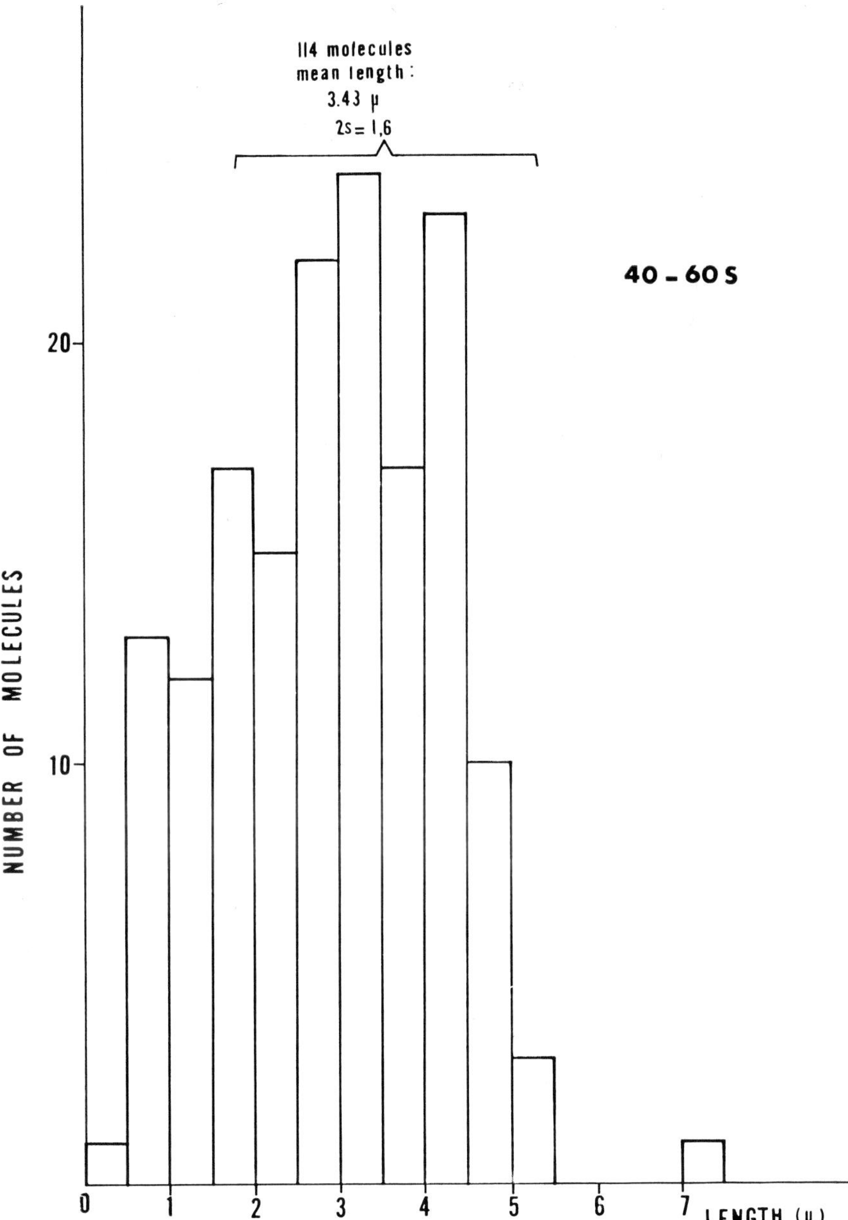

Chart 4. Distribution of the lengths of 158 molecules and fragments of the 40-60S RNA fraction, spread in the presence of urea. It appears to be heterogeneous. Mean length for the 114 molecules between 2 and 5.50 μ = 3.43 $\pm$ 0.80 μ.

II. Viral RNA

1) Double-stranded Viral RNA from Reovirus. This double-stranded viral RNA appears to be especially susceptible to breakage, as is evident from the ultrastructural studies made on phenol-extracted samples (28, 13). Even when phenol extraction is avoided and the RNA is released by disrupture of the capsid, long exposure of the viral suspension to urea (18) or sodium perchlorate (7) resulted in fragmented RNA molecules. Preferential breakage points have been described resulting in molecules of length of 0.35 μ, 0.60 μ and 1.10 μ (28, 13, 7).

But after gentle disruption of the capsid with urea at 4°C, it was possible to demonstrate long molecules of double-stranded RNA from reovirus (Fig. 3). The distribution centered around 5 to 5.50 μ with a peak at 5.10 — 5.20 μ (Chart 5). The mean length was 5.14 $\pm$ 0.10 μ (18). The distance between two successive base pairs in double-stranded RNA from reovirus and wound tumor virus was determined to be 3.00 — 3.05 Å by x-ray diffraction analysis (30, 43). The average value of the molecular weight of one base pair was calculated to be 710 for the RNA from reovirus (14, 13). Taking these data into account, the molecular weight of the double-stranded reovirus RNA molecule having a mean length of 5.14 μ is 12 x 10^6. This value fits well with that obtained from the chemical analysis of reovirus which showed that the amount of RNA per virion is greater than 10^7 daltons (14, 15). The viral genome appears, therefore, to be contained in a single molecule.

2) Single-stranded Viral RNA from Avian Myeloblastosis Virus. The samples of phenol-extracted RNA contained very long molecules of single-stranded RNA but also numerous smaller molecules (20). The long molecules were linear (Fig. 4) with a model length of 8.70 μ and a mean length of 8.30 μ. The analysis of the cumulative distribution of the lengths of all of the molecules and fragments showed that, besides the population of entire molecules, there exists a population of short fragments corresponding to quarter molecules and a population of fragments having a mean length of 5.1 μ which does not exactly correspond to half molecules (Chart 6). The significance of this last population is not yet known.

Figure 3. Double-stranded viral RNA molecule from reovirus, treated with urea for 5 min at 4°C before spreading. Length of this molecule = 5.60 μ. M = 29,500 X.

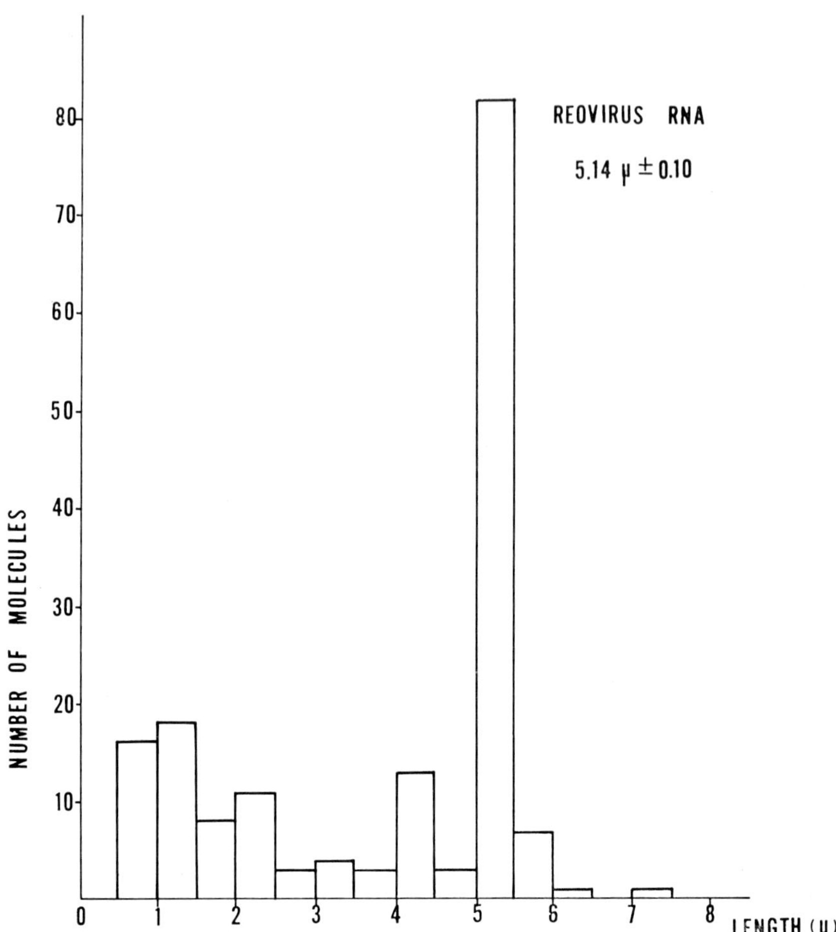

Chart 5. Distribution of the lengths of 170 molecules and fragments of RNA from reovirus treated with urea, peak between 5-5.50 μ with a mean length of 5.14 $\pm$ 0.10 μ (from *J. Microscopie, 6:* 1967) .

From the base composition of this RNA (25), the average value of the molecular weight for one nucleotide is 341. The number of nucleotides would be 29,350 if the sedimentation coefficient of 67S does correspond to a RNA molecule of molecular weight 10^7 daltons. Then, for the modal length of 8.70 μ, the distance between two successive bases should be 2.96 Å (8.7 μ divided by 29,350) . This value is close to that of 3.00 — 3.05 Å determined for the double-stranded RNA from reovirus and wound tumor virus (30, 43). We can therefore conclude that the high sedimentation coefficient does result from the predicted size of this long linear molecule of single-stranded RNA and is not due to any special configuration of the molecule. The latter conclusion must be taken with some reservations since the morphology was observed on molecules spread in the presence of urea.

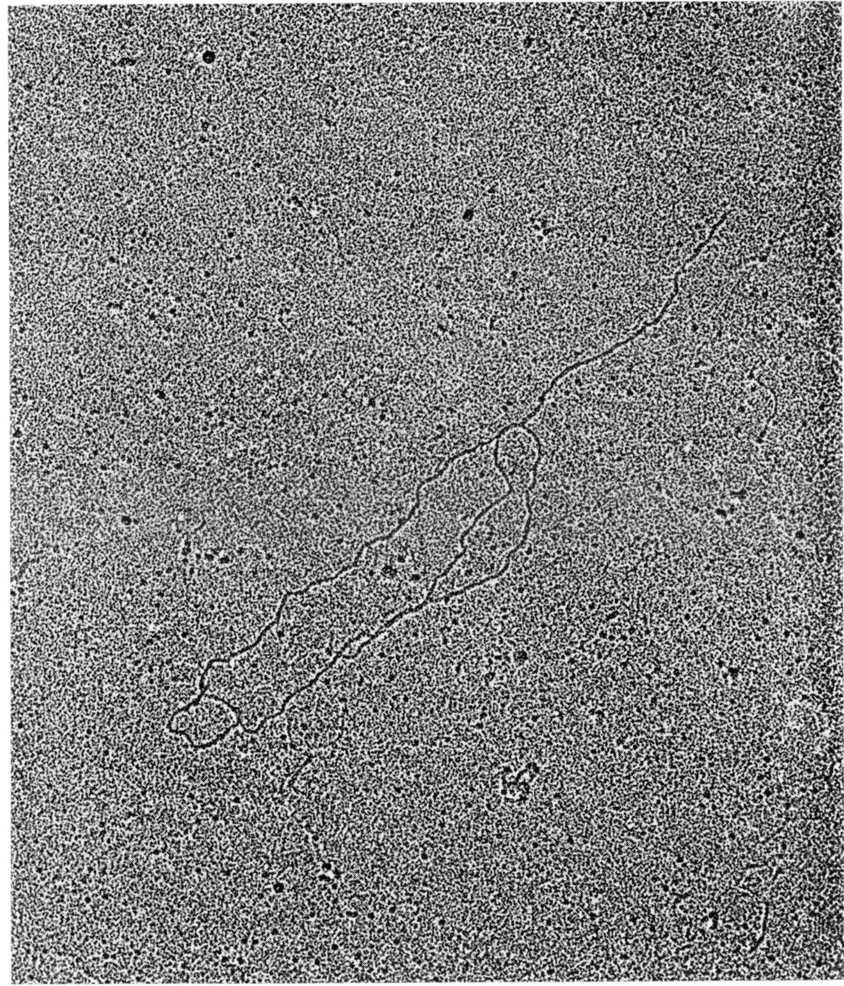

Figure 4. Long single-stranded RNA molecule extracted from avian myeloblastosis virus (AMV), spread with urea. M = 38,800 X.

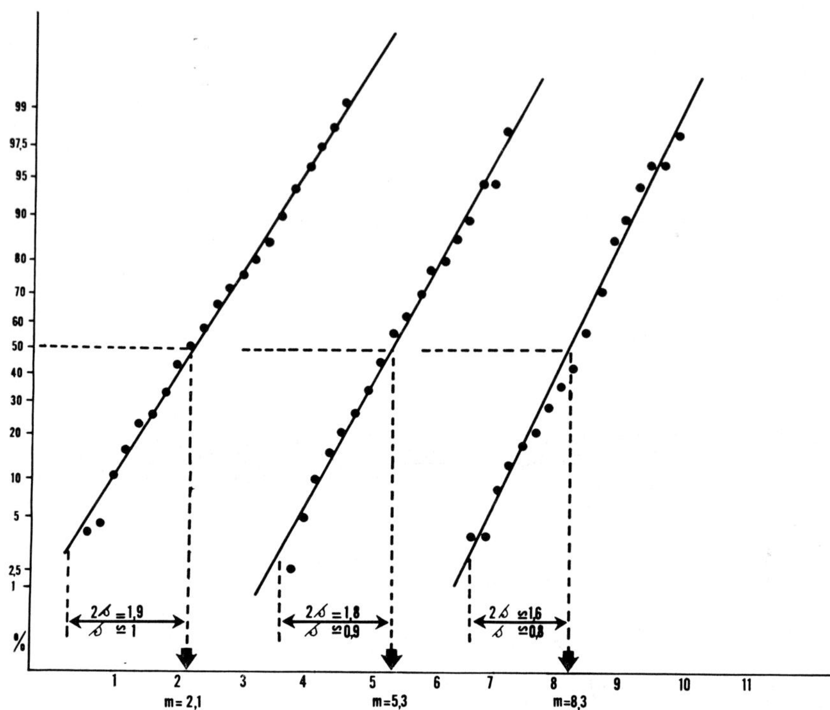

Chart 6. Cumulative distribution plotted on a gaussian plot of the lengths of 250 molecules and fragments of RNA extracted from avian myteloblastosis virus (AMV), spread in the presence of urea (from *J. Mol. Biol., 16:*571, 1966).

3) Single-stranded Viral RNA, Replicative Form, and Replicative Intermediate of the Bacteriophage R17. This material offered the best opportunity to compare the respective morphological features of the unique types of RNA appearing during the replication of a small RNA virus. Indeed, we can study the single-stranded RNA or replicative form (1) acting as template for the synthesis of single-stranded RNA, and the replicative intermediate or double-stranded template with nascent single strands (8, 10, 11).

Figure 5. Single-stranded viral RNA, double-stranded replicative form, and replicative intermediate from bacteriophage R17:

a and b) Single-stranded RNA molecules extracted from the virus, spread in the presence of urea. Length = 1.06 μ. M = 37,000 X.

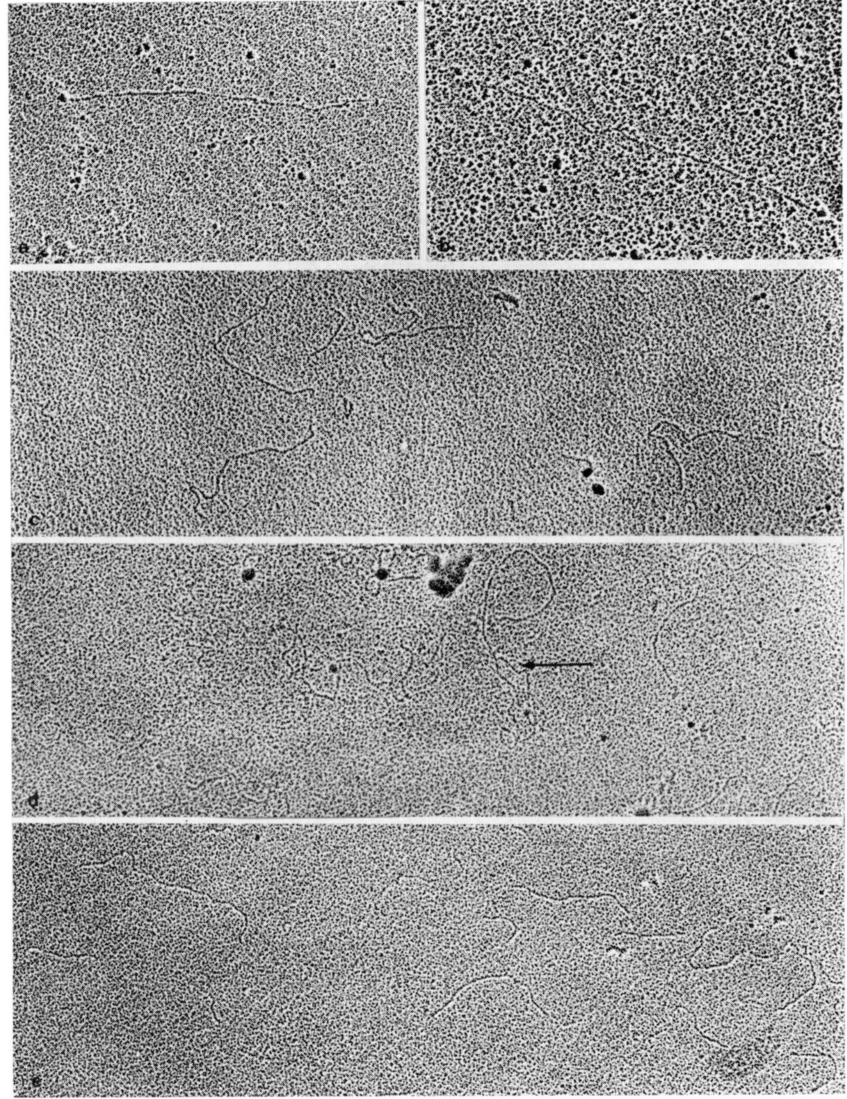

c) Double-stranded replicative form, spread in the presence of urea. M = 32,600 X.

d) Replicative intermediate spread in the absence of urea. The arrow shows one single-stranded branch attached to the double-stranded backbone molecule. M = 37,000 X.

e) Replicative intermediate treated with RNase showing removal of single-stranded branches. These molecules cannot be distinguished from the replicative form. M = 32,600 X.

The single-stranded viral RNA from R17 bacteriophage was
seen as short linear molecules (Fig. 5 a, b) with a mean length of
1.06 $\pm$ 0.06 μ (16) (Chart 7). Its molecular weight is 1.1 x 10^6
(31, 42) with 3,342 nucleotides (42). The distance between two
successive bases would therefore be 3.17 Å, a value which is close
to that of the double-stranded RNA.

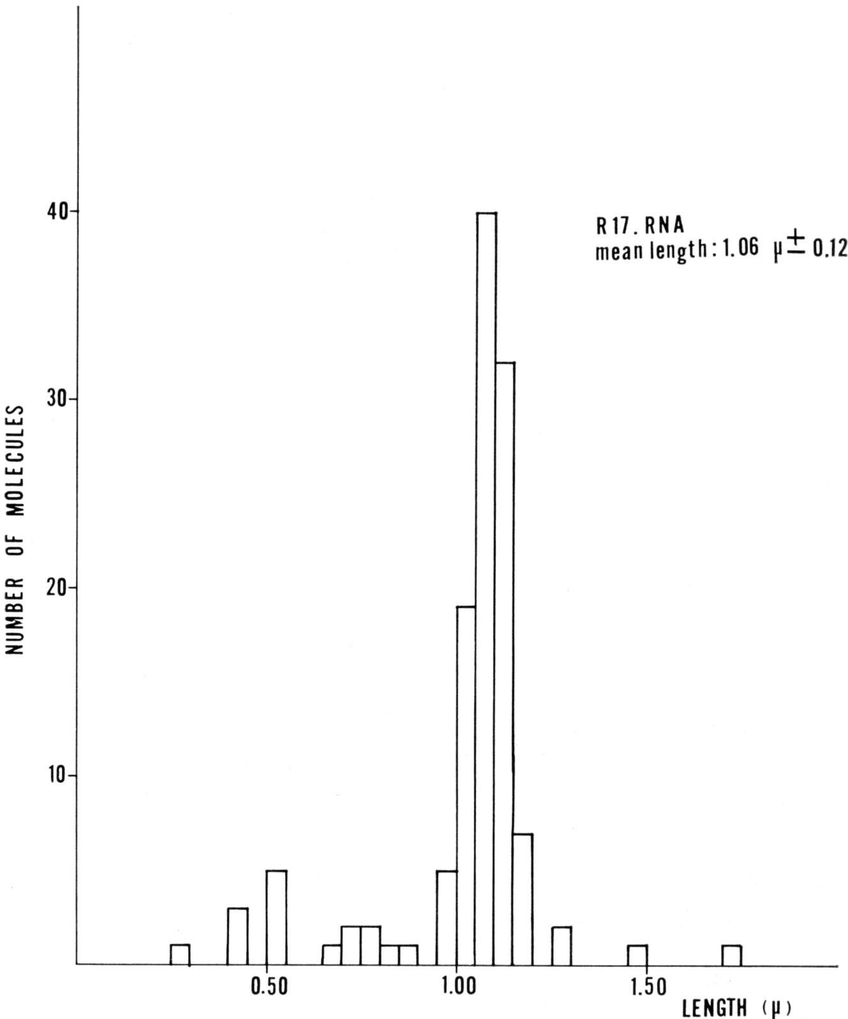

Chart 7. Distribution of the lengths of 123 molecules of single-stranded
RNA from bacteriophage R17, spread in the presence of urea. Mean length
= 1.06 $\pm$ 0.12 μ.

The double-stranded replicative form, spread with or without urea, was constituted of linear molecules having the same length as the single-stranded viral genome (Fig. 5c). It appeared, however, to be much more susceptible to breakage into fragments which have half and quarter lengths. Therefore, it breaks in the same manner as double-stranded DNA submitted to shearing forces (24, 2). It seems, however, to be more sensitive to this breakage than DNA, for reasons still unknown.

The length of the molecules of replicative form (1.07 μ) fits well with the molecular weight of 2.0 x 10^6 (9). Its linear configuration and its length were similar to the shape and length of the replicative form of the RNA phage M12 (1).

The replicative intermediate, spread with or without urea, had the structure of a branched polymer (Fig. 5d) (16). The entire molecules had the same length as those of replicative form. Fragments, corresponding to half and quarter molecules, were also present (16) (Chart 8). Usually, the double-stranded backbone molecule possessed one single-stranded branch. The length of the branch and the position of the branch point along the molecule were variable. There was a 1:1 correlation between the length of the branch and that of the corresponding portion of the backbone molecule (17). About 40% of the intact molecules of replicative intermediate were found to have branches in the presence or absence of urea (16). This value fits moderately well with predictions from biochemical and biophysical considerations (11, 16).

In the sample of replicative intermediate treated with pancreatic RNase, the structure of the molecules was similar to that of the replicative form, i.e., the single-stranded branches had disappeared (Fig. 5e) (16).

On the basis of our observations of branched molecules found only in preparations of replicative intermediate, removal of these branches by RNase, and the lack of influence of urea on replicative intermediate "backbone" molecules, we can give strong morphological support to biochemical and biophysical data. We can therefore conclude that the replicative intermediate is characterized by single-stranded nascent branches attached to the double-stranded template (8, 10, 11, 16). These morphological studies on the new species of RNA molecules appearing during the repli-

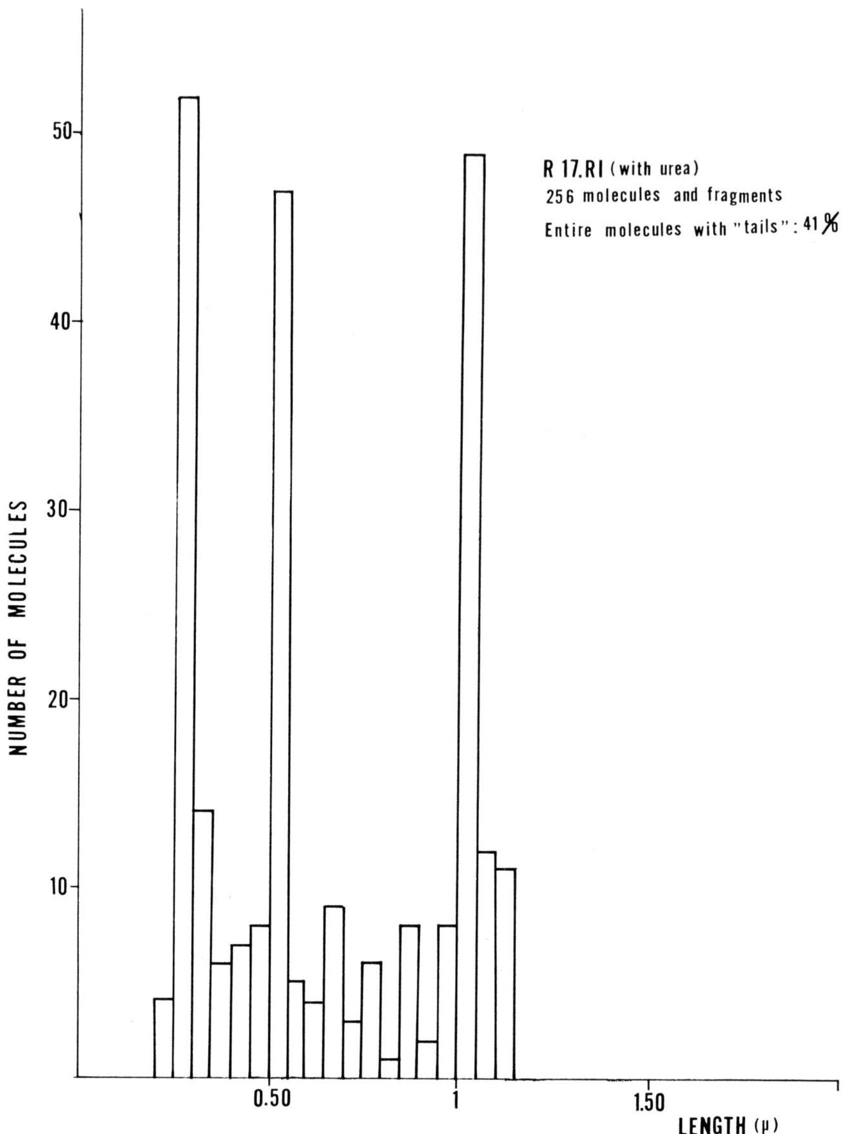

Chart 8. Distribution of the lengths of 256 molecules and fragments of replicative intermediate of bacteriophage R17, spread in the presence of urea. Presence of numerous fragments with half and quarter the length of the entire molecules (from *J. Mol. Biol., 22*:173, 1966).

cation of a RNA phage cannot, however, distinguish between molecules undergoing conservative or semi-conservative modes of replication.

The availability of single-stranded RNA with widely different sedimentation constants has afforded an opportunity to study the correlation between the sedimentation coefficient, the molecular weight, and the length of the molecules. A good correlation was found in plots of ln S (Svedberg units) versus ln L (microns) (Chart 9) or of molecular weight calculated from the Spirin equation versus length in microns (Chart 10). From Chart 9 the values of A and B have been calculated in the equation $S = AL^B$. Since there is obviously a direct relationship between M and L, i.e.,

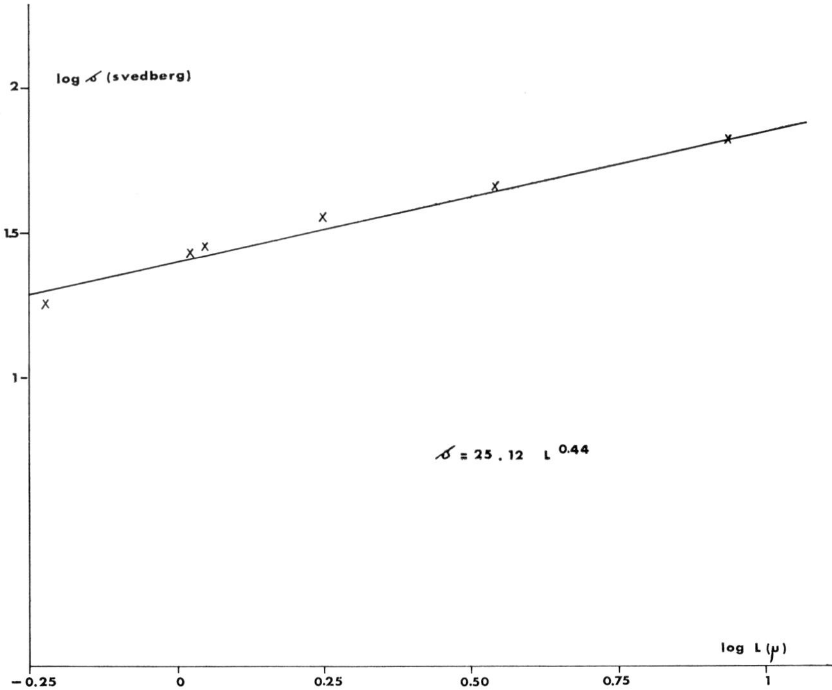

Chart 9. Plot of ln S (Svedberg units) versus ln L (microns) for six types of single-stranded RNA molecules. These are successively: 18S ribosomal RNA, 27S RNA from bacteriophage R17, 28S ribosomal RNA, 35S precursor of ribosomal RNA, 45S precursor of ribosomal RNA, 67S RNA from AMV. The slope of the straight line was used to obtain the exponent in the equation $S = 25.12\ L^{0.44}$.

 Medical and Applied Virology

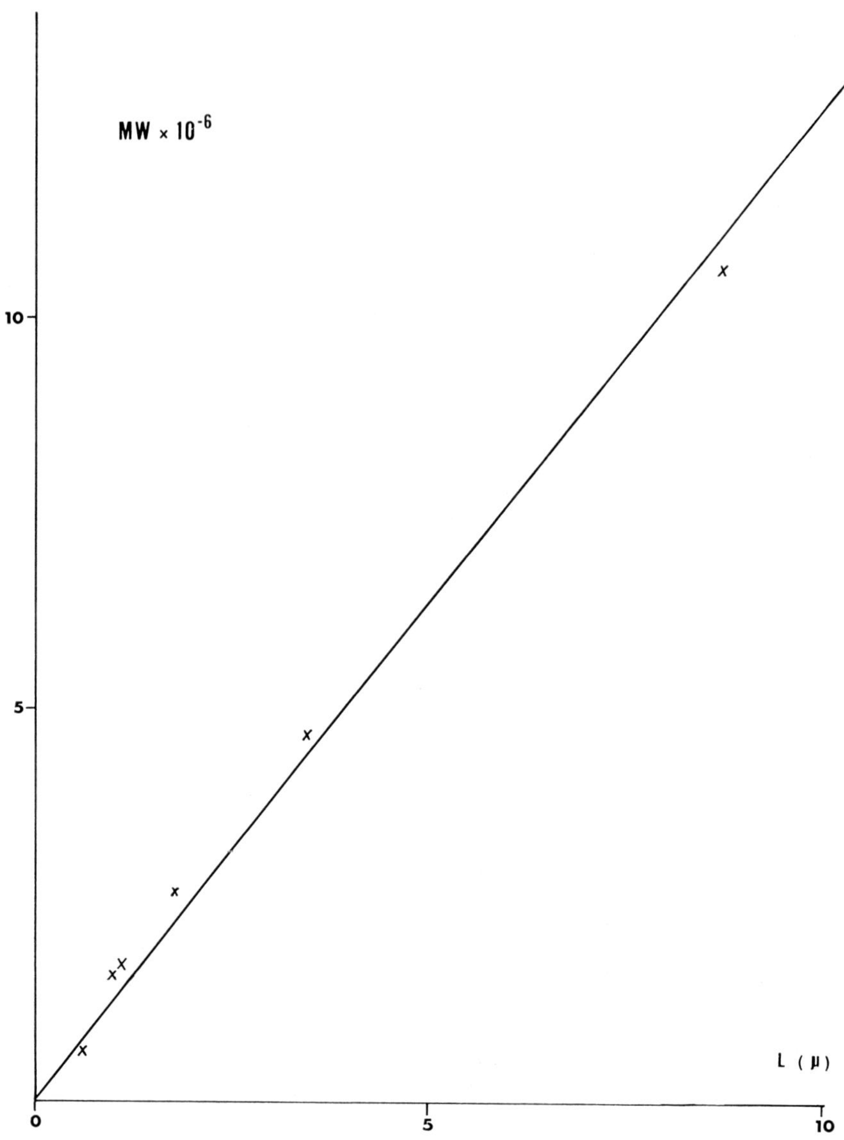

Chart 10. Correlation between molecular weight calculated from the Spirin formula ($M = 1550\ S^{2.1}$) and length of single-stranded RNA molecules. The points correspond successively to 18S r-RNA, 27S RNA from bacteriophage R17, 28S r-RNA, 35S precursor r-RNA, 45S precursor r-RNA, 67S RNA from AMV. Abcissa = length in microns; ordinate = M.W. x 10⁻⁶.

$M = \alpha L$, where α is a constant, then the exponent B must be related to the exponent in the equation relating S to M. In the Spirin equation $M = 1550\ S^{2.1}$ and therefore B would be $1/2.1$ and in the Gierer equation $M = 1100\ S^{2.2}$ and therefore B in this case would be $1/2.2$. The best value of B was $0.44 = 1/2.2$ and A was 25.1. Thus the relationship between S and length of molecules can be nicely correlated to the relationship between S and molecular weight, considering that data on only six types of single-stranded RNA are available. This relationship between S and L, and therefore between S and molecular weight, extends at present up to 67S. The direct measurement of lengths of RNA molecules thus provides strong support to the extension of the Spirin relationship between S and molecular weight up to molecular weights of 10^7 (25, 23, 33, 34, 22, 12, 32, 4, 5).

REFERENCES

1. AMMAN, J., DELIUS, H., and HOFSCHNEIDER, P. H.: Isolation and properties of an intact phage-specific replicative form of RNA phage M 12. *J. Mol. Biol., 10:* 557, 1964.

2. BURGI, E., and HERSHEY, A. D.: A relative molecular weight series derived from the nucleic acid of bacteriophage T2. *J. Mol. Biol., 3:*458, 1961.

3. DANON, D., MARIKOVSKY, Y., and LITTAUER, U. Z.: A comparative electron microscopical study of RNA from different sources. *J. Biophys. Biochem. Cytol., 9:*253, 1961.

4. DUESBERG, P. H., and BLAIR, P. B.: Isolation of the nucleic acid of mouse mammary tumor virus. *Proc. Natl. Acad. Sci., Wash., 55:*1490, 1966.

5. DUESBERG, P. H., and ROBINSON, W. S.: Isolation of the nucleic acid of Newcastle Disease virus (NDV). *Proc. Natl. Acad. Sci., Wash., 54:*794, 1965.

6. DUESBERG, P. H., and ROBINSON, W. S.: Nucleic acid and proteins isolated from the Rauscher mouse leukemia virus (MLV). *Proc. Natl. Acad. Sci., Wash., 55:*219, 1966.

7. DUNNEBACKE, T. H., and KLEINSCHMIDT, A. K.: Ribonucleic acid from Reovirus as seen in protein monolayers by electron microscopy. *Z. Naturforsch.,* 22 b, 159, 1967.

8. ERIKSON, R. L., and FRANKLIN, R. M.: Formation and properties of a replicative intermediate in the biosynthesis of viral ribonucleic acid. *Bact. Rev., 30:* 267, 1966.

9. ERIKSON, R. L., FENWICK, M. L., and FRANKLIN, R. M.: Replication of bacteriophage RNA: Some properties of the parental-labeled replicative intermediate. *J. Mol. Biol., 13:*399, 1965.

10. FRANKLIN, R. M.: Purification and properties of the replicative intermediate of the RNA bacteriophage R 17. *Proc. Natl. Acad. Sci., Wash., 55:*1504, 1966.

11. FRANKLIN, R. M.: Replication of bacteriophage RNA: Some physical properties of single-stranded, double-stranded, and branched viral RNA. *J. Virology, 1:*6, 1967.

388   *Medical and Applied Virology*

12. GAMIBERT, F., BERNARD, C., CHENAILLE, B., and BOIRON, M.: Acide ribonucléique de haut poids moléculaire isolé du virus leucémogène de Rauscher. *Compt. Rend. Acad. Sci., Paris, 261*:1771, 1965.

13. GOMATOS, P. J., and STOECKENIUS, W.: Electron microscope studies on Reovirus RNA. *Proc. Natl. Acad. Sci., Wash., 52*:1449, 1964.

14. GOMATOS, P. J., and TAMM, I.: The secondary structure of Reovirus RNA. *Proc. Natl. Acad. Sci., Wash., 49*:707, 1963.

15. GOMATOS, P. J., and TAMM, I.: Animal and plant viruses with double helical RNA. *Proc. Natl. Acad. Sci., Wash., 50*:878, 1963.

16. GRANBOULAN, N., and FRANKLIN, R. M.: Electron microscopy of viral RNA, replicative form and replicative intermediate of the bacteriophage R 17. *J. Mol. Biol., 22*:173, 1966.

17. GRANBOULAN, N., and FRANKLIN, R. M.: In preparation.

18. GRANBOULAN, N., and NIVELEAU, A.: Etude au microscope électronique du RNA du Reovirus. *J. Microscopie, 6*:23, 1967.

19. GRANBOULAN, N., and SCHERRER, K.: Examen au microscope électronique de RNA monocaténaires. *J. Microscopie, 5*:54a, 1966.

20. GRANBOULAN, N., HUPPERT, J., and LACOUR, F.: Examen au microscope électronique du RNA du virus de la Myéloblastose Aviaire. *J. Mol. Biol., 16*:571, 1966.

21. HALL, C. E.: In *Proc. IVth International Congress of Biochemistry, 9*:90, 1958.

22. HAREL, L., GOLDE, A., HAREL, J., MONTAGNIER, L., and VIGIER, P.: Isolement d'un acide ribonucléique de haut poids moléculaire de deux souches différentes de virus de Rous. *Comp. Rend. Acad. Sci., Paris, 261*:4559, 1965.

23. HAREL, J., HUPPERT, J., LACOUR, F., and HAREL, L.: Mise en évidence d'un acide ribonucléique de très haut poids moléculaire dans le virus de la myéloblastose Aviaire. *Comp. Rend. Acad. Sci., Paris, 261*:2266, 1965.

24. HERSHEY, A. D., and BURGI, E.: Molecular homogeneity of the deoxyribonucleic acid of phage T2. *J. Mol. Biol., 2*:143, 1960.

25. HUPPERT, J., LACOUR, F., HAREL, J., and HAREL, L.: High molecular weight RNA from Avian Myeloblastosis Virus. *Cancer Res., 26*:1561, 1966.

26. JOHNSON, M. W., and HILLS, G. J.: Electron microscopy of turnip yellow mosaic virus ribonucleic acid. *Virology, 21*:517, 1963.

27. KISELEV, N. A., GAVRILOVA, L. P., and SPIRIN, A. S.: On configurations of high-polymer ribonucleic acid macromolecules as revealed by electron microscopy. *J. Mol. Biol., 3*:778, 1961.

28. KLEINSCHMIDT, A. K., DUNNEBACKE, T. H., SPENDLOVE, R. S., SCHAFFER, F. L., and WHITCOMB, R. F.: Electron microscopy of RNA from Reovirus and Wound Tumor Virus. *J. Mol. Biol., 10*:282, 1964.

29. KLEINSCHMIDT, A. K., LANG, D., JACHERTS, D., and ZAHN, R. K.: Darstellung und Längenmessungen des gesamten desoxyribonucleinsäure Inhaltes von T2-Bakteriophagen. *Biochim. Biophys. Acta., 61*:857, 1962.

30. LANGRIDGE, R., and GOMATOS, P. J.: The structure of RNA. *Science, 141*:694, 1963.

31. MITRA, S., ENGER, M. D., and KAESBERG, P.: Physical and chemical properties of RNA from the bacterial virus R 17. *Proc. Natl. Acad. Sci., Wash., 50*:68, 1963.

32. MORA, P. T., McFARLAND, V. W., and LUBORSKY, S. W.: Nucleic acid of the Rauscher leukaemia virus. *Proc. Natl. Acad. Sci., Wash., 55*:438, 1966.

33. ROBINSON, W. S., and BALUDA, M. A.: The nucleic acid from avian myeloblastosis virus compared with the RNA from the Bryan strain of Rouse sarcoma virus. *Proc. Natl. Acad. Sci., Wash., 54*:168, 1965.

34. ROBINSON, W. S., PITKANEN, A., and RUBIN, H.: The nucleic acid of the Bryan strain of Rous sarcoma virus: Purification of the virus and isolation of the nucleic acid. *Proc. Natl. Acad. Sci., Wash., 54*:137, 1965.

35. SCHERRER, K.: Information transfer in a differentiated animal cell: m-RNA pattern in avian erythroblasts. In *Morphological and Biochemical Aspects of Cytodifferentiation.* Basel/New York, Karger, 1966.

36. SCHERRER, K., and DARNELL, J. E.: Sedimentation characteristics of rapidly labelled RNA from HeLa cells. *Biochem. Biophys. Res. Comm., 7*:486, 1962.

37. SCHERRER, K., and GRANBOULAN, N.: In preparation.

38. SCHERRER, K., and MARCAUD, L.: Remarques sur les ARN messagers polycistroniques dans les cellules animales. *Bull. Soc. Chimie Biol., 47*:1697, 1965.

39. SCHERRER, K., LATHAM, H., and DARNELL, J. E.: Demonstration of an unstable RNA and of a precursor to ribosomal RNA in HeLa cells. *Proc. Natl. Acad. Sci., Wash., 49*:240, 1963.

40. SCHERRER, K., MARCAUD, L., ZAJDELA, F., BRECKENRIDGE, B., and GROS, F.: Etude des RNA nucléaires et cytoplasmiques a marquage rapide dans les cellules erythropoiétiques aviaries difféenciées. *Bull. Soc. Chimie Biol., 48*:1037, 1966.

41. SCHERRER, K., MARCAUD, L., ZAJDELA, F., LONDON, I. H., and GROS, F.: Patterns of RNA metabolism in a differentiated cell: A rapidly labelled, unstable 60S RNA with messenger properties in duck erythroblasts. *Proc. Natl. Acad. Sci., Wash., 56*:15H, 1966.

42. SINHA, N. K., FUJIMURA, R. K., and KAESBERG, P.: Ribonuclease digestion of *Sci., Wash., 56*:157, 1966.

43. TOMITA, K. I., and RICH, A.: X-Ray diffraction investigations of complementary RNA. *Nature, 201*:1160, 1964.

44. VASQUEZ, C., GRANBOULAN, N., and FRANKLIN, R. M.: Structure of the RNA bacteriophage R 17. *J. Bact., 92*:1779, 1966.

45. WARNER, J. R., SOIERO, R., BIRNBOIM, H. C., GIRARD, M., and DARNELL, J. E.: Rapidly labeled HeLa cell nuclear RNA. I. Identification by zone sedimentation of a heterogeneous fraction separate from ribosomal precursor RNA. *J. Mol. Biol., 19*:349, 1966.

VIRAL INHIBITORS — DISCUSSION

DR. LEVY: The *in vitro* cell-free work that we presented with Dr. Marcus would at least indicate in case of the RNA viruses the alteration in the ribosomes is responsible for the action of interferon. I don't think that it is really necessary to bring in the possibility of an altered messenger RNA of the virus.

DR. JOKLIK: There is clearly a difference of experimental evidence here. There are two possibilities. Either vaccinia virus behaves differently from the RNA viruses as Dr. Levy has indicated, or there is the other possibility that the *in vitro* system does not reflect conditions *in vivo*. This is a possibility and must be checked. I think that is all one can say. I personally would prefer to believe that all viral messengers act the same way rather than that there is a difference between vaccinia messenger RNA and RNA virus messenger. It is, of course, possible, but this is just my viewpoint.

DR. LEVY: I didn't mean to suggest for one minute that the RNAases act differently. The point of conversion to the two systems is in the altered ribosome rather than any alteration of the RNA, because I think that your work is certainly consistent with an altered ribosome, and the work that Dr. Marcus has and that we have, also indicated altered ribosome both *in vivo* and in a cell-free system.

DR. JOKLIK: There is no way yet of telling whether it is an altered ribosome or an altered messenger RNA. However, it appears to be rather dangerous for interferon to mingle with the normal uninfected cell. I would rather believe that what interferon does is more specifically directed towards the invader rather than towards the factory.

DR. BALTIMORE: I am struck by the last point. Apparently, the virus can break out of host cell polyribosomes at the same time yet cannot code for anything, can't make any viral products. You suggest that it may be part of the virus which is in fact causing the destruction. Will that happen in the presence of Actinomycin? In other words, if you add virus in the presence of Actinomycin where there is only a slow decay of host cell polyribosomes, are these rapidly broken down?

Dr. JOKLIK: With the Mengovirus system, no.

Dr. BALTIMORE: With the vaccinia system?

Dr. JOKLIK: We haven't tested any viruses of that system.

Dr. BALTIMORE: You are predicting that in the presence of Actinomycin you get this rapid breakdown.

Dr. JOKLIK: Yes, there should be, but we haven't tested it yet. We are just struck by the complete absence of the polyribosomes at a very early time after infection.

Dr. BALTIMORE: I am just wondering when you talk about complete absence, is it the 10% absence you couldn't see?

Dr. JOKLIK: You are absolutely right. I call 10% residual complete absence.

Dr. BALTIMORE: Yes, but 10% of a protein may be enough to cause an acid induction.

Dr. JOKLIK: It is conceivable, yes, but once again, taking the wider view, I prefer to believe that if one viral messenger can't attach, no viral messenger RNA can attach.

Dr. BALTIMORE: Does the interferon treated cell with 95% virus inhibition have a 5% breakthrough?

Dr. JOKLIK: Yes, there is 5% breakthrough.

Dr. BALTIMORE: This must be in a small amount of cells making normal yield.

Dr. JOKLIK: I don't know.

Dr. BALTIMORE: Because most of your cells are gone.

Dr. Joklik: Yes, right. One has to draw a line somewhere as to how much interferon one wants to put in. No doubt if one added more interferon, it would be complete.

Dr. Lockart: I am struck also by the fact that the polyribosomes of the cells are broken down very rapidly in the interferon treated cells. Yet, you get a rapid increase of messenger formation of your viral messenger. This says, then, that the transcribing enzyme is probably increasing either in amount or in availability. Also, this system in this case is the cellular enzyme. Is it your suggestion, then, that just the availability of the enzyme probably from the nucleus in the cell is leaking faster into the cytoplasm to work or how do you explain this?

Dr. Joklik: Yes. There is, of course, in the vaccinia-infected cell the problem as to the source of the enzyme which transcribes the vaccinia messenger RNA which must be a DNA dependent, RNA preliminaries. This is a process that takes place in the cytoplasm and the question is — there are two possibilities, either it is the host cell enzyme which also exists in the cytoplasm or the vaccinia virus brings into the cell the formation for the code or the synthesis of a DNA dependent RNA preliminaries in a form which is signing off DNA. This is most unlikely. I think that the host enzyme which is used and your suggestion that there may be more of this enzyme leaking out of the nucleus is a good one. This could also be the case. We have not actually tested the level of the preliminaries in the cytoplasm. I don't really know what the answer is. There is the possibility of a control mechanism as well. There is further work which time did not permit to discuss, but the one thing I will say is that the pattern of early and late viral messenger RNA synthesis in HeLa cells and L cells is completely different. Here we have two cells, a HeLa cell and an L cell and the relative rates of early and late vaccinia messenger RNA synthesis is completely different in the two cells. In L cells, there is a tremendous burst in early viral messenger RNA synthesis and so little late viral messenger RNA synthesis that one can scarcely detect it. So, there are control mechanisms which say how much the early and late viral messenger is being copied. I don't think that can be explained solely by the availability of the transcribing enzyme. There are other factors involved.

Dr. Gordon: I think that Dr. Joklik has very clearly made his point. The interesting thing, then, is the recognition system that exists in common between viruses susceptible to the action of interferon and which must be rather widely distributed among viruses. If not universal, and by the same token, viruses that do shut down cellular macromolecular synthesis are able to recognize a recognition point presumably in the cell genome, perhaps by messenger, which isn't available to them in the virus messenger. If one wants to speculate then, these are major and fundamental biological differences.

QUESTION: What is the nature of the residual virus formed even if in the presence of high concentrations of IBT? Is it possibly mutant virus, or why is there a certain amount of replication?

DR. JOKLIK: Because the half-life of the messenger RNA is increased from 30 to 60 minutes not to 0 but to 5 minutes so that a small amount of protein is formed. We have isolated an IBT resistant mutant and thought at first that would help us greatly in understanding the mechanism of the action of interferon on IBT but unfortunately we haven't found the handle for it. This mutant is just resistant to IBT and that's all there is to it.

DR. DONOVIC: I am wondering whether this disruption is characteristic of all cells and I am talking about the interferon. In animals you expect a greater amount of pathology than in, for instance, mice protected from infection with interferon. You should be able to note the difference in the pathology from the protected mice.

DR. JOKLIK: Well, presume that the number of cells infected would be very, very small compared to all the cells in the whole mouse. I mean, when you have a cell treated with interferon which is infected with virus, there may be, say one thousand or even a million cells affected which is very small as compared to a whole mouse. I wouldn't think that you'd be able to pick that up at all. It depends again, on how massive the infection is. If you can infect under such condition say 10% of the cells, that would represent a tremendous virus inoculum. We have had mouse interferon made available to us and we have used L cells. We haven't been able to locate a chick system or HeLa system so I couldn't answer that, although I prefer to believe that it is a universal phenomenon.

DR. KABARA: Do these compounds have any contact inhibition on your virus? Has this been studied?

DR. KAPLAN: IUDR for example? No, cells will grow in the presence of IUDR, depending on the concentration. I don't know about actual contact inhibition, that is, we don't see any piling up if that's what you mean by contact inhibition.

DR. SHATKIN: Is the reovirus RNA extracted by urea infectious?

DR. GRANBOULAN: We have not determined the infectivity of the RNA after urea because we worked to prepare a trial microscopy and obtained a very small quantity. I think that it could be possible to take the preparation by mixing in a quart bottle pure reovirus preparation and urea for five minutes controlling the action of urea by relative thinning. Certainly, using large quantities we could measure the infectivity.

DR. BOXACE: After treatment with Guanidine, is it completely accountable for in terms of double-strand release RNA or do you have accumulations within the Guanidine?

DR. BALTIMORE: You do have further accumulation with Guanidine. The amount of RNA that is partitioned is about 50-50 single strands and double-strands. About 50% of the RNA that is made is identical to viral RNA by any criterion that we can use and there aren't many, but size is the best one. Fifty per cent is double-strand RNAases which is equivalent to the normal double-strand RNA.

We have evidence that the replicate of intermediate functions much less often than it does in the normal cell. The normal cell replicate of intermediate functions with a half-life of 10 minutes or thereabout. Since a molecule takes 45 seconds to be made, and there are about four molecules per structure, it functions to make about 40 or 50 molecules before it does. Here, it is not functioning anywhere near that rate. It is functioning something like four moulecules and then it dies so that you see a large amount of many single-strand synthesis going on but at a much lower rate than normal. It just stays in there presumably. I don't know if inactive preliminaries are being released. Maybe this is building up the number of sites which is also increasing aside from the activity of any one site being decreased.

MR. ARMSTRONG: I understood you to say that magnesia would release the RNA from the Guanidine.

DR. BALTIMORE: No, the lack of magnesium. It will release the single strands of RNA from the Guanidine but will not release the replicate of the intermediate.

MR. ARMSTRONG: Do you feel that the Guanidine is acting as a molecule in replacement of magnesium in some sort of RNA protein complex? Second, with the availability of the labeled Guanidine do you know its fate in this system? Can you pick it up in the Guanidine, or in the virus, or where?

DR. BALTIMORE: I have never tried that because Guanidine being such a strong base would bind to all sorts of things. I am not sure that I would trust any results I got. As far as Guanidine acting as a replacement for magnesium, Guanidine shouldn't be DPA sensitive whereas magnesium should, I would guess but maybe I am wrong. It seems to me that the magnesium is necessary and Guanidine is not necessary to hold it together so I think it's actually magnesium.

DR. FRANKLIN: On the basis of your scheme, can you account for the specificity of Guanidine in that it acts only on certain RNA viruses and not on others?

DR. BALTIMORE: I doubt that the specificity is related to the scheme. Since you can get mutants which are insensitive to Guanidine poliovirus, type I mutants which are insensitive to Guanidine but still presumably making RNA the exact same way, I would just guess that the picorna viruses that are not sensitive are naturally mutants, in one sense or another.

DR. PAYNE: There is one RNA tumor virus which you did not mention in a reference as appeared by Lyons and others in *Nature* some month or two ago. The Bittner virus has been observed by electromicroscopy and a cell line developed by site from a mammary tumor some years ago and this cell line has been in continuous culture for perhaps 100 or 110 passages. The addition of Actinomycin D to these cultures stimulated the production of the Bittner agent approximately five-fold as evidenced by electromicroscopy.

DR. ROBINSON: Some people have found similar things with the Rous agent. I think the difficulty is that they are not really measuring *de novo,* in a sense, replication of Rous. It is possible in a Rous system to add Actinomycin D and have formation of virus from viral components that were in the cell before you added the Actinomycin D, and still find production of infectious Rous following addition of Actinomycin D. If you use a measure of Rous production that really measures *de novo* synthesis of Rous components, like measuring the incorporation of treated uradine into the virus or into the viral RNA that you find in the virus in the medium, then you find an immediate complete inhibition of Rous at any time after infection. I think that explains the discrepancy in the Rous results, so I really can't comment on the Bittner findings. They may have been looking at virus that really was formed from the viral components in the cell that were made before they added Actinomycin D.

DR. BALTIMORE: I wonder if it is possible to use the kind of trick that Dr. Joklik uses in studying the RNA of vaccinia virus by using only cytoplasm after short times of labeling? Have you tried anything like that specifically with the Rous system to label let's say for 10 minutes with a large amount of treated uradine and then look to see what's in the cytoplasm that's bigger than 4S?

DR. ROBINSON: Yes, we have a number of experiments like that fractionating the cell. The fact that Rous RNA seems to have a sedimentation of 71S, you ought to be able to find an RNA like that in the cell. We have done it by fractioning the cell and looking only in the cytoplasm after very short pulses and still we see nothing that looks like Actinomycin D inhibits virus production. We've done it without Actinomycin D and it is very difficult. You know Rous RNA is made at an estimate from reovirus at less than 1% production. If you look at the rate of virus production, then it look like viral RNA synthesis. If you look at the cytoplasm, you still have a background out at 71S, even RNA from the cytoplasm does.

DR. FRANKLIN: I believe, according to Graham, Actinomycin can be added to cells even before infection and also he uses higher concentrations of Actinomycin than you do without affecting the synthesis of reovirus RNA. Is there a true discrepancy between your experiments and Graham's?

DR. SHATKIN: No, I think he only used a higher concentration of Actinomycin and finds RNA synthesis when he adds it later rather than earlier. In other words, if he adds 5 gammas per ml very early, he doesn't see any virus specific RNA synthesis and he gets a progressive resistance as he adds it later and later. Even by later, I mean three hours after infection or even four, so that there seems to be some early process that is Actinomycin sensitive. Once by that process, the system is resistant to Actinomycin suppression.

DR. FRANKLIN: I take it that at the present time you have no direct evidence for the viral specific enzymes?

DR. SHATKIN: Well, let's just say that I have looked hard but I have no evidence for it.

DR. BALTIMORE: I was struck by the possibility in the scheme that I presented earlier that the reovirus actually multiplies the same way poliovirus does except possibly a little more slowly and with the exception that it wraps up what poliovirus throws away and all the single-strand RNA goes into the polyribosomes and the double-strand gets wrapped up. The only thing it wouldn't explain is how does it get started again in the next cycle which you obviously don't have an explanation for either? Is there anything to contradict the idea that actually multiplication is the same for the two viruses?

DR. SHATKIN: I think that maybe it is the same except for the initiation point. In fact, we have done a few experiments looking for replicative intermediate in reovirus infected cells with some promise that there may be such a molecule. We don't have any data that we can really present. On the other hand, this system obviously is different, in that there probably are several enzymes, at least more than one enzyme involved in the process, whereas I am not sure that's the case with the polio system. I am not sure anybody knows.

INDEX*

---

*Certain terms have been used in almost every presentation and discussion. It is not practical to index these in toto, e.g., DNA, RNA, etc.